eries, 1880–1940

Chapter 6 Materia Medica	Chapter 7 Nutrition	Chapter 8 Microbiology

Cocaine
  by Koller
	Polished rice
	  and beri-beri
	  by Eijkmann	

Novocain
  by Einhorn
	Synthetic diets
	Amino-acids
	  by Hopkins
		Cell free enzymes
		  by Buchner

	Ship beri-beri
	  by Holst
		Adaptable microbes
		  by Twort and
		  Massini

	Vitamins
	  by Hopkins, Funk,
	  and Mellanby
		Bacteriophage
		  by Twort and
		  d'Herelle

Ephedrine by Chen and Schmidt	Sunlight Pure vitamins	Microbial chemistry
Prostigmine Ergometrine Tubocurarine	Synthetic vitamins	Essential metabolites Culture of viruses Prontosil

comprehensive record of the important discoveries during the period. See respective chapters for

In Search of a Cure

IN SEARCH OF A CURE

A History of
Pharmaceutical Discovery

M. WEATHERALL

Oxford New York Tokyo
OXFORD UNIVERSITY PRESS
1990

Oxford University Press, Walton Street, Oxford OX2 6DP

Oxford New York Toronto
Delhi Bombay Calcutta Madras Karachi
Petaling Jaya Singapore Hong Kong Tokyo
Nairobi Dar es Salaam Cape Town
Melbourne Auckland

and associated companies in
Berlin Ibadan

Oxford is a trade mark of Oxford University Press

Published in the United States
by Oxford University Press, New York

British Library Cataloguing in Publication Data
Weatherall, M.
In search of a cure.
1. Drugs, history
I. Title
615.109
ISBN 0—19—261747—8

Library of Congress Cataloging in Publication Data
Weatherall, M. (Miles)
In search of a cure: a history of the pharmaceutical industry/
M. Weatherall.
Includes bibliographical references.
Includes index.
1. Pharmacology —History.   2. Therapeutics—History.
3. Pharmaceutical industry—History.   I. Title.
[DNLM: 1. Drug Industry—history.   2. Drug Therapy—history.   QV 711.1 W362i]
RM45.W43 1990   615'.1″09—dc20
90-7426   CIP
ISBN 0—19—261747—8

Typeset by Joshua Associates Limited, Oxford
Printed in Great Britain by
Bookcraft Ltd, Midsomer Norton, Avon

Preface

This book is about the discovery of materials which prevent or cure diseases. It describes the steps which people have taken to understand how diseases occur and how medicines, vaccines, dietary materials, anaesthetics, antiseptics, and other medical materials came to be invented and used. Everyone should know something about these discoveries and their impact on society, and an unfamiliarity with science need not be a barrier to understanding them. The subject is approached historically and the ideas of scientists and doctors are seen as they developed. The story is not a comprehensive but a personal view, influenced by the experiences of the author as both an academic and an industrial scientist in Great Britain.

New medicines are not designed like a new aircraft or a new oven. If ever we come to know as much about life as we do about machinery and engineering, we may be able to design drugs on the drawing board, but for now our knowledge is inadequate. Nor does progress depend as much on dramatic 'breakthroughs' as some stories suggest. Breakthroughs occur only after a lot of valuable work has paved the way for them. Furthermore, once a 'breakthrough' has occurred an immense amount of 'applied' research is necessary to convert a promising discovery into a useful medicine. Brilliant possibilities are apt to fade into insignificance because they cannot be applied practically. Neither the extensive background to, nor the technical development of, 'promising' ideas receive as much credit as they deserve, nor is the nature and extent of the resources needed for either activity usually understood.

Humans are often only too ready to believe in the virtues of new drugs. Innumerable medical treatments have been adopted enthusiastically, only to be discarded when experience and careful investigation showed that they did no more good than faith and time, which are themselves great healers. Proving that a treatment works better than anything else, and defining the circumstances in which it works, is very difficult, but it is an essential part of completing a discovery. The development of clinical trials and other ways of judging whether treatments really work has had a major, beneficial effect on therapeutics and so is included in this book.

A few definitions may be helpful. Medicines usually include one or more active ingredients, called *drugs*, as well as other substances which are there because the active ingredient has not been completely purified, or because the medicine must be made stable and convenient to use. The use of the word *drug* only for the substances used and abused for their psychological effects is too narrow, and here the word is used with the wider meaning.

Another word which has recently been used in an unduly narrow sense is *organic*. The chemistry of materials originating from living organisms has been studied for the last two centuries under the name *organic chemistry*, and the substances with which this branch of science deals are known as *organic* compounds. The recent use of the word organic to refer to unpurified fertilizers of animal or vegetable origin and to materials grown only with their aid is unduly limited. The word is used in this book in its historically accepted sense.

The naming of drugs themselves is complicated. There are at least three kinds of name: proprietary names which apply to the product of a particular manufacturer; official names which are established on a national and, usually, international basis; and chemical names which follow highly technical agreements about nomenclature and are mostly too elaborate for everyday use. Making drugs available for medical use often depends on the activities of a particular manufacturer; proprietary or trade names, indicated by a capital letter, are used here when it seems historically relevant, otherwise official and, rarely, chemical names are adopted.

The discovery of drugs involves many sciences, including those of disease (pathology), of the healthy living body (physiology), of microbes (microbiology), of the chemistry of living organisms (biochemistry), and those of drugs themselves (organic chemistry, pharmacology, and pharmaceutical science). We cannot consider all of them simultaneously, and so Chapters 3 to 9 each deal with different facets of the subject. Each starts in about the middle of the nineteenth century and ends at about the end of World War II. The sequence of chapters is arranged in chronological order of the major advances they cover; for instance, insulin and steroid hormones (Chapter 5) come after the first achievements in chemotherapy (Salvarsan, Chapter 3) and before vitamins (chapter 7) and antibiotics (Chapter 9). Chapters 10 to 13 are mostly concerned with events after 1945, but include some earlier historical background. The synopsis of discoveries (1800–1940) will help readers to appreciate the chronology of different discoveries.

Experts in every science are increasingly apt to use their own jargon, often with scant regard for *any* reader's understanding. I have kept technical terms to a minimum, explained them, and included references in the index to the places where they are explained. To anyone working in the field who is troubled by over-simplification or omission, I can but apologize for not getting nearer to the truth while aiming first for intelligibility.

The notes to each chapter contain references, which will help anyone who wishes to know more about particular discoveries. Most of the references are to the papers or books in which the first accounts of discoveries were made public; or to reviews written within a few years of a discovery being recognized as important and chosen so as to give an impression of the scientific climate at the time; or to more recent articles which analyse the historical evidence about particular discoveries and the circumstances in which they were made. Books which will provide background information are listed at the end of the references quoted in Chapter 12.

Most of my working life has been spent in seeking and judging new medicines, and I am grateful to many scientists with whom I have discussed the subject. I am particularly indebted to my colleagues and friends who have most generously found time to read and comment on drafts of some or all of this book, especially Professor George Brownlee, Dr William Bynum, Revd L. A. Garrard, Mr John Griffith, Professor J. A. H. Lee, Mrs E. Penning-Rowsell, Dr Charles Phelps, and Professor Duncan Vere. Dr Tim Crow, Dr Gertrude Elion, Dr Norman Finter, and Professor Michael Shepherd have given me expert advice about particular chapters. Dr R. A. Maxwell and Dr E. M. Tansey kindly allowed me to read unpublished manuscripts which were very informative on particular points. Dr H. Blaschko and Dr B. Hazelman gave me useful references to published papers. I am deeply obliged to all of them for helping. Whatever errors remain are entirely my own responsibility and I shall be grateful to anyone who points them out to me.

Part of the material in this book was published in a series of articles in the *Pharmaceutical Journal* (1986–87) and now appears by agreement with the editor. Preparation of the text would have been impossible without help readily given by the library staff of the Royal Pharmaceutical Society, the Royal Society of Medicine, and the Wellcome Institute for the History of Medicine. I am indebted to Mr William Schupbach and staff of the Wellcome Institute for the History of Medicine for finding and providing illustrations, and for permission to reproduce plates 1–23. Plate 24 was kindly provided by Dr Elion. I am grateful to the Royal Society for permission to quote Sir Henry Dale's account of the creation of an insulin standard; *The Lancet* for permission to quote from the obituary notice of Sir Frederick Hopkins; The Johns Hopkins University Press for permission to quote a passage by Professor D. W. Richards on the clinical use of vitamins; and Dr R. Schwartz and Burroughs Wellcome USA for permission to quote from his account of the discovery of the immunosuppressant properties of some drugs. I have always been particularly indebted to my secretaries, but on this occasion preparation of the text depended entirely on my word processor, which appears to be indifferent to gratitude. Members of the staff of the Oxford University Press have been most helpful in editing the book for publication, and my best thanks are due to them.

1989   M.W.

Contents

List of Plates   xiii

1 The Beginnings of Medicine and Medicines   3

Aesculapius to Paracelsus. Therapeutic sceptics. Treatment in the seventeenth and eighteenth centuries. The beginning of immunization. New science in revolutionary Paris. 'Modes of ascertaining the effects of medicines'. A science of drugs

2 The Chemistry of Medicines   27

The new chemistry. Medicines from new chemicals. New light on medicines. Growth of an industry. The many achievements of Pasteur. The germ theory of disease. Lister and antiseptics. Science becomes acceptable in medicine

3 Immunotherapy and Chemotherapy   51

Defence against germs. Producing the remedies. Standardizing the materials: the genius of Paul Ehrlich. Ehrlich's receptor theories. The magic bullet. Progress and decline of arsenical drugs

4 The Physiological Basis of Medicine   69

English physiology. Dale's mess of pottage. A National Institute for Medical Research. Resolution of conflicting theories

5 Replacement Therapy   83

Early ideas about ductless glands. Hormones. The thyroid problem. The discovery of insulin. Steroid hormones and the contraceptive pill. Drugs and hormones

6 Scientific Study of Drugs   103

*The formative forces.   A science in its infancy.   A pattern for
inventing drugs.   Scientists, doctors, and the ergot problem solved.
Limitations*

7 Deficiency Diseases   117

*Essential foods.   Scurvy and beri-beri.   Synthetic diets.
Accessory food factors and vitamins.   Resolving the enigma of
rickets.   From scepticism to credulity.   Subtler deficiencies*

8 War on Germs   141

*Biological remedies.   Microbial chemistry.   Chemotherapy
despised.   Chemotherapy prevails: the defeat of streptococci.   From
achievement to understanding: how sulphanilamide works*

9 Antibiotics   161

*Microbes against microbes.   Fleming.   Florey.   A new tool for
bacteriologists.   Trials and apathy in diverse places.   Breaking
through.   Penicillin arrives.   Publicity.   Streptomycin and
onwards*

10 Towards the Conquest of Infectious Diseases   189

*Malaria: discovery in the laboratory.   Malaria: discovery by
committee.   Tuberculosis.   At the frontiers of life: viruses.
The new look in science*

11 Cancer   209

*The background.   The beginning of radiotherapy.   Hormones and
cancer.   Swords into ploughshares: war gases and the arrest of
cancer.   First whisperings of DNA.   Converging discoveries.
A search with many facets.   Progress in cancer research*

12 Cause Unknown   235

*Internal medicine.   Blood pressure.   Transmitters and
mediators.   Indian snake-root.   Receptor for adrenaline.
Salt and water drugs.   Evaluation*

13 Drugs and the Mind 251

Poppy or mandragora. Problems of insanity. The devious paths to tranquillizers. Relief of depression. Rescued from the remnants: a new range of tranquillizers. Brain mechanisms and drugs. New light on old drugs. Body and mind

14 The Present and the Prospects 271

Some lessons of history. The safety of medicines. Alternatives. The problems of success

Index 287

Plates

(see between pp. 146 and 147)

1 Paracelsus (1493–1541), from a Dutch seventeenth century line engraving.

2 James Lind (1716–1794). Portrait by J. Wright after Sir G. Chalmers, 1783.

3 William Withering (1741–1799). Portrait by W. Bond after C. F. v. Breda.

4 Edward Jenner (1749–1823). Portrait by J. R. Smith, 1800.

5 François Magendie (1783–1855). Lithograph portrait by N. E. Maurin, probably between 1842 and 1858.

6 Pierre-Joseph Pelletier (1788–1842). Portrait by Catherine Buisson, 1870, after Eliza Dérivieères.

7 Joseph Caventou (1795–1877). Portrait by Catherine Buisson, 1870, after Eliza Dérivieères.

8 Claude Bernard (1813–1878). Undated photograph.

9 Friedrich Wohler (1800–1882). C. Cook, after design by C. L'Allemand. William Mackenzie, not dated.

10 Louis Pasteur (1822–1895) in his laboratory. From an engraving published in *The Graphic*, November 1885.

11 Robert Koch (1843–1910) in his laboratory at Kimberly, South Africa, *c.* 1896–97.

12 Emil von Behring (1854–1917), *c.* 1909.

13 Joseph Lister, later Lord Lister (1827–1912), about the time of his experiments in antiseptic surgery, *c.* 1862.

14 Sir Thomas Fraser (1841–1920). Photographic portrait by A. Swan Watson, Edinburgh, not dated.

15 Paul Ehrlich (1854–1915), not dated.

16 H. H. Dale, later Sir Henry Dale (1875–1968), *c.* 1918.

17 F. G. Banting (1891–1941).

18 C. H. Best (1899–1978).

19 Sir (Edward) Charles Dodds (1899–1973).

20 Sir Frederick Gowland Hopkins (1861–1947) in 1938.

21 Sir Edward Mellanby (1884–1955).

22 Discoverers of antibiotics. Back row: left to right, S. Waksman, H. Florey, J. Trefouel, E. Chain, A. Gratia. Front row: P. Fredericq, M. Welsch.

23 Sir Henry Dale (1875–1968) in 1959.

24 Dr G. H. Hitchings and Dr Gertrude Elion at the Wellcome Research Laboratories, New York, about 1948.

In Search of a Cure

Chronology of Chapter 1

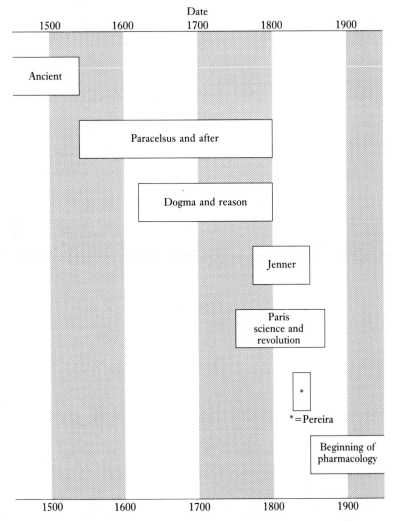

NB The blocks show periods discussed in the text, and do not mean that there was no activity about the subject at other times.

1

The beginnings of medicine and medicines

Aesculapius to Paracelsus

The idea that something can be done to avert illness has ancient beginnings. It is much too old for us to know how it first arose. Perhaps it was a natural part of the evolution of human curiosity and ingenuity. Perhaps it started from a simple wish to survive and, rightly or wrongly, a belief that a particular action would actually preserve health or expel disease.[1] But such attempts have always been limited by ignorance.

The early treatment of illness appears to have been inseparable from religious and magical practices. The mythological Asklepios, latinized as Aesculapius, and now a sort of patron saint of healing, was the head of a cult of temple-medicine to which the sick could resort for treatment. Homer, in the *Odyssey*, uses the word pharmakon to refer to a drug or a charm, and the derived words pharmacy and pharmacology are used today for a place where medicines are supplied and for the science of drugs or medicines.[2,3]

Hippocrates of Cos (born 460 BC) introduced a more rational outlook by basing his practice on observation and experience instead of on appeals to the gods or to magic. We owe to him ideas of different temperaments based on four 'humours' supposed to permeate the body: fluid essences named the sanguine, the phlegmatic, the choleric, and the melancholic. All these terms influenced medical thought for centuries and persist today, though no longer with much influence. Aristotle (384–322 BC) dissected animals and formed more refined concepts about the special properties of living organisms. His idea, that matter was hot, cold, wet, and dry, combined in different proportions to form the four elements earth, air, fire, and water, can be related to the humours of Hippocrates. Like them, the four elements persisted until the growth of modern chemistry created new ideas about the elements of matter.

In the following centuries, many medicinal plants were discovered or recognized in the civilizations which existed round the Mediterranean. Most of them can be identified more or less accurately, and a certain amount learnt about the diseases for which they were used, but it is impossible to find out how they were first discovered. After the imaginative and speculative civilization of Greece had given place to the organized and regulated civilization of Rome, the Roman physician Galen (AD 129–199) wrote extensively on medical matters. Some of his statements were at variance with easily observed facts, but this did not prevent their wide acceptance. In the troubled centuries which followed, originality, and even the Roman Empire, declined, and for 1200 years the study of medicine in Europe stagnated under the weight of authority and tradition.

Other parts of the world were developing independently, especially China and the Arabic-speaking countries (Table 1.1). Chinese medicine had almost no impact on European knowledge and practice, but contact between the Middle East and other countries around the Mediterranean allowed the passage of medical ideas and medicines. The encyclopaedia of the Persian physician Avicenna (AD 980–1037) was thus added to the standard authority of Galen. However, once again there is little information about how the drugs recommended by Avicenna were first discovered.

Table 1.1. Some potent vegetable drugs known before 1800

Drug	Plant	Discovery/ Recognition	Uses
Opium	*Papaver somniferum*	Ancient Greece	Relief of pain
Hemlock	*Conium maculatum*	Ancient Greece	State poison
Mandragora	*Mandragora officinatum*	Ancient Greece	Soporific: magic
Belladonna	*Atropa belladonna*	By *c.* 1500	Cosmetic: poison
Ipecacuanha	*Cephaelis ipecacuanha*	Brazil *c.* 1600	Bloody fluxes
Jesuit's Bark	*Cinchona* species	Peru *c.* 1630	Tertian and quartan agues
Coca Leaves	*Erythroxylon coca*	Bolivia and Peru *c.* 1688	Chewed to prevent fatigue
Ma-Huang	*Ephedra* species	China, BC USA *c.* 1924	Asthma: stimulant
Foxglove	*Digitalis* species	England, 1775–85	Dropsy: heart disease

At the end of the fifteenth century, the shackles of authority began to be weakened. Columbus (*c.* 1451–1506) sailed to America and widened the boundaries of the known world, while Copernicus (1473–1543) expanded our view of the heavens. Luther (1483–1546) challenged the domination of the Catholic Church and began the religious reformation by the dramatic act in 1520 of burning a papal bull publicly. In medicine, Theophrastus Bombastus von Hohenheim (1493–1541), who assumed the name Paracelsus, rapidly achieved the distinguished position of town physician and university professor in Basel. In imitation of Luther he started his lectures on therapeutics by setting fire to the works of Galen and Avicenna. He was naturally unpopular with his more reverent colleagues. After a few years he was involved in a dispute about fees with a canon of the cathedral whom he had treated. He lost his case, and left Basel for good, to become a wandering practitioner.[4,5]

Paracelsus was as contemptuous of orthodox alchemists as he was of priests and physicians. He was a pioneer of extensive chemical experiments. Many of his observations were accurate and some of his inferences were far ahead of their time but fundamentally correct. He noted that air was necessary for wood to burn and stated that without air all living things would die. He favoured the use of minerals as well as plants in medicine. He may have introduced mercury for the treatment ot syphilis. He extolled antimony and found medicinal uses for iron, copper, arsenic, and lead. He made much of a secret remedy which he named laudanum. It is not clear what he put into it, but opium was probably the main constituent, and preparations of opium continued to be called laudanum until well into the present century. Although in some ways he was very much of a realist, he had as much capacity for belief as for scepticism. He had a mystical philosophy of gnomes, sylphs, nymphs, and salamanders which were the 'Elementals' of earth, air, water and fire, whilst recognizing the chemical 'elementals' combustible sulphur, volatile mercury, and residual salt. Despite his imitation of Luther's incendiarism, he had no use for Lutheran beliefs and distressed his friends by his disregard for religion. He was one of the first doctors to separate completely the material treatment of patients from attention to their beliefs and feelings, a separation which has again become prominent and which raises interesting philosophical problems.

Paracelsus is credited with having achieved some remarkable cures, but it is difficult to judge how effective his remedies really were. It has always been too easy to be influenced by the results of single cases and to attribute recovery to the treatment given, just as it has been (and still is) too easy to blame medicines for ill-effects. Mercury, arsenic, and antimony are powerful poisons and, in generous doses of uncertain purity, may well have killed or permanently harmed his patients. Laudanum at least had the advantage of giving a pleasanter death than metallic poisoning. From what we know of the character of Paracelsus, we can be sure that he did not fail to exploit any credit he could get from apparent successes, and it does not look as though his zeal for experiment led him to put his discoveries to the test by what would now be called a 'controlled' trial, that is, an experiment in two sections, one of which included some treatment of particular interest and the other left it out.

Therapeutic sceptics

When Paracelsus burnt the texts of Galen and Avicenna he showed a healthy disrespect for authority, but he gave no real grounds for dismissing the orthodox remedies as worthless. Nor was he entirely original as a sceptic. Petrarch (1304–1374), in a letter to Boccaccio (1313–1375), suggested that if he took 1000 sick people and let 500 be treated by doctors while the rest were kept away from medicines, the untreated would fare better.[6] How informative it would be if the experiment had actually been done! It is very difficult to collect facts about the advantages and disadvantages of a particular treatment, but exceptional circumstances are sometimes quite revealing.

In 1537, a great figure in Renaissance surgery, Ambroise Paré (1510–1590), was responsible for treatment of the wounded after the capture of the castle of Villaine. The casualties were numerous, and supplies of oil (for the standard method of cauterizing the wounds with boiling oil) ran out. In its place, Paré improvised a cold dressing of egg yolk and turpentine.

That night I could not sleep at my ease, fearing that by lack of cauterization I would find the wounded upon which I had not used the said oil dead from the poison. I raised myself early to visit them, when beyond my hope I found those to whom I had applied the digestive medicament feeling but little pain, their wounds neither swollen nor inflamed, and having slept through the night. The others to whom I had applied the boiling oil were feverish with much pain and swelling about their wounds. Then I determined never again to burn thus so cruelly the poor wounded by arquebuses.[7,8]

So much for the merit of a treatment which caused Paré to fear for the suffering of those who were deprived of it. Petrarch might have said 'You see what I mean'.

In 1600 four ships set out on the first expedition from England to India by the newly formed East India Company. One ship alone, that of General James Lancaster, carried stocks of lemon juice. On this ship, and only on this ship, did the crew remain largely free of scurvy, at that time an almost inevitable consequence of long voyages. The Company was impressed, and provided lemon water for subsequent voyages.[9] No one knew why the remedy worked, but the benefit was established by sound evidence.

However, its use did not become universal. It was evaluated again in a famous experiment by the British naval surgeon James Lind (1716–1794). His *Treatise of the scurvy* gave stronger evidence that scurvy was prevented by a diet which contained sufficient vegetables or fruit juices.[10] The Admiralty took a long time to act on the evidence, but after 20 years or so the issue of lime juice to British sailors was established; during most of the nineteenth century ships were able to remain at sea for prolonged periods without casualties from scurvy. The ability to keep crews healthy without periods ashore was crucial for the naval blockade of Europe during the Napoleonic wars and significantly contributed to the survival of an independent England.

Later in the century the navy began to obtain limes. The new limes happened to be less effective against scurvy, and for a time the treatment fell into disrepute (see Chapter 7).[11] So a good treatment was temporarily lost, because the exact identity of the cure was not known. In other trials, the difficulty lay in ignorance about the diseases themselves. Scurvy was easy to recognize; the bleeding gums, the easy bruising, the general weakness, and death, especially in such well-defined circumstances as long sea voyages, sieges, and famine, together made a recognizable condition. But many illnesses did not fall into recognized patterns. It was necessary to study diseases before trying to invent or to judge remedies.

Thomas Sydenham (1624–1689), sometimes regarded as the English Hippocrates, contended that most forms of ill-health could be ranked in definite categories. He gave clear accounts of the common diseases of his day, including plague, smallpox, gout, and malaria. Acute fevers and inflammations he regarded as wholesome reactions of the body to injurious influences from without, but to understand chronic diseases he went back to the ancient Hippocratic idea of bodily humours. According to the doctrine of humoral pathology, the blood, the phlegm, or the yellow or the black bile might dominate the temperament, and disorders could be remedied by adjusting their influence. However, the doctrine gave no lead towards finding medicines which would influence these humours. Nor could much be done about injurious influences from without until they were identified. Sydenham himself was mostly sceptical of the value of medication, but commended the use of laudanum and of the newly introduced Peruvian bark for quartan agues (recurrent fevers, probably malarial).

For centuries, with increasing momentum, disease was further studied by examining the bodies of patients after death, internally as well as externally, and relating what was found to the symptoms experienced during life. Public dissection had been practised in mediaeval Italian medical schools as early as the fourteenth century, but mainly for instruction in anatomy and for legal purposes in cases of suspected poisoning. Gradually, doctors went to the post-mortem room to see what internal disorders had accounted for the illness and death of their patients. Hermann Boerhaave (1668–1738) of Leyden was noted for teaching students in this way, and Battista Morgagni (1682–1771) of Padua for beginning to classify medical knowledge on the basis of what came to be called 'morbid anatomy'.

Information for medical use also began to be obtained by examining the records of births and, particularly, deaths in a community. John Graunt (1629–1674), a self-educated London merchant, studied the parish registers and bills of mortality and wrote about factors influencing the expectation of life and the age of childbearing. When infectious diseases were rampant and epidemics of plague or smallpox swept periodically through the civilized world, evidence obtained from such sources showed how epidemics spread and suggested methods that could be used to check their course. Even without a formal analysis of the records kept by the community or by hospitals, differences in mortality were well enough known to suggest thoughts about the cause and management of illness. John Aiken (1747–1822), a distinguished non-conformist physician of Warrington in Lancashire, wrote in his tract *Thoughts on hospitals*:

Every surgeon attending a large and crowded hospital, knows the very great difficulty of curing a compound fracture in them. This is so universally acknowledged that the most humane and judicious of them have been obliged to comply with that dreadful rule of practice, immediate amputation in every compound fracture. Mr Pott, than whom there is not in the whole profession a more unprejudiced patron of improvement, or a warmer advocate for humanity, has recommended this general rule upon the grounds of accurate and impartial observation. Yet, that it is not founded upon the nature of the case, but on the added malignancy of hospital air, is evident from the different successes in private practice and the country infirmaries.[12]

So the science of epidemiology began to emerge, though it made little progress until the middle of the nineteenth century, when the recording of births and deaths began to be established on a secure national basis in England and other countries.[13]

Treatment in the seventeenth and eighteenth centuries

For a long time, medical treatment was based on reasoning, but the reasoning started from the dogmas of the time, which had little basis in fact.[14] In a world in which man thought of himself as the centre, it was easy to believe that there must be a remedy for every disease and that the remedy could be seen by anyone who looked for it. So the doctrine of signs or signatures was invoked. The yellow colour of saffron showed that it was good for jaundice. The red colour of rust or wine that they were good for bloodlessness or anaemia, and the leaves of the plant lungwort, which had some resemblance to the appearance of the lungs, showed that the plant was good for lung disease.[15]

Less absurd by present standards were procedures based on the limited physiological knowledge of the time. The circulation of the blood had been demonstrated by William Harvey (1578–1657), but the old ideas of 'humours' still had much influence. Excess of these humours could be relieved, it was thought, by bleeding, which quietened patients and was indeed beneficial in certain circumstances. Other physical procedures, which let real or imaginary humours escape from the body, or reduced the pressure in the tissues, included dry and wet cupping, and the application of leeches. In dry cupping a heated bowl was pressed on the skin, which was sucked into the bowl as it cooled. In wet cupping the skin was incised under the applied cup to release tissue fluid and blood. All these practices lasted for a long time. Because the use of leeches was increasing at the Middlesex Hospital, in London, in 1821, it was suggested, as an economy, that each leech should be used twice.[16] In 1940, dry cupping was demonstrated to me and other medical students by an elderly physician: the patient did not complain, and, as far as I recollect, got better.

Medicinal plants have always been the mainstay of therapeutics. Purgation was another way of getting material out of the body, and the ritual use of purgatives became notorious. Strong purgatives caused colic, which could be relieved by the leaves of deadly nightshade, *Atropa belladonna*. So there were reasons for using two drugs at once, and many other kinds of reasoning led to complicated prescriptions containing many supposedly active drugs. Most of them were probably innocuous. Deadly nightshade was not; it caused mental disorders and death if too much was used. The Latin name of the plant referred to Atropos, one of the three classical Greek fates who control human destiny. It also referred to the fact that the drug dilated the pupils of the eyes so making ladies more beautiful. Like all potent drugs its merits depended on getting the dose right. When lesser drugs failed to give relief, opium or laudanum gave comfort, perhaps more desirable than the prolongation of life. The range of *materia medica* was increased by plants with medicinal reputations brought from distant lands by the growing number of explorers. Tea, and later coffee, came from the East Indies and were not only welcomed as new social and invigorating beverages but also reviled as destructive to health and wealth.[17] Two centuries passed before their medicinally-active ingredients, caffeine and theophylline, were extracted and identified.

Some of the new imports were potent medicines. Peruvian bark, also called Jesuit's bark, the source of quinine, was brought to Europe between 1630–1640, perhaps by Jesuit missionaries and perhaps by the viceroy to Peru, whose wife, the Countess Anna del Cinchon, had been cured by it.[18]

It was a specific cure for the easily recognized fever which recurred every third or fourth day and which is now called malaria. This disease was widespread in Europe until large tracts of marshland were drained in the nineteenth century. The demand for Jesuit's bark had important consequences in promoting new exploration, in starting new cultivation in different lands, and in developing new drugs. Ipecacuanha (which means 'the little plant which grows by the wayside and causes vomiting') also came from South America. It was sometimes effective in dysentery and was used indiscriminately for a long time. It was necessary to discover more both about dysentery and about the plant before proper use could be made of it, as is described later.

Among plants indigenous to England, the bark of the willow was recorded as successful in curing ague by the Revd Mr Edmund Stone, of Chipping Norton in Oxfordshire. Its effect was reported to the Royal Society in 1763. Its bitterness suggested to its discoverer that it might have the properties of Peruvian bark. Also, he wrote:

As this tree [the willow] delights in moist or wet soil, where agues chiefly abound, the general maxim, that many natural maladies carry their cures along with them, or that their remedies lie not far from their causes, was so very apposite to this particular case, that I could not help applying it; and that this might be the intention of Providence here, I must own had some little weight with me.[19]

The 'general maxim' and faith in Providence, so characteristic of a human-centred view of the universe, has not been a fruitful source of discovery of drugs. Willow bark was less of a therapeutic success than the Peruvian material. However, a century later, active principles were isolated from willow bark and became known as salicylates. They were found to be of considerable importance, not in malaria, but in another 'ague', rheumatic fever.

Twenty or so years later William Withering, physician to the General Hospital at Birmingham, wrote *An account of the foxglove and some of its medical uses etc; with practical remarks on dropsy and other diseases*. He referred to accounts of the plant from 1542 onwards, usually mentioning that the plant caused purging and vomiting. He wrote:

In the year 1775, my opinion was asked concerning a family receipt for the cure of the dropsy. I was told that it had long been kept a secret by an old woman in Shropshire, who had sometimes made cures after the more regular practitioners had failed. I was informed also, that the effects produced were violent vomiting and purging; for the diuretic effect seemed to have been overlooked. This medicine was composed of twenty or more different herbs but it was not very difficult for one conversant in these subjects, to perceive, that the active herb could be none other than the Foxglove.[20] [The punctuation is Withering's.]

It is not clear why Withering selected this one of the dozen or so ingredients of the 'receipt', but he began using it in his practice in 1775.

Ten years later he published an analysis of his cases and an account of the virtues and deficiencies of the remedy. He described its use mainly as an agent which increased the flow of urine and so reduced the fluid accumulation which appeared as dropsy. He recognized the powerful action on the heart and suggested that the 'power might be converted to salutary ends'.

Withering's treatise is rightly held to be a model of careful clinical observation and it brought digitalis into widespread use. But his interpretation was handicapped by the general lack of knowledge of bodily mechanisms at this time. The circulation of the blood was known and accepted, but the role of the circulation in transporting water was far from clear, nor was dropsy due to heart failure distinguished from dropsy due to disease of the kidneys. Progress in converting the power of digitalis to salutary ends, and in seeking better medicines, necessitated knowledge of the body on which the medicine acted. At the time when Withering was making his advances, the experimental approach, essential for acquiring such knowledge, was beginning to be adopted to great effect in France (p. 16).

Digitalis and other plants had obvious effects on the human body, but whether the effects were curative or not was less obvious. The intellectual climate which promoted proper trials did not begin to develop until the nineteenth century, and neither the diseases nor the remedies were sufficiently well defined for such trials to have much prospect of being profitable. Some of the remedies used, especially the violent purgatives, can be assumed to have made the patient uncomfortable, but they were credited unjustifiably with benefits supposed to warrant their use. In excessive dosage all the potent remedies were capable of being lethal. If a drug interferes powerfully with any bodily system an excess is likely to be fatal, so the science of medicines is intimately linked with the science of poisons. Many other plants were used which had the great merit of being innocuous even when the largest doses were administered. Later knowledge has suggested that their therapeutic effect depended on giving confidence to the patient directly or through his doctor.

Mineral substances, as advocated by Paracelsus, had their place, as had various materials of animal origin, sometimes suggesting the traditional ingredients of a witch's cauldron. Exotic remedies based on the tales of travellers thrived and there were few ways of distinguishing between genuine experimenters, quacks, and charlatans, especially as the therapeutics of orthodox practitioners rested on uncertain foundations. Inevitably practitioners of the same techniques or school of thought, banded together and formed societies for promoting their ends.

Organized medicine in England depended on the physicians, who had a rigid code of practice and professional standards, and founded a college, which received royal patronage in 1518. Lower down the social hierarchy, apothecaries treated patients and also prepared medicines: their activities included the functions of general practitioners of medicine and of pharmacists or 'chemists' today. The apothecaries became a livery company in the city of London in 1617. Membership of the Society of Apothecaries was recognized as a medical qualification and remains so today. Further down the scale barber-surgeons undertook treatments that their name suggests: the surgeons gradually dissociated themselves and, after some difficulties, acquired a Royal Charter for their College in 1843.

Less orthodox practitioners included herbalists, whose activities were based on the available plant remedies, and homeopaths, who developed interesting theories which related the medicine used to the exact symptoms displayed by the individual patient. The doctrine of 'like cures like' resembled the old doctrine of signs, but now it was the effects of the medicine which were supposed to show what it was good for. Homeopaths departed completely from the general advance of scientific knowledge by a further doctrine of diluting their medicines until it was reasonable to suppose that the dose given contained none of the original therapeutic material. The safety of homeopathy had great merit, especially in contrast with the gross overdosage practised by the more flamboyant of the physicians and apothecaries. Attempts to judge the efficacy of homeopathy by ordinary criteria are of recent origin.[22] For those who sought alternatives to any medication or to surgery, osteopaths provided a different sort of treatment. And childbirth was still mainly a domestic matter attended by midwives, until the growth of towns brought more women to be delivered in the dangerous environment of malignant hospital air. Until the Medical Act of 1858, there was no legal recognition of qualifications which could distinguish the more from the less authentic practitioners.

The beginning of immunization

A quite different approach to disease began to be applied in Europe during the seventeenth century. Its aim was preventative, not curative. The approach depended on the well known fact that one attack of certain diseases was sufficient to protect a person from further attacks. The protection was very valuable, particularly in epidemics of smallpox, because people who had already had the disease could go about their business securely and even nurse new victims with impunity. This fact appears to have been recognized in China by, at latest, the sixth century AD. So people were deliberately infected with smallpox, under conditions believed to favour a mild attack.[23]

Originally the disease was transferred in various ways, often using the dried scabs from a patient's pustular rash and feeding them, or pricking them into the skin of the recipient. The practice, widespread in the Near East by the eighteenth century, was described favourably in the vivacious letters of Lady Mary Wortley Montagu, wife of the British ambassador at Constantinople in 1717. It spread to England, and was called variolation, after variola, the Latin name used for smallpox. A proportion of successes, some confirmed by subsequent deliberate attempts at reinfection, encouraged variolation both in England and in America during the eighteenth century. But the risks were great. There was no certainty that the induced attack of smallpox would be mild and it was quite possible that it would be fatal.

In some country districts of England, and probably elsewhere, people thought that infection with cowpox, contracted from cattle, gave immunity to smallpox. Edward Jenner (1749–1823), a country physician at Berkeley in Gloucestershire and a personal friend of the great surgeon and naturalist John Hunter (1728–1793), investigated the belief by careful experiment, such as Hunter advocated in the famous words 'Why think? Why not do the experiment?' Jenner inoculated material from cowpox vesicles into a boy and later inoculated him with smallpox to see whether he caught the disease. The risk was considerable and the experiment was, by modern ethical standards, probably reprehensible, but conditions in the eighteenth century were very different and we have profited greatly from the advance of knowledge made in a bolder time. In fact, the inoculation produced immunity and the boy suffered no ill effects. Similar inoculations of persons who had contracted cowpox directly from cattle showed the same immunity. As material from cowpox vesicles was seldom available, Jenner devised a method of arm to arm inoculation, by which large numbers of people could be immunized with a very limited amount of material. The Royal Society rejected his account of his work and it was published privately in 1798.

Jenner continued his studies on 'vaccination', i.e. inoculation with material from lesions of vaccinia or cowpox, and developed methods of preserving vaccine lymph. Supplies of vaccine lymph were subsequently organized by the Royal Jennerian Institute, which was established under his presidency in 1803. Vaccination aroused much opposition, some of it blindly unreasonable. It was modified in various ways in the light of new knowledge and became the subject of legislation in various countries, first to make vaccination compulsory and later to allow refusal to be vaccinated permissible on grounds of conscience.[24] Nevertheless, its value was shown repeatedly, often in the different mortalities of the vaccinated and unvaccinated. The objectors were not damagingly numerous and so the disease became rare where preventive measures were well applied. With more recent advances in making and using the vaccine smallpox appears to have been eliminated as an active disease.[25]

Jenner's work, invaluable though it was, threw no light on the cause of smallpox and cowpox, though it clearly implied some physical means of spread from person to person. One must realize that at this time the idea of 'germs' or microbes as a cause of disease was quite unknown. Disease of all kind was often regarded as an act of divine retribution and those who sought more immediate causes talked about miasmas and contagions, with little idea of their nature. Early microscopists with their primitive instruments detected organisms invisible to the naked eye but seldom associated them with disease; naturally enough because the great majority of microbes are innocent and often beneficial creatures. Speculations about such physical means had been made before. Richard Bradley (died 1732), professor of botany at Cambridge, attributed putrefaction (which might include putrefaction of living organs, i.e. destructive diseases) to 'insects', which in his usage meant microscopic organisms, and plague to a living contagion. His contemporary, Benjamin Martin, wrote of 'A new theory of consumptions' and invoked 'animalculae'. But they did not devise experiments to test their ideas.[26] The evidence of Aiken and others about corrupt air and avoiding overcrowding in hospitals gave some support to the doctrine. The success of variolation and vaccination was more positive evidence about the physical existence of 'something' which caused or carried a particular disease.

Still more positive evidence was discovered by an Italian physician, Agostino Bassi (1773–1856), who was losing his sight and turned to farming in his native Lombardy. He cultivated silkworms and investigated the contagiousness of certain silkworm diseases. He showed that infectious material spread from efflorescent, dead, and decaying silkworms and could be made harmless by treatment with suitable chemical agents or disinfectants. His account was published in 1835. It interested botanists and agriculturists, but had little influence on medical ideas of the time. Bassi himself wrote further papers on contagions in general. He claimed that smallpox, spotted fever, plague, and syphilis were due to living parasites, but he did no more experiments, perhaps because of his loss of sight and perhaps because of his isolation from medical activity. His doctrine of 'facts before judgements' shines brightly in a time when speculation was rife and experiments were still rare, but his isolation did not help the spread of his conviction. Nor were many minds receptive to the idea that diseases of anything as insignificant as silkworms could possibly be related to the important problems of human health.[27]

The hazards of putrid effluvia were recognized fitfully, but ideas were confused and not clear enough to be allowed to change established practices. Childbed fever became a serious problem in the early part of the nineteenth century, when industries had developed which brought many people to live in towns, and the delivery of women in hospitals instead of their own cramped and confined homes became common. The hospitals provided a splendid opportunity for the spread of infection, particularly when doctors proceeded straight from the post-mortem room to conduct deliveries without changing their outer clothes or washing. Criticism was rare and not welcome. In Boston, Oliver Wendell Holmes (1809–1894) carefully compiled records of the spread of the disease, but made little headway in changing the conduct of New England lying-in hospitals. In Vienna, Ignaz Philipp Semmelweiss (1818–1865), less prudent than Holmes, was subjected to bitter abuse and hostility for observing that infection was carried by nursing and particularly by medical attendants. He prescribed washing with chlorinated lime water and achieved a spectacular reduction of mortality in certain lying-in wards. The implication of faults in current medical practice was intolerable to the medical authorities, especially coming from a Jew, and Semmelweiss had to go.[28,29,30]

Another disease of epidemic proportions in the nineteenth century was cholera. Careful investigation of its spread again gave evidence about the 'something' which caused the disease. In England and Wales a system of civil registration of births, marriages, and deaths was established in 1836, and in time came to include recording the supposed cause of death.[13] The work centred on the General Register Office housed in London, and, from the beginning its statistical superintendent William Farr (1807–1883) made great use of this information to throw light on the causes and associations of disease. In conjunction with Dr John Snow[31] (1813–1858), the medical practitioner who administered the first anaesthetic to Queen Victoria in 1853, Farr produced detailed tabulations of the 1849 cholera epidemic. The tabulations gave Snow a basis for his hypothesis that cholera was caused by a self-reproducing organism excreted by victims of the disease and spread by contaminated water supplies. The hypothesis was brilliantly confirmed by meticulous identification of the sources from which water was supplied to the houses where cholera occurred in the 1854 epidemic. Nevertheless, another decade was to pass before the Italian physician Filippo Pacini (1812–1883) described his microscopic observations of the responsible organisms.[32] Further advances in bacteriology made the facts crystal-clear, but as late as 1912 it was not universally accepted, even by experts, that infected drinking water was the usual source of cholera.[33]

So evidence accumulated, for those who cared to look, about the 'germs' which caused disease, but it had negligible influence on practical medicine until microbes were actually demonstrated in a positive way. By the end of the century the role of microbes was widely understood, but only because of work in quite different fields. The chief exponent of the germ theory of disease was not, as might be naively expected, a medical doctor but a chemist, Louis Pasteur (1822–1895). But first, a much wider view of the emergence of natural science is necessary.

New science in revolutionary Paris

Sometimes advances in knowledge have been made in isolation. The work of Jenner, a physician in what was then remote rural England, is an example of such progress. But science grew, particularly in centres where many active minds stimulated and supported each other. The time when Jenner was making his rural experiments on vaccination was a period of intellectual activity in France. Alongside the ferment of political ideas which flourished among the injustices of the crumbling *ancien regime*, scientific facts and ideas were being developed by deliberate experimental investigations. We note, particularly, Antoine Lavoisier (1743–1794), often called the father of modern chemistry, whose achievements ranged from geology to investigating the processes of life, and who saw the living body as, at least partly, an elegant piece of chemical machinery. His study of the production of heat by animals and its connection with breathing was a fundamental step towards understanding living processes and their conformity with physical laws.[33]

The questioning spirit of the times led to the revolution, to the appalling sequel of the terror, and to the restoration of civil stability with the rise of the Napoleonic regime. Lavoisier died under the guillotine as retribution for his activities in maintaining the revenue of the state. But in spite of all the hazards of civil turmoil, science and medicine developed rapidly.[35] The new scepticism was expressed in an essay *Du degré de certitude de la médecine* written in 1788 by a rising Parisian physician–philosopher–politician Pierre Cabanis (1757–1808). He joined those who were critical of theories in medicine and doubted the efficacy of any treatment: nevertheless, he noted that if physicians did little good, or even some harm, they prevented much evil because they displaced grossly incompetent and ignorant practitioners who undoubtedly did harm. His near contemporary Xavier Bichat (1771–1802) gave practical expression to the spirit of enquiry in his experiments in physiology and in post-mortem dissections to discover the immediate causes of death. The importance of observation, measurement, and counting received its strongest support from Pierre Charles Alexandre Louis (1787–1872), sometimes, not uniquely, described as the 'first modern clinician, who made bedside medicine a science as well as an art'. he conducted a trial of the merits of bleeding which undermined the classical belief in its efficacy and contributed to the decline of the practice. Louis's writing on 'the numerical method' was greatly welcomed by his followers, but gradually became overlooked and had to be re-discovered from time to time.[36,37]

In the general reorganization of the revolutionary period medicine was further influenced by the emergence of the profession of pharmacy, distinct from apothecaries, botanists, and chemists, but combining the skills of all three in advancing the knowledge of drugs. At the same time, the value of laboratory experiments on animals was recognized in a more detailed analysis of the mechanisms of health and disease. Bichat's successor, Francois Magendie (1783–1855), is often regarded as the founder of modern experimental physiology. He is most famous for work identifying which nerves take messages into the spinal cord and brain and which ones lead away from the central nervous system to the muscles which carry out voluntary movements. Magendie was perhaps the first man to recognize that poisons were invaluable tools in the armoury of a physiological investigator. He also, clear-sightedly, saw that such tools were most reliable when they were purest. The collaboration of an outstanding pharmacist, Pierre Joseph Pelletier (1788–1842), led to the study of plants with a reputation as strong poisons. From the roots of the South American plant ipecacuanha they isolated a substance which caused vomiting, and named it emetine, and from the seeds of the small Indian tree *Strychnos nux-vomica* a convulsant poison, strychnine, which became a cornerstone in Magendie's analysis of reflex activity of the spinal cord.[38]

Pelletier and his colleague Joseph Bienaimé Caventou (1793–1877) also developed the discovery of 'Morphium', or morphine, a substance was first isolated from opium by an obscure Westphalian apothecary's assistant, Friedrich Sertürner (1783–1841), who precipitated the material with ammonia from an acid extract of the raw material. He showed by experiments on himself and his friends that the material which he had isolated was highly active as a sedative or narcotic. His material was much more potent than anything which had previously been isolated from opium and he unwittingly took an excessive dose in one of his experiments. He was fortunate to survive. He is generally credited with first making known the existence of drugs which, like morphine, formed salts with acids.[39] Minerals with the same salt-forming property were already called alkalies, so the vegetable substances became known as alkaloids.[40]

Pelletier and Caventou adapted Sertürner's methods and, between 1817 and 1821, isolated a remarkable range of alkaloids. They included a close relative of strychnine, which was named brucine; the febrifuge and anti-malarial quinine; caffeine from coffee beans; and veratrine from certain hellebores, which has complex physiological effects and limited medicinal use. Isolation of the active principles from any therapeutically-active material became an established basic procedure. It required collaboration between chemist and physiologist in order to show that the substances had the same actions as the parent material. This required trials in man, such as the one that Sertürner had rashly conducted, or, less hazardously, experiments on animals, in which some anlaysis of the mode and site of action was also possible.[41]

Anaesthetics had not yet been introduced into medicine or into physiological experimentation, and Magendie acquired an unwelcome reputation for being indifferent to the pain which was inflicted on animals in some experiments. At a time when surgery on humans was performed with no more relief of pain than could be afforded by alcohol and opium, Magendie's experiments did not appear so shocking. The popular and quite incorrect notion that experiments on animals were *necessarily* painful, or that experimenters enjoyed inflicting pain, perhaps had its origins at this time. As will become apparent, these misconceptions have often delayed the discovery of drugs and their use for human benefit.[42]

The study of poisons was extended by the Majorca-born physician Joseph Orfila (1787–1853), who came to Paris and published a famous textbook, his *Traité des Poisons*[43] in 1814. He does not appear to have been a great experimenalist himself, but his encyclopaedic work was an obvious source-book for scientists as well as for physicians and lawyers. From a scientific point of view there is little to distinguish poisons from potent drugs, except the matter of dosage. The investigation of poisons, or toxicology, developed, as part of experimental pharmacology, inseparably with experimental physiology. Magendie's pupil and successor to the chair of physiology at the Institut de Paris, Claude Bernard (1813–1878), made some of the earliest experiments which showed that drugs act at specific, identifiable bodily sites and not in some diffuse way all over the system. His studies on the South American arrow poison curare are classical and many later discoveries are based on them. Bernard showed that the paralysis caused by curare was not the result of poisoning muscles, because muscles still responded to 'galvanic' or electrical stimulation in frogs paralysed with curare. But curare did not act on the leg muscles if it was prevented from reaching them by a ligature tied round the leg. So it must act somewhere between the nerves, which it could still reach, and the muscle, i.e. at the point between nerve and muscle, on the specialized tissue called the neuromuscular junction. This specific identification of the site of action of a drug, and simultaneous recognition that the site was one with special physiological properties which were worth further study, was profoundly important in the evolution of the new sciences of physiology and pharmacology.[44,45]

Modes of ascertaining the effects of medicine

Innovation was in the air, but few people had much idea of how it should be done or what it would lead to. We can feel the flavour of the time by looking at a contemporary textbook *Elements of material medica and therapeutics*, published in 1839–40 for Jonathan Pereira (1804–1853).[46] Its author[47] was a physician at the London Hospital in Whitechapel and professor of chemistry at the Royal College of Surgeons. He was also one of the first teachers at the School of Pharmacy, which had just been established by the newly-formed Pharmaceutical Society, now the Royal Pharmaceutical Society of Great Britain.[48-50] Most of Pereira's book deals with the remedies then in general use, but six of its 1800 pages are titled *Modes of ascertaining the effects of medicine*. The modes he recognized were:

1. The sensible qualities of medicines.
2. Natural historical properties.
3. The chemical properties.
4. The dynamical properties.

Today, these terms seem a little strange. By sensible qualities, Pereira referred to the colour, taste, and odour of the medicine, and observed that it was a waste of time to dwell on this subject. The old doctrine of signs or signatures was 'out'. One did not decide that lungwort was good for lung diseases because its leaves were patterned so that they looked somewhat like lungs, though of a most unnatural colour, nor that red rusty particles and red wine were, because of their colour, good for relieving the bloodless pallor of anaemia. At last ideas were progressing beyond the simple assumption that the world had been made for human convenience and that all could be solved by casual observation and naive reasoning.

Under 'natural historical properties', Pereira noted that the medicinal powers of minerals couild not be deduced from their crystalline form and structure. For two pages he discussed the natural orders of plants and the similarity or otherwise of medicinal properties of plants in the same natural order. Much progress had been made in the classification of plants and it must have seemed reasonable to Pereira to hope that such classification, based on botanical characteristics, would bear some relationship to their therapeutic properties. If particular plants had particular active principles, surely they would follow the same orderly principles that were recognized in plant structure. The isolation of numerous alkaloids from plants was quite recent (Table 1.2) and their exact chemical identity was still shrouded in mystery, as his next section showed, but one feels that our modern ideas of plant chemistry and plant evolution would have appealed greatly to him. The section on medicines of animal origin is very limited, though there is a curious mention of the toxicity of bear's liver, a sign of the growth at that time of arctic exploration and the fur-trapping trade. The function of glands such as the thyroid had not yet been conceived and the therapeutic powers of extracts of these glands were unknown.

With reference to chemical properties, Pereira wrote that sulphuric, nitric, and hydrochloric acids acted very much alike, as did potash and

Table 1.2. Discoveries about some alkaloids

Source	Alkaloid	Isolated	Uses
Opium	Narcotine	Derosne 1804 (a)	Relief of cough
Opium	Morphine	Sertürner 1806	Relief of pain: addictive
Ipecacuanha	Emetine	Pelletier & Magendie 1817	Treatment of amoebic dysentery
St. Ignatius's beans	Strychnine	Pelletier & Caventou 1818	Tonic (obsolete): (b)
Autumn crocus	Colchicine	Pelletier & Caventou 1819	Relief of gout
Jesuit's bark	Quinine	Pelletier & Caventou 1820	Treatment of malaria: (c)
Coffee beans	Caffeine	Robiquet 1821 Runge 1821	Mental stimulant
Hemlock	Coniine	Geiger 1831	(d)
Opium	Codeine	Robiquet 1832	Relief of cough
Nightshade	Atropine	Geiger & Hess 1833	Reduction of secretions etc.
Opium	Papaverine	Merck, 1850	Relief of spasm of involuntary muscles
Coca leaves	Cocaine	Niemann 1860	Local anaesthetic
Calabar beans	Physostigmine	Jobst & Hess 1873	Treatment of glaucoma

(a) Derosne's material, a minor constituent of opium, was identified some years later. (b) Strychnine has important uses in the physiological analysis of spinal reflexes. (c) Quinine was synthesized for the first time in 1944. (d) Coniine was the first alkaloid to be synthesized, in 1886. Its use as a relaxant of muscles is limited and obsolete.

For details see:

Fluckiger, F. A. and Hanbury, D. (1879). *Pharmacographia. A history of the principal drugs of vegetable origin met with in Great Britain and British India.* Macmillan, London.

Manske, R. H. F. and Holmes, H. L., eds (1950–). *The alkaloids: chemistry and physiology.* Academic Press, New York. (35 volumes published by October 1989; edited from vol. 21 by A. Brossi).

soda, all being corrosive poisons. On the other hand, the alkaloids quina and morphia (quinine and morphine) appeared to be similar chemically, but medically quite different. The state of chemical knowledge at the time meant that complex organic compounds could not be distinguished from each other easily; advances in this field were essential for all progress, as we shall discuss in the next chapter.

By 'dynamical properties' Pereira meant the effects caused by giving medicines to animals, effects which we would now call physiological or pharmacological. He noted that not much could be found out by applying medicines to dead animals and that the study of living animals was much more valuable and important. He recognized the problem of differences between species, though he discussed only the nervous system, the structure of the digestive organs, and the skin. The heart and blood vessels were not mentioned. Finally, he referred briefly to the study of the effects of medicines on humans.

So, although at this time animal experiments and clinical research were in their infancy, Pereira, at least, was in no doubt that they were essential to progress.

A science of drugs

New editions of Pereira's textbook were soon needed, and a fourth, edited by colleagues, appeared soon after his death at the early age of 49. The book was translated into German by an able young doctor, Rudolf Buchheim (1820–1879), who had recently graduated from Leipzig. In preparing the translation, Buchheim learnt much pharmacology and acquired a reputation for his knowledge so that he was invited to teach the subject at the University of Dorpat (now Tartu) in Estonia. Dorpat was an old university in an outlying part of the Russian empire, but it had strong German connections and, for a time, attracted many outstanding individuals in science and medicine. Buchheim's appointment was as 'ausserordentlicher Professor', but after 2 years he became 'ordentlicher' or established professor of pharmacology, so filling the first chair established under that name in Europe and probably in the world. At first he had no laboratory and studied in his own home the effects of the familiar drugs of the time—metals, purgatives, alkaloids, and alcohol among others—until he succeeded in having a pharmacological institute built where he continued research and trained younger men who became eminent in the subject. Buchheim remained at Dorpat for nearly 20 years before returning to Giessen in Germany to further his children's education. Again he planned new laboratories, but his later years were spoilt by ill health and he did not live to work in them.[51,52]

Buchheim made no great discovery, but did much to establish the science as a distinct discipline. One of his pupils, Oswald Schmiedeberg (1838–1921), succeeded him at Dorpat and then moved to Strassburg in 1872 to set up a new institute. Germany had just won the Franco–Prussian war; Alsace and Lorraine were annexed, and the formerly French city of Strasbourg, capital of Alsace, was made a showplace of German culture.[53] The University received much support and Schmiedeberg's laboratory became a great centre for the development of experimental pharmacology. Here Schmiedeberg, with the clinician Bernhard Naunyn (1839–1925), founded the earliest journal specifically on the subject, the *Archiv für experimentelle Pathologie und Pharmakologie*. In his laboratory, the medicines then in common use were investigated to discover how they acted on bodily tissues and their supposed therapeutic benefits were related to observable physiological facts. Young investigators came to be trained in the new methods of enquiry and left to create departments of pharmacology both in Germany and abroad. By the time Schmiedeberg retired in 1918 over 100 pupils had been trained in his laboratory.[54]

The science of pharmacology also took strong roots in Edinburgh, where there was a long tradition of teaching materia medica. Gradually, as the chemistry of drugs became better understood, the attachment to botany diminished, and the ways in which medicines acted were studied rather than their origins. Toxicology was included with pharmacology, although it was also associated with the teaching of forensic medicine.[55] The long political association of Scotland with France led academically minded young doctors from Edinburgh to go to Paris and see for themselves the exciting scientific advances which were being made there. Among the travellers was (Sir) Robert Christison (1797–1882), who was appointed to the chair of medical jurisprudence at the age of 22. He studied in Paris under Magendie, Pelletier, Caventou, and Orfila, and went back to Edinburgh eventually to take the chair of materia medica. Like Orfila, Christison wrote a textbook on poisons.[56] He also conducted experiments not only on frogs and rabbits, but also on himself. He wrote an absorbing account of his investigation of the bean-like seeds of *Physostigma venenosum*, a tropical plant from Calabar in West Africa. The beans were used by natives as an ordeal poison administered to persons accused of witchcraft: if the victim survived it was taken as evidence of his innocence. Samples were forwarded by a missionary to Edinburgh, where Christison observed their effects on animals and on his own heart and blood vessels, and noted the muscular weakness or paralysis that they gradually induced. His interest in what he called 'this powerful and subtle poison' was well justified: it continued to play a key part in many investigations in the next 80 years.[57]

He was succeeded by (Sir) Thomas Fraser (1841–1920), who occupied the chair for 41 years. Buchheim was among the many names who supported his candidature for the chair. Fraser's graduation thesis was on the Calabar bean from which he isolated or alkaloid which he named eserine, after the native name of the plant. His results were not published until another account of the alkaloid had appeared in which it was given the name physostigmine, derived from the Latin name of the plant, and so physostigmine became the proper name of the drug (although it is easier to call it eserine). Fraser made the important discovery that another alkaloid, atropine, blocked some of the actions of eserine. Such interaction between two drugs had not been carefully analysed before, and later threw much light on the ways in which drugs worked. Fraser considered the chemical relationship between the two drugs, and collaborated with Alexander Crum Brown (1838–1922), who later occupied the university chair of chemistry, in studies of antagonism between drugs and the relation of chemical constitution to the action of drugs. Their papers are a landmark in the evolution of pharmacology. The vision of Paracelsus, in applying ideas of chemistry to medicine, was vindicated, but it was a very different chemistry to anything which Paracelsus conceived. We must go back two centuries to see how it evolved.

Notes

1. Singer, C. and Underwood, E. A. (1962). A short history of medicine. 2nd edn. Clarendon Press, Oxford.
2. Temkin, O. (1964). Historical aspects of drug therapy. In *Drugs in our society*, ed. P. Talalay, p. 29. The Johns Hopkins Press, Baltimore; Oxford University Press, Oxford.
3. Leake, C. D. (1975). *An historical account of pharmacology to the 20th century.* Charles C. Thomas, Springfield, Illinois.
4. Temkin, O. (1964). Historical aspects of drug therapy. In: *Drugs in our society*, ed. P. Talalay, p. 5. The Johns Hopkins Press, Baltimore; Oxford University Press, Oxford.
5. Pagel, W. (1982). *Paracelsus. An introduction to philosophical medicine in the era of the renaissance.* 2nd, rev., edn. Karger, Basel.
6. Lilienfeld, A. M. (1982). *Ceteris paribus*: the evolution of the clinical trial. *Bulletin of the History of Medicine* 56, 1–18.
7. Paget, S. (1897). *Ambroise Paré and his times 1510–1590*, p. 34. Putnam's, New York.
8. Packard, F. R. (1926). *Life and times of Ambroise Paré*, p. 163. Hoeber, New York.
9. Drummond, J. C. and Wilbraham, A. (1939). *The Englishman's food.* London, Cape.
10. Lind, J. (1753). *A treatise of the scurvy.* Sands, Murray and Cochran, Edinburgh. Reprinted, 1953, eds C. P. Stewart and D. Guthrie. Edinburgh University Press, Edinburgh.
11. Henderson Smith, A. (1919). A historical enquiry into the efficacy of lime-juice for the prevention of scurvy. *Journal of the Royal Army Medical Corps* 32, 93–116; 188–208.
12. Aikin, J. (1771). *Thoughts on hospitals*, p. 25. J. Johnson, London. Quoted by J. F. Fulton (1933). The Warrington Academy (1757–1786) and its influence upon medicine and science. *Bulletin of the Institute of the History of Medicine* 1, 50–80.
13. Nissel, M. (1987). *People count. A history of the General Register Office.* Her Majesty's Stationery Office, London.

14. Earles, M. P. (1961). Early theories of the mode of action of drugs and poisons. *Annals of Science* 17, 97–110.
15. Leake, C. D. (1975). *An historical account of pharmacology to the 20th century*, p. 27. Charles C. Thomas, Springfield, Illinois.
16. Rivett, G. (1986). *The development of the London hospital system 1823–1982*. King Edward's Hospital Fund for London.
17. Sigerist, H. E. (1943). A literary controversy over tea in eighteenth century England. *Bulletin of the History of Medicine* 13, 185–199.
18. Haggis, A. W. (1941). Fundamental errors in the early history of cinchona. *Bulletin of the History of Medicine* 10, 417–59; 568–92.
19. Stone, E. (1763). An account of the success of the bark of the willow in the cure of Agues. *Philosophical Transactions of the Royal Society* 53, 195–200.
20. Aronson, J. K. (1985). *An account of the foxglove and its medical uses 1785–1985.* Oxford University Press, Oxford.
21. Cushny, A. R. (1925). *The action and uses in medicine of digitalis and its allies.* Longmans Green, London.
22. Turner, P. (1980). Clinical trials of homeopathic remedies. *British Journal of Clinical Pharmacology* 9, 441.
23. Hopkins, D. R. (1983). *Princes and peasants. Smallpox in history.* University of Chicago Press, Chicago.
24. Porter, R. and Porter, D. (1988). The politics of prevention: antivaccinationism and public health in nineteenth century England. *Medical History*, 32, 231–52.
25. Hopkins, note 23, pp. 301–10.
26. Williamson, R. (1955). The germ theory of disease. Neglected precursors of Louis Pasteur. *Annals of Science* 11, 44–57.
27. Major, R. H. (1944). Agostino Bassi and the parasitic theory of disease. *Bulletin of the History of Medicine* 16, 97–107.
28. Lesky, E. (1979). *The Vienna medical school of the 19th century.* Johns Hopkins University Press, Baltimore.
29. Nuland, S. B. (1979). The enigma of Semmelweiss—an interpretation. *Journal of the History of Medicine* 34, 255–72.
30. Carter, K. C. (1985). Ignaz Semmelweiss, Carl Mayrhofer, and the rise of germ theory. *Medical History* 29, 33–53.
31. Ellis, R. H. (1989). Dr. John Snow. His London residences and the site for a commemorative plaque in London. In *The History of Anaesthesia*, International Congress and Symposium Series, No. 134, eds R. S. Atkinson and T. B. Boulton, pp. 1–7. Royal Society of Medicine Services.
32. Howard-Jones, N. (1972). Choleranomalies; the unhistory of medicine as exemplified by cholera. *Perspectives in Biology and Medicine* 15, 422–33.
33. Howard-Jones, N. (1973). Gelsenkirchen typhoid epidemic of 1901. Robert Koch and the dead hand of Max von Pettenkofer. *British Medical Journal* i, 103–5.
34. McKie, D. (1952). *Antoine Lavoisier, scientist, economist, social reformer.* Constable, London.
35. Lesch, J. E. (1984). *Science and medicine in France. The emergence of experimental physiology, 1790–1855.* Harvard University Press, Cambridge, Mass.
36. E.B. (reviewer) (1836). Researches on the effects of blood-letting in some inflammatory diseases, and on the influence of tartarized antimony and

vesication in pneumonitis. By P. C. A. Louis. Translated by C. G. Putnam, M.D. With preface and appendix by James Jackson, M.D., physician of the Massachusetts General Hospital. 1 vol., p. 171. Boston, 1866. *American Journal of the Medical Sciences* 18, 102–11.

37. Bollet, A. J. (1973). Pierre Louis: the numerical method and the foundation of quantitative medicine. *American Journal of the Medical Sciences* 266, 92–101.

38. Olmstead, J. M. D. (1944). *François Magendie. Pioneer in experimental physiology and scientific medicine in XIX century France.* Schuman's, New York.

39. Hanzlik, P. J. (1929). 125th anniversary of the discovery of morphine by Sertürner. *Journal of the American Pharmaceutical Association* 18, 375–84.

40. Henry, T. A. (1913). *The Plant Alkaloids.* Churchill, London.

41. Lesch, J. E. (1984). *Science and medicine in France. The emergence of experimental physiology, 1790–1855*, p. 125 f. Harvard University Press, Cambridge, Mass.

42. Paton, W. (1984). *Man and Mouse. Animals in Medical Research.* Oxford University Press, Oxford.

43. Orfila, M. P. (1814). *Traité des poisons tirés des regens mineral, végétal et animal, ou toxicologie generale.* Crochard, Paris.

44. Olmstead, J. M. D. (1938). *Claude Bernard, Physiologist.* Harper and Brothers, New York.

45. Holmes, F. L. (1974). *Claude Bernard and animal chemistry.* Harvard University Press, Cambridge, Mass.

46. Pereira, J. (1839–40). *Elements of Materia Medica and Therapeutics.* Longman, London.

47. Shellard, E. J. (1981). The life and work of Jonathan Pereira 1804–1853. *Pharmaceutical Journal* 227, 631–3.

48. Wallis, T. E. (1961). The Society's house and school. *Pharmaceutical Journal* 186, 225–8.

49. Matthews, L. G. (1962). *History of Pharmacy in Britain*, pp. 112–141. Livingstone, Edinburgh.

50. Holloway, S. W. F. (1987). The orthodox fringe: the origins of the Pharmaceutical Society of Great Britain. In *Medical fringe and medical orthodoxy 1750–1850*, eds W. F. Bynum and R. Porter, pp. 129–57. Croom Helm, London.

51. Kuschinsky, G. (1968). The influence of Dorpat on the emergence of pharmacology as a distinct discipline. *Journal of the History of Medicine* 23, 258–71.

52. Habermann, E. R. (1974). Rudolf Buchheim and the beginning of pharmacology as a science. *Annual Review of Pharmacology* 14, 1–8.

53. Alsace and Lorraine were restored to France in 1919 and Strassburg again became Strasbourg.

54. Koch-Weser, J. and Schechter, P. J. (1978). Schmiedeberg in Strassburg 1872–1918: the making of modern pharmacology. *Life Sciences* 22, 1361–72.

55. Comrie, J. D. (1932). *History of Scottish medicine.* Ballière, Tindall and Cox, London.

56. Christison, R. (1829). *A treatise on poisons in relation to medical jurisprudence, physiology and the practice of physic.* Adam Black, Edinburgh.

57. Gaddum, J. H. (1942). The development of materia medica in Edinburgh. *Edinburgh Medical Journal* 49, 721–35.

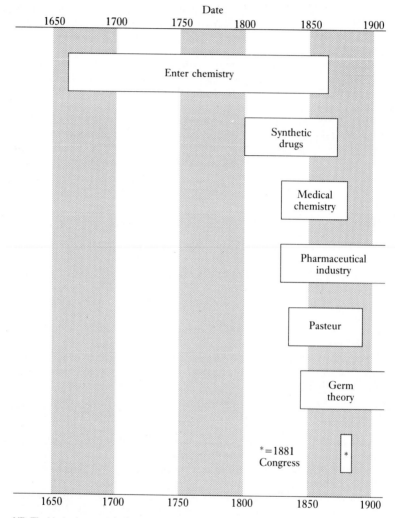

Chronology of Chapter 2

Date

NB The blocks show periods discussed in the text, and do not mean that there was no activity
about the subject at other times.

2

The chemistry of medicines

The new chemistry

A great expansion of scientific thought in the seventeenth century was marked by the founding of the Royal Society in London in 1660 and the Académie des Sciences in Paris in 1666. It involved fundamental changes in ideas about the elements from which the universe is made. The old 'principles', earth, water, and air might be equated with solids, liquids and gases, and fire with energy, but notions that all kinds of matter were obtained by compounding these principles in different ways led nowhere. Robert Boyle (1627–1691), one of the founders of the Royal Society and author of *The sceptical chymist*, was perhaps the first to make a crucial distinction between mixtures, which were of variable composition and easy to separate, and compounds, which were assumed to be of constant composition (proof came much later) and were decomposed only by specific chemical procedures. Substances which could not be resolved into any other distinct substances he called elements.[1] His definition served chemistry well, until the discovery of radioactive elements (see Chapter 11) complicated the scene.

Naturally these ideas were not immediately accepted, and the old principles of earth, water, air, and fire continued to influence chemical thought for most of another century. The supposition that matter existed as aggregates of unitary particles or atoms was gradually adapted to fit facts provided by new experiments. This was not difficult for elements, but more complicated theories were necessary to deal with substances made up from more than one element. Clearly the smallest particle of a 'compound' must contain more than one atom, and so could not itself be an atom. Today we call the smallest particle of a compound a 'molecule'. The difficulties were evident in the terminologies of the time. The English natural philosopher John Dalton (1766–1844) used the term 'compound atoms' for what we now call molecules, while the Italian physicist Amedeo Avogadro (1776–1856) wrote of 'solitary elementary molecules', meaning atoms. But, while their ideas developed, neither was consistent in his usage.[2,3]

Such difficulties did not deter chemists from tackling problems of great complexity, especially those related to living organisms. The idea that life depended on the presence of some vital substance or vital force was widespread and much research was pursued in the hope of finding it. Substances were isolated and purified, and found to contain carbon, hydrogen, oxygen, often nitrogen, and sometimes other elements, but an element of life was elusive and illusory. Thus the science of organic chemistry, the chemistry of carbon compounds, grew up. The belief that these compounds possessed a vital force hampered progress, because it suggested that they could not be synthesized in laboratories, and so an important method of investigation was neglected.

However, there are always iconoclasts. The substance urea, the chief nitrogenous constituent of urine, was undeniably 'organic', because all urine came from animals. In 1828, the young German chemist Friedrich Wöhler (1800–1882) prepared urea from the inorganic materials ammonium sulphate and potassium cyanate, 'without' (as he triumphantly reported) 'requiring a kidney of an animal, either man or dog'. His former teacher, the great Swedish chemist Jons Jakob Berzelius (1779–1848), remained unconvinced, but further transformations and syntheses were achieved until the fact that a substance had been synthesized by living tissues ceased to have any philosophical importance. The concept of vitalism at a chemical level dwindled gradually as older chemists in whose minds the notion was deeply established died or faded away.[4,5]

As more and more substances were isolated from plant and animal sources it became clear that they all contained the elements carbon, hydrogen, and oxygen, often nitrogen, and sometimes phosphorus, sulphur, or other elements. With improved methods of analysis, it was possible to find out what weights of each element were combined in a given weight of the compound. But in order to calculate the number of atoms of each kind in a molecule of the compound it was necessary to know the relative weights of the atoms of different elements. Then one could state its exact composition and start to form notions of how the atoms might be arranged in each molecule.

Basic advances in organic chemistry were far from complete when a young English chemist, William Henry Perkin (1838–1907), inspired by his teacher August von Hofmann (1818–1892) at the newly founded Royal College of Chemistry in London, attempted the synthesis of quinine. Quinine was isolated from cinchona bark by Pelletier and Caventou in 1820, and its composition was eventually established as $C_{20}H_{24}N_2O_2$. It was in much demand for the treatment of malaria and other fevers, and supplies, which at that time came only from mountainous parts of South America, were very limited. Much benefit might accrue if the drug could be made synthetically in a laboratory or factory.

Perkin attempted the synthesis by oxidizing a compound available to him, allyltoluidine. The procedure seemed rational on a simple count of atoms in the molecules and looked all right on paper.[6]

$$2(C_{10}H_{13}N) + 3O = C_{20}H_{24}N_2O_2 + H_2O$$
allyltoluidine oxygen   quinine   water

Structural knowledge, acquired later, shows that this attempt was bound to fail, because the atoms are put together quite differently in allyltoluidine. There is no way by which one compound could be converted directly to the other. Perkin obtained a dirty looking precipitate, which was evidently not quinine. However, it inspired him to related experiments, from which in due course he isolated the first synthetic dye, mauveine or aniline purple. With astonishing enterprise, he not only found support to undertake the commercial development of this dye, but virtually founded the synthetic dyestuff industry.

The synthesis of quinine had come to nothing, but the development of new dyes on an industrial scale had widespread consequences both in increasing the amenities of life and in more technical processes such as photography. New dyes found a special application in microscopy, because they could be used to stain specimens and reveal structures that would otherwise be invisible. The fact that a particular structure was selectively stained was practically convenient. Also it raised the question in some alert minds, 'Why does the dye react with this structure and not with others?' Profound consequences followed from the pursuit of this idea (see Chapter 3).

Parts of the dyestuff industry widened its interests to other fine chemicals. Some firms, particularly in Germany, began to embark on the production of drugs, either by purifying them from natural sources or by synthesis. Without the existence of a major and capable chemical manufacturing industry, pharmaceutical development would have been very different and much more meagre. So Perkin's therapeutic aims were, in a very roundabout way, achieved, and humanity reaped the benefit on a scale far beyond what the synthesis of quinine alone could have offered.

Medicines from new chemicals

With the rise of synthetic chemistry substances never previously known came into existence. There was no reason to expect them to have any effects on man, but some, especially if they were volatile, had obvious and even drastic effects. Just before 1800, Humphry Davy (1778–1829) discovered by personal experience the stupefiant power of nitrous oxide and suggested that it might be used to relieve pain in surgical operations. The idea was too strange to be taken up. The stupefying effects of ether were described in 1814 by Orfila in his classic text on toxicology,[7] and occasionally promising experiments in anaesthesia were carried out with nitrous oxide or with ether during the early part of the nineteenth century. Nitrous oxide became known as 'laughing gas', because of its curious effects, and it was used by entertainers for parties and stage shows. In Connecticut, a practising dentist, Horace Wells (1815–1848), observed during such an entertainment that a subject who received an accidental injury noticed no pain. In 1844 he began to employ the gas in his dental practice. With the primitive equipment available nitrous oxide was difficult to use, and it was not adopted widely for another 20 years. But the climate of opinion was being changed by such attempts. William Morton (1819–1868), a former associate of Wells, experimented with ether and in October 1846 gave a successful demonstration of its use at the Massachusetts General Hospital in Boston. The results were sufficiently dramatic and the audience sufficiently distinguished for this event to be very influential. Oliver Wendell Holmes (1809–1894), then about to become professor of anatomy and physiology in the medical school of Harvard University, coined the words 'anaesthesia' and 'anaesthetic' to describe what he had seen.[8-10] News of the event spread and ether was administered for the first time in Britain to a patient in University College Hospital, London, in December 1846.[11]

Another novel volatile substance, chloroform, was used on animals by Pierre Flourens (1794–1867), a colleague of Magendie, in 1847, and after self-experimentation as an anaesthetic for man by the adventurous (Sir) James Young Simpson (1811–1870) in Edinburgh in the same year.[12] The process of anaesthesia, like many innovations, was roundly condemned as flying in the face of nature and opposing the will of God. However, the benefits were more unmistakable than the available estimates of the will of God. After Queen Victoria had been given chloroform during childbirth[13] in 1853 and 1857, the use of nitrous oxide, ether, and chloroform extended widely and led to discoveries of safer and more effective agents.[9]

Other volatile substances had quite different effects. Antoine Jerome Balard (1802–1876) of Montpellier, who had already become famous by discovering the element bromine in sea water, came across an aromatic liquid in 1844 during his investigation of materials which spoilt the odour of eau de vie de marc, the brandy distilled from pressed grape pulp after wine has been made. He identified it as amyl nitrite and noted that it caused violent headaches, but does not appear to have pursued its physiological properties.[14] The English chemist Frederick Guthrie, who also studied the compound, wrote:

One of the most prominent of its properties is the singular effect of its vapour, when inhaled, upon the action of the heart. If a piece of bibulous (blotting) paper, moistened with two drops of the nitrite of amyl, be held to the nostrils, through which the breath is exclusively drawn, after the lapse of about fifty seconds, a sudden throbbing of the arteries is felt, immediately followed by flushing of the neck, temples and forehead, and an acceleration in the action of the heart. These symptoms last for about a minute and then cease as suddenly as they began.[15]

Guthrie described also some experiments on rabbits, and in a footnote suggested that Balard's material had not been freed from hydrocyanic acid, to which he attributed Balard's headaches.

From a chemist's point of view, this meant that it was better to avoid inhaling amyl nitrite. From a medical point of view, it raised all sorts of questions about how the effects were produced. In the next 5 years it was investigated by Sir Benjamin Ward Richardson (1828–1896), who described its properties to the British Association for the Advancement of Science at its annual meeting in 1864. He noted its ability to preserve flowers and dead animal tissues, reported on its effects when applied to the skin (tingling and flushing, but less than with chloroform or turpentine), its action by inhalation in man, and its effects by mouth and after injection into frogs, rabbits, and cats. Physiology was not sufficiently advanced to give him much help in understanding how the substance produced its effects. He discussed possible uses, but concluded, 'I have been too closely and intently occupied in the task of obtaining elementary facts to divert time to the practical elucidation of this important point'.[16]

It remained for another London physician, Sir Thomas Lauder Brunton (1844–1916), to study the flushing which was regularly observed after amyl nitrite was inhaled. He deduced that small blood vessels in the skin were being dilated and guessed that the blood vessels to the heart might also be dilated and if so, that it might relieve the pain of angina pectoris. Brunton tried treating patients and achieved considerable clinical benefit. His report is often quoted as establishing amyl nitrite as a valuable palliative medicine.[17]

The physiological properties of a related substance, glyceryl trinitrate or nitroglycerine, were recognized independently, not because the substance was volatile but because the chemist who made it, Ascanio Sobrero (1812–1888) of Turin, was adventurous enough to taste his new substance. He reported his observations on the oily liquid, in 1847

sa saveur est douce, piquante, aromatique. Il faut toutefois être sur ses gardes en faisant cet essai, car il suffit d'en tenir une tres petite quantité (ce qu'on peut en prendre en y mouillant legèrement le bout du petit doigt) sur la langue pour en éprouver une migraine assez forte pendant plusieurs heures.[18]

(Its taste is sweet, pungent, aromatic. One must always be careful in trying this, because a very small quantity (such as one gets in lightly wetting the end of the little finger) on the tongue can bring on a powerful migraine lasting for several hours.)

It does not appear that he made further studies of its physiological properties.

The substance attracted the attention of American homeopaths, in whose hands there was little risk of overdosage, and it acquired the name, later discarded, of 'glonoin' or 'glonoine'.

In due course it reached England. A. G. Field, a surgeon resident in Brighton, wrote:

On the evening of 2nd February, 1858, I was conversing with a homeopathic practitioner, when he mentioned a medicine which possessed peculiar and extraordinary qualities, some of which he described as having affected himself, though he had taken it in very minute quantities. I laughed at his credulity, and offered to take as much as he pleased, upon which he let two drops of what he called the first dilution of glonoine fall on my tongue. . . . In about three minutes I experienced a sensation of fulness in both sides of the neck, to this succeeded nausea, and I said 'I shall be sick'. The next sensation of which I was conscious was, as if some of the same fluid was being pushed down my throat and then succeeded a few moments of uncertainty as to where I was . . . My intellects returned, however, almost immediately, and I remember saying, 'This has nothing to do with homeopathy, but it has to do with a very powerful poison; there are more things in heaven and earth than are dreamt of in the philosophy of some of us'.[19]

Field interpreted his experiences as direct sedation of the nervous system (though now they would be regarded as the result of disturbances of blood flow). He experimented on cats, rabbits, and other animals, and, being disappointed by the results, studied its effects on some patients. His observations prompted a brisk correspondence in the journal in which he had published his results, and brought forth some comments, interesting for the period, on the quality control of the material and uncertainty of dosage. Field modestly replied, 'It affords me great satisfaction to find the action of nitroglycerine has engaged the attention of those who are so much more capable than I am of doing justice to the subject; and I hope that we shall soon be furnished with more precise information on this curious medicine'.[20] In fact 20 years passed before W. Murrell 'did justice to the subject', and introduced the drug as a longer acting alternative to amyl nitrite.[21]

Sometimes new substances were tried as drugs because they were related chemically to familiar medicines. The element bromine, Balard's first discovery, was evidently related to iodine. Salts of iodine were highly respected for a variety of purposes, including the treatment of syphilis, so salts of bromine were tried against this disease and various other conditions including eczema, scrofula, and enlargement of the liver. However, bromides proved to have quite different properties and uses. A distinguished London physician-accoucheur, Sir Charles Locock, who had attended Queen Victoria at the first confinement in which chloroform was used, and was evidently not averse to innovation, noted in 1857 an account 'some years past, in the *British and Foreign Review*' of experiments by 'a German, on himself' who found that ten grains (about 640 mg) of potassium bromide three times a day for 14 days caused temporary impotence. (What else it may have caused is not mentioned). Locock found the drug 'of the greatest service' in hysteria in young women, and in epilepsy.[22,23] As epilepsy was then seen as a consequence of onanism or masturbation, this observation was pleasingly rational and a gratifying confirmation of a strongly held theory. It was not accompanied by any data more substantial than authoritative assertion.

T. S. Clouston, then physician to a Westmorland asylum, gave admirably objective evidence of the efficacy of potassium bromide in reducing the number of fits in epileptics. His paper shows a critical attitude rare at the time and an interesting display of the state-of-the-art of evaluation before the rise of statistically sound methods for the design and analysis of experiments.[24] Bromides were widely used for many years also to promote sleep, for which they were effective only in prolonged and near toxic doses, and they were reputed to be used by army authorities to reduce the sexual activities of their troops. But the original idea that bromides might imitate or be an improvement on iodides found no support. They belong to the large class of remedies which turned out to be useful for quite other purposes than they were intended. However, they had many disadvantages and better agents superseded them. In 1941, when the now celebrated text by Louis Goodman and Alfred Gilman first appeared, bromides occupied more than ten of the 1300 pages.[25] In recent editions they receive no space at all.

New light on medicines

The empirical discoveries of anaesthetics, of drugs which dilated blood vessels, and of the unforeseen properties of bromides, all raised questions about how drugs worked and gave little clue to the answers. Not enough was known about physiology and about the chemistry of the body. More progress was made at this time by investigating the chemistry of natural products, especially the alkaloids[26] which Pelletier and others had isolated so successfully. Once their chemical constitution was known they became invaluable models with specific bodily actions, which could be imitated and sometimes simplified to provide useful alternative drugs. But the alkaloids were complex substances, mostly containing at least 30 atoms in each molecule and often many more. As Perkin discovered in 1856, it was not enough to know how many atoms of each kind were present and to proceed accordingly. When the structure of most of the familiar alkaloids had been worked out 20 or 30 years later, the situation was more

favourable, though quinine itself was among the most difficult. Establishment of its structure was one of the classics of organic chemistry and
depended on many workers, mostly in Germany, between 1880 and 1907.
Quinine was not synthesized until 1944, and its production on a
manufacturing scale has never been practicable.[27]

However, much could be done in elucidating the chemical properties of
alkaloids without knowing the whole structure of their molecules. In
Edinburgh, Fraser and Crum Brown[28] investigated the antagonism
between physostigmine and atropine and compared the properties of
various alkaloids. They knew the approximate composition of these
alkaloids and were aware that almost identical formulae in terms of the
number of atoms present did not lead to similar pharmacological actions.
Crum Brown modified various alkaloids, including strychnine, codeine,
morphine, and atropine, by attaching an additional methyl (-CH$_3$) group to
the nitrogen atom which each of them contained. The addition involved a
change in the number of atoms with which the nitrogen combined,
resembling that which happens when ammonia gas reacts with water,

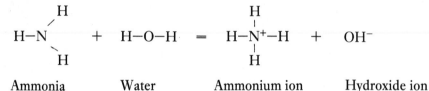

| Ammonia | Water | Ammonium ion | Hydroxide ion |

and had subtle chemical consequences, especially on the electrical charge
on the nitrogen atom. Fraser showed that it also had striking pharmacological results. The original alkaloids differed widely in their actions:
strychnine was convulsant, morphine soporific, and atropine relieved colic
(among many other actions). All the new compounds had a new action, like
that of curare, in causing paralysis. Sometimes they also lost their own
typical actions. The presence of the ammonium-like structure, later known
as an onium compound, consistently conferred this kind of action on any
drugs to which it could be introduced. Crum Brown and Fraser were less
successful than they hoped in applying their discoveries to the invention of
useful drugs. Curare and curare-like drugs did not come into medicinal use
for another 70 or 80 years. But Brown and Fraser's experiments were
fundamentally important. They were the first major demonstration of a
connection between chemical structure and pharmacological action, and
are still often quoted in scientific papers today.[29,30]

A different kind of observation was made by the physician Alexander
Ure, who used benzoic acid as a medicine and observed that after its use
the urine contained crystals of a material which he identified as hippuric
acid. Hippuric acid is made when benzoic acid combines with glycine, a

constituent of proteins widespread in bodily organs, but the reaction does not usually take place when the components are put together at body temperature. Evidently chemical combination, of a kind which did not happen spontaneously, had taken place in the living organism.[31] Ure looked for uric acid in the urine following the administration of benzoic acid, and failed to find it, so he claimed that the uric acid had been 'superseded' by hippuric acid and that benzoic acid was an effective treatment for gout. The experiment was repeated and confirmed, apart from the uric acid aspect, in Wohler's laboratory.[32] It was one of the first examples of what is now called the 'biotransformation' of drugs. Very few drugs pass through the body unchanged, and study of the chemical changes which they undergo is often essential to understanding the actions they have and the time for which they act. Commonly changes of this kind make drugs less active or inactive, and are known as detoxication. But some biotransformations have the opposite effect, with advantages if the new formed material is present in therapeutic amounts. An inactive substance which slowly releases potent material can be a very useful medicine: several instances of such 'pro-drugs' appear later. However, the potent materials released are as likely to be harmful as beneficial, depending on their actions and the quantities released. The impressive phrase 'lethal synthesis' has been applied to such dangerous biotransformation.[33,34]

A theoretically possible biotransformation was envisaged both by Buchheim at Dorpat and by Oscar Liebreich (1839–1908), working in the chemical division of the pathological laboratories in Berlin. The substance chloral hydrate had been prepared by Liebig in 1832, about the same time as chloroform had been first made. The two substances were fairly closely related. Perhaps the body would convert chloral to chloroform and release the anaesthetic gradually and so produce anaesthesia. Buchheim and Liebreich each tried the idea and found that choral hydrate did indeed render people insensible. The depth of unconsciousness resembled sleep rather than anaesthesia, but that made the substance all the more useful. Buchheim appears to have been rather intimidated by the implications, but Liebreich published an account of his discovery in 1869, and chloral hydrate became established as a useful drug. The precise hypothesis on which it was tried was at most half right: chloral does not break down to chloroform but it does undergo 'biotransformation' (as was shown in 1944), to a compound which produces sleep.[35] So chloral is an early example of a pro-drug. But the general idea was completely right, and many drugs have been introduced which act not directly but after being transformed into an active principle, often with the benefit that the active substance is made available gradually over a long period and without the dangers of a single large dose acting all at once or the inconvenience of frequent repeated administration of smaller doses.

Growth of an industry

Discovery that certain new 'chemicals', such as ether, chloroform, and amyl nitrite, were valuable drugs made it both necessary and attractive to make them on a large scale. For this task, the established suppliers of drugs were largely unequipped. Neither the individual apothecaries and pharmacists nor the merchant-importers had resources for chemical manufacturing, which indeed were largely unknown at the time. However, some pharmaceutical wholesale businesses developed the skills of applied chemistry. In the family business founded by the apothecary E. Merck of Darmstadt in 1668, H. E. Merck in 1827 organized the large scale extraction from plant sources of several recently discovered alkaloids, and began to market the purified drugs. His son, G. F. Merck, acquired up to date experience by working in the laboratory of Justus von Liebig (1803–1873), pioneer in the chemical analysis of foods and in studying the cycles of carbon and nitrogen between plants and animals. He acquired sufficient technical skill to isolate a new alkaloid, papaverine, from opium. Shortly afterwards, in 1848 he became head of the business, in which role he continued to promote applications of the new chemistry in the firm's activities.[36,37]

The greatest development of a chemical industry occurred in Germany, where academic chemists, especially Liebig, vigorously promoted industrial applications of their subject. Early in the second half of the century substantial chemical factories were being built, in order to extract essential ingredients for the synthesis of a variety of products. Dyestuffs, though small in tonnage, were particularly valuable, and required more refinement than other products. Naturally, by-products began to accumulate, for which uses were sought, with prospects that the medical field would be particularly profitable. The newly founded businesses, including Agfa, Bayer, and Hoechst, began making and selling new kinds of drug. Industrial secrecy was paramount and little is known about the steps by which the medicinal activity of particular compounds such as 'antifebrine' (acetanilide) and phenacetin was discovered, or about the circumstances in which trials were conducted. It was an age of self-experimentation, and the field was too novel to be in any way submitted to governmental regulation.[38]

Analogy with existing drugs of known or partially known composition gave some guidance. Part of the structure of quinine was being discovered, and the drug was as much in demand as when Perkin tried to synthesize it 30 years earlier. So compounds which bore some resemblance to parts of the quinine molecule, as far as it was then known, received a considerable impetus because they could be made easily from products of coal tar distillation. They relieved headaches and minor pains, and they reduced body temperature in fevered patients. In the 1870s, clinical thermometers were introduced.[39,40] Temperature charts were being introduced and the new drugs had a good start. Indeed the precipitously descending lines on the graphs of fevered or 'pyrexial' patients showed how effective the new medicines were. Consequently they became known as antipyretics (against fever) and analgesics (against pain).

These drugs included phenazone ('antipyrine'), acetanilide ('antifebrine'), phenetidin, phenacetin, and amidopyrine, most of which appeared between 1884 and 1894.[41-43] Studies on toxicity in animals were, by modern standards, almost non-existent, and several compounds were withdrawn because they caused ill-effects in too many patients. Several had very much longer lives, but more exacting standards of safety took their toll. Amidopyrine was falling into disrepute in the 1930s and was eventually superseded by the related compound phenylbutazone, introduced in 1949 and in turn coming into increasing disfavour especially in the 1980s. Whether the reduction of fever had any long-term benefit, or whether it suppressed a natural defence mechanism remained unknown.

While these 'antipyretics' were being discovered and put to use, the most familiar of all was lying unappreciated on a chemist's shelf. Aspirin or acetylsalicylic acid was synthesized by Charles Gerhardt (1816–1856) in 1853, but its medicinal properties were not recognized until 1899. Its virtues were probably suspected because of its chemical resemblance to salicylic acid, derived from willow bark (see Chapter 1) and used to relieve fever and rheumatic pains.

The lapse of nearly half a century before aspirin was recognized as a useful medicine may seem surprising. But many compounds have lain for a long time on chemists' shelves before their medicinal properties were appreciated. It is difficult, tedious, and not without risk to discover whether a substance has unpredicted medicinal properties, and the odds against any particular compound being useful are enormous. Aspirin became probably the most widely used of all synthetic medicaments in the western world, and competes for favour with paracetamol, the chief survivor of the coal tar antipyretic analgesics.[44-46]

New hypnotics appeared in this period, perhaps because sleep, like anaesthesia, was easy to recognize, both in animals and in man. The success of chloral was a stimulus to seek drugs more or less like it, and several were found. Most of the early agents have been superseded, though the barbiturates, of which the first (Veronal)[47] appeared in 1903, continue to have restricted uses. Drugs of other kinds followed, many based on finding simple analogues of plant alkaloids. The German chemical, and by this time pharmaceutical, industry continued to dominate the field until World War I, when problems of supply compelled many nations to establish their own manufacturers.[48-51]

The many achievements of Pasteur

The advance of chemistry as a practical study left untouched some deep philosophical issues. Chemists continued to investigate and imitate processes of living matter either as if they had no distinctive and valued qualities or in the hope of finding the true essence of life. But, as usual, the frontier between living and non-living was not well defined. Belief in the spontaneous generation of living organisms was widespread. Stories of phoenixes rising from their ashes or salamanders emerging from fire were no longer credible, but growing moulds appeared from nowhere on jam and maggots arrived in decaying meat. If life could arise spontaneously, how did it achieve vitality? But the idea of spontaneous generation was open to experimental test, and several investigators carried out more or less satisfactory experiments, with conflicting results. The most convincing evidence came from Louis Pasteur (1822–1895), though his research in this field was only a tithe of the astonishing range of his achievements.[52,53]

Pasteur, the son of a tanner in Arbois, near the Jura mountains, made his way with difficulty to become a student in Paris. His teachers thought well of him, and, to his delight, he obtained a place in Balard's laboratory. His work for a doctoral thesis was a valuable example of the use of physical methods in chemistry, and is worth a little attention. It involved the way in which polarized light was affected by certain organic compounds. Polarized light is formed by passing a beam of light through a prism, from which it emerges with its waves all in one plane. If such light is shone through a solution of certain organic compounds, the beam rotates, by a certain amount depending on the kind of compound, the strength of the solution, and the distance through it. A particular compound interested Pasteur because it behaved anomalously. It was called racemic acid, and its chemical reactions were identical with those of tartaric acid, obtained from the deposits of tartar in old wine casks. However, tartaric acid from wine casks rotated the beam of polarized light, but racemic acid had no effect on the plane of the beam. Why not? What was the difference between racemic and tartaric acids? Pasteur observed that the crystals of the racemic acid laboratory material were of two kinds, mirror images of each other. He separated the two kinds by hand, dissolved each of them separately, and showed that one rotated polarized light to the left and the other rotated it to the right. A mixture in equal proportions did not rotate it at all, so that was why the racemic acid appeared to have no optical activity.[54]

The two kinds of tartaric acid crystal were later shown to differ in the arrangement of their atoms, which resembled each other like a left and a right hand, or like an asymmetric object and its mirror image. Many other organic compounds were found to exist in left-handed and right-handed forms, always with the difference in their effect on polarized light and often with striking differences in their biological properties. In many drugs, of which atropine from belladonna was a notable example, the form which rotated light in one direction was frequently a much stronger drug than its mirror image. Such a surprising fact was very valuable in developing ideas about how drugs acted on the cells of the body (see Chapter 6).

The study of the optical activity of organic compounds was left to others. Pasteur moved to Strasbourg, then a French city,[55] where he married, and in 1854 he moved to Lille, as Dean of the Faculty of Sciences. His famous remark, that chance favours only the prepared mind, is said to have been made in his inaugural address. His admirable record as a conscientious administrator at Lille and later at the Ecole Normale in Paris might suggest that he had no more time for research, but the facts are very much otherwise. The industrial problems of Lille turned his work towards alcoholic fermentation and the prevention of sour wine. The fermentation of sugars (from grapes or malt) appeared to require yeast, but it was not clear why. Distinguished chemists, including Wohler and Liebig, maintained that some substance in the organisms acted as an inert catalyst, which promoted the processes of fermentation without taking part in them. If so, it did not matter whether the yeast was alive or not.

To establish whether yeast cells were necessary, experiments were essential in which they were excluded, and evidence was required that they could not generate spontaneously. Pasteur did a variety of experiments which showed that living organisms could be prevented from appearing, i.e. simply did not generate, if appropriate measures were taken to exclude them. He demonstrated convincingly that when living organisms were found, a source of infection could always be traced or at least not excluded. This was most important in studying both fermentation which produced alcohol and souring which produced vinegar: both processes were prevented if live organisms were excluded. On a philosophical plane, the evidence undermined traditional beliefs to a most disturbing extent, and controversy raged, not least because Pasteur was uncompromising and fortified by carefully conducted experiments and carefully observed facts.[56]

The germ theory of disease

From this time on, Pasteur's career was more like that of a biologist, a veterinarian, or even a medical man than of a chemist. He was invited to investigate a disease of silkworms which was rampant and ruining the silk producing regions of France. While he was acting on the invitation, he suffered a stroke and paralysis of one side, and was confined to bed more or less completely for 2 months. Within another 2 months, still partly paralysed, he was back at his work. With his growing knowledge of microscopic organisms and rejection of spontaneous generation, he was well equipped to recognize the disease as a result of microbial infection. Like Bassi, who had come to a similar conclusion earlier (see chapter 1), Pasteur saw the analogies with diseases of higher animals and man, and began to formulate what became known as the 'germ theory of disease'.

Later, he observed the role of doctors and nurses in spreading puerperal sepsis (childbed fever). He was in a position to make a stronger case than Semmelweiss had been able to do. Pasteur had seen, under the microscope, chains of microbes isolated from the blood of women dying of the disease, and similar microbes in contaminated linen and on the hands of their attendants. With a changing climate of opinion, the inescapable facts began to be generally accepted. From then on, much could be done to prevent childbirth fever. But another 60 years passed before a drug was discovered which arrested the course of the disease.

After the horrors of the Franco–Prussian War (1870–1871), Pasteur was able to study intensively both fowl cholera and 'charbon', *alias* anthrax or splenic fever of sheep, which, like the silkworm disease, caused heavy economic losses. Pasteur was encouraged and impressed by Jenner's discovery of vaccination. It showed that whatever caused immunity was not inescapably bound to whatever caused virulence. A fortunate observation came when, after a 2-week vacation, he injected an old culture of the fowl cholera bacillus into some chickens and found that it did not kill the birds. With the genius for exploiting what looked like an experiment gone wrong, Pasteur did not throw the cultures away, but injected some fresh culture into these birds and, as controls, into some birds which had not previously been exposed. The controls died, so the fresh culture was undoubtedly virulent. The previously inoculated birds survived. This was a real discovery: Pasteur had immunized his birds when he inoculated them with the organisms which had lost their virulence. In other words, he had found how to make an 'attenuated' strain of organism which could be used, like vaccinia against smallpox, to protect from serious infection, and a method, which might well work also with other organisms, of producing non-virulent but immunity-producing strains.

Discovery of the complex life cycle of the anthrax bacillus by Robert Koch[57] (1843–1910) helped Pasteur further. The organism presented special difficulties because of its capacity to form spores which were resistant to ordinary methods of sterilization and could persist with dangerous consequences. But an attenuated culture was produced by various means, and its efficacy proved in a famous experiment conducted at Pouilly-le-Fort, near Melun[58] in 1881. Virulent anthrax cultures were injected into 48 sheep, half of which had been previously vaccinated with attenuated anthrax culture. After 2 days, 22 unvaccinated sheep were dead and 2 dying. All the vaccinated sheep were alive.

In honour of Jenner (though with disregard for etymology; vaccine is derived from Latin *vacca*, a cow) Pasteur called his attenuated culture a vaccine. (The word is still used for all such preparations, although none are now associated with cows). Finally, he and his associates produced a vaccine against rabies. The achievement was all the more remarkable because rabies is caused by a virus and is not visible under the ordinary microscope, and because the vaccine was effective after the patient had been infected by the bite of a rabid dog; the long latent period of the disease allowing time for immunity to develop provided the vaccine was given early enough.

The course of Pasteur's research is astonishing in its variety, and yet consistent in its pursuit of practical ends. Trained as a chemist, a science which at that time was much more demanding of exact technique than biology, Pasteur became a magnificent experimenter and did not allow himself to be constrained by the frontiers of chemical thought at the time. He broke away from orthodoxy into a more biological approach to fermentation, and advanced the science of microbiology so far that it could be applied to provide new cures for animal and human diseases. That he was born a countryman, that he was a devout Catholic, that he had a strong sense of duty, that he worked ceaselessly in the face of disability, that he was a pugnacious exponent of the truths discovered by experiment; all these contributed to his success. Above all, he was not restricted by his own past; he prepared his mind and accepted what chance offered; and he opened new frontiers which all who were able could expand.

Lister and antiseptics

The first application of Pasteur's work to the practical treatment of disease was made as soon as his evidence against the spontaneous generation of microscopic organisms became known. Pasteur's report *Récherches sur la putréfaction* was shown to Joseph Lister (later Lord Lister) (1827–1912) while he was professor of surgery in Glasgow during the 1860s.[59] Lister, much concerned with the putrefaction of wounds in his wards, explored the idea that it was the result of contamination with microbes and devised means of keeping them out of surgical and open wounds.[60] Without techniques for detecting the presence of microbes until putrefaction had actually occurred, it was very difficult to choose the best procedures. For a long time, Lister relied on diluted carbolic acid applied to wounds, dressings, and as a spray in the operating theatre. The practice of antiseptic surgery, like most other innovations, was regarded in every way from immense enthusiasm to intense hostility, but the results were obvious and gradually convinced everyone. As usual, more credit was given to the chosen substance, carbolic acid, than to the principle of excluding germs. Lister himself was clear about the primary objective, but the search for agents which would destroy germs, 'disinfectants', was naturally attractive and developed in many ways (see Chapter 3).

Within a few years bacteriological techniques were developing rapidly, and the efficacy of germicides could be tested, crudely, in laboratories. Lister's notebooks record that in 1871 he observed the failure of growth of bacteria in a culture medium contaminated by a mould, *Penicillium glaucum*.[61,62] The mould was preserved, and cultures were used by Lister for the treatment of an abscess in 1884. Similar experiments were made several times early in the twentieth century before penicillin became a universally known drug, but Lister was one of the first persons to recognize the activity of some moulds against bacteria (see Chapter 9).

In 1876, Lister was invited to the chair of surgery at King's College in London. For his inaugural lecture, he chose to speak on the nature of fermentation. He described his work on putrefaction of blood *in vitro* and on milk, which was an easier substance to study. 'The audience gave the address a patient hearing, but they wondered what all this milk research had to do with surgery. A lecture of such striking originality, dealing with an obscure problem in bacteriology at a time when bacteria were hardly known, and delivered by a professor of surgery, was a bewildering phenomenon to the majority. Only a few realized the connection between the fermentation of milk and that chief blight of surgery, the putrefaction of wounds'.[63] Other accounts of the same event describe a more lurid picture, with jeering students and remarks such as 'Shut the door or one of Mr. Lister's microbes will come in'.[64] Whatever the exact facts, evidently it was hard to come to terms with the new ideas, and even more to recognize that advances in the treatment of patients could come from experiments done in laboratories to which no patients went, on materials which had no human origin. Many people still find it difficult to recognize that *new* ideas are likely to come from *unfamiliar* sources.

All over Europe concepts of disease were becoming broader and deeper. The practice of performing post-mortem examinations was widespread and 'morbid anatomy' was recognized as a specific subject. Medical education was more formalized and chairs in morbid or pathological anatomy were founded in several universities, including London and Paris. The most notable advances were made by Rudolf Virchow (1821–1922), first occupant of the first chair in Germany, at Würzburg, and later at Berlin. Virchow was one of the earliest pathologists to progress from naked eye pathology, to microscopic examination of diseased tissues, and he developed theories of the causes of disease based on disturbed function of specific cells in the body's tissues.[65] He founded the first journal which dealt with the subject. Later it became entitled *Virchow's Archiv* and it is still published under that name.

The enquiring spirit of Virchow and others about bodily changes in disease had a great influence on methods of treatment. The terrible destruction of internal organs seen on the autopsy table and the microscopic evidence of damage to organs which looked outwardly healthy all suggested the futility of hoping to repair them with elixirs and extracts. Therapeutic scepticism developed into therapeutic nihilism, and the emphasis grew towards methods of treatment which promoted the normal working of the body. Then as now, nothing could altogether quell the human thirst for medicine, but the advance of science was doubly discouraging to medication. Pathologists showed that ravages of disease were too profound to be amenable to simple remedies, and pharmacologists showed that medicines which had demonstrable effcts in laboratories were few and far between, and those which did were potentially poisonous and suitable only for desperate situations or very careful application.[66]

The germ theory of disease gave new hope of finding ways of controlling these ravages, so many of which resulted from the destruction of tissues by micro-organisms. Men of wider vision took microbes seriously and set out to discover all that they could about them. Koch made discoveries about anthrax with the aid of a primitive laboratory which he set up while working as a district physician in East Germany.[57] Like Pasteur, he grew microbes in liquid soups and broths. Later, with better facilities, he started to grow microbes on the surface of solid culture media, where the organisms flourished as moulds do at the top of a jam pot. It was then possible to spread a sample of infected material so sparsely over a layer of jelly that each microbe grew its own separate colony of descendents. From such colonies, pure growths of individual species could be obtained, and examined separately. Koch also made use of the newly discovered aniline dyes to stain single cells. Bacteria, a few micrometers in diameter or length, could now be seen because a more powerful microscope lens had been invented and the device of immersing the lens and the object studied in oil made higher magnification possible. These fundamental advances in technique greatly helped the identification of the causal organisms of different diseases. Further progress depended especially on the discovery of conditions under which the organisms could be grown *in vitro*, a study which in turn led to striking advances in the discovery of drugs (see Chapters 5 and 8).

Science becomes acceptable in medicine

These advances in physiology, toxicology, pharmacology, and microbiology, were part of the revolution of thought which was gathering momentum in the nineteenth century. The men who made these advances rejected traditional authorities as a source of knowledge, and relied on what they could see and measure for themselves, and what they could demonstrate to their fellows. Most of them were probably not conscious of any philosophical method underlying their actions. Some, notably among early German physiologists,[67] were explicitly anxious to clarify their philosophical position. Inevitably, the nature of 'life' was a stumbling block. Could all the phenomena of life be reduced to explanations given by physics and chemistry? Resort to philosophy did not alter what appeared to be the practical facts of life; first, that the advances of knowledge involved rejecting, or escaping from, currently held and often deeply and sincerely felt ideas, and second, that intellectual conflicts were an inevitable consequence of such escape. One may add, with hindsight, that the advances in knowledge made in this way were essential for modern advances in the treatment of disease.

By 1880 the revolution of thought was becoming familiar to the leaders of the medical profession in Europe and America. Advances both in medical ideas and in methods of transport had led to a series of international medical congresses held in different centres in Europe. The early congresses, held in Paris, Vienna, Brussels, Florence, and Amsterdam, were comparatively modest. The seventh congress was held in London in 1881.[68] The organizing committee, which included Gull (Chapter 5) and Lister, chose the distinguished surgeon Sir James Paget as President of the congress, and supported his insistence that the congress 'should be firstly and chiefly for the purposes of true scientific work and for the advancement of our profession'.[69] Moreover, the fashionable world was to see what great developments were taking place. Distinguished guests included the Prince of Wales, the Crown Prince of Germany, and the Archbishop of York. This congress marks a watershed in the acceptance of scientific contribution to medicine. Doctors could no longer assert knowledge only on the grounds of their experiences and eminence; the evidence of proper experiments must be studied, however much they conflicted with the authorities of the day, and the eminent doctors wanted to be seen to be listening to those who had made the experiments.

A great deal of organizing skill went into the congress, and it attracted over 3000 participants. Among them were such outstanding innovators as Pasteur, Virchow, Koch, Lister, and Huxley. The major scientific developments were widely debated and reported. Pasteur spoke on 'vaccination in relation to chicken cholera and splenic fever'. Koch demonstrated his method of solid culture media for the isolation of pure bacterial colonies, and, in spite of the lingering antagonisms of the Franco–Prussian conflict, received from Pasteur the famous praise '*C'est un grand progrès, Monsieur*'. Virchow spoke formidably on the value of science and on the inescapability of experiments on animals if advances were to be made. 'None of those who attack vivisection as an aid to science have any conception of the true importance of science, nor of the value of this means of acquiring knowledge'.[70] His remarks came at a critical time, because in England feelings had run high in the previous decade about vivisection, and had led to an act of parliament, the Cruelty to Animals Act of 1876, which made experiments on animals illegal except by licensed persons in licensed places.[71,72] A statement on the subject by a distinguished foreign scientist who also had much political experience[65] was of great value. Among other matters, the theory of spontaneous generation, which (as an anonymous commentator remarked)[73] 'had become almost a thing of the past, was again revived by Dr. Bastian and M. Béchamp, but no new argument of importance was brought forward'. Pasteur, not unreasonably, was enraged that the subject should still gain attention.

Nevertheless, none of this brought forward any discovery which actually provided new means of treating patients and saving their lives. Pharmacology and materia medica were represented by Fraser (see Chapter 1), by this time Sir Thomas Fraser, of Edinburgh, but he modestly did little more than draw on Magendie and Bernard to illustrate discoveries about the way in which drugs worked. It was left to the formidable T. H. Huxley (1825–1895) in a *General address on the connexion of the biological sciences with medicine* to point out the implications.[74] He quoted some recent German and Scottish discoveries and went on:

There can surely be no ground for doubting that, sooner or later, the pharmacologist will supply the physician with the means of affecting, in any desired sense, the functions of any physiological element of the body. It will, in short, become possible to introduce into the economy a molecular mechanism which, like a very cunningly contrived torpedo, shall find its way to some particular group of living elements, and cause an explosion among them, leaving the rest untouched. The search for the explanation of diseased states in modified cell life; the discovery of the important part played by parasitic organisms in the etiology of disease; the elucidation of the action of medicaments by the methods and the data of experimental physiology—appear to me to be the greatest steps which have ever been made towards the establishment of medicine on a scientific basis. I need hardly say they could not have been made except for the advance of normal biology'.[74]

Events of the next 10 years vindicated Huxley, though it was not yet pharmacologists who brought forward the first dramatic and unequivocal life-restoring substance. The path from Huxley's cunningly contrived torpedo to Ehrlich's magic bullet lay through bacteriology and immunology, with life-saving achievements on the way.

Notes

1. Maddison, R. E. W. (1969). *The life of the honourable Robert Boyle F.R.S.*, p. 104. Taylor and Francis, London.
2. Greenaway, F. (1966). *John Dalton and the atom.* Heinemann, London.
3. Morselli, M. (1984). *Amedeo Avogadro Scientific Biography.* Reidel, Dordrecht.
4. Hopkins, F. G. (1928). The centenary of Wöhler's synthesis of urea (1828–1928). *Biochemical Journal* 22, 1341–8.
5. McKie, D. (1944). Wöhler's 'synthetic' urea and the rejection of vitalism: a chemical legend. *Nature* 153, 608–10.
6. Read, J. (1958). The life and work of Perkin. In *Perkin centenary London. 100 years of synthetic dyestuffs*, pp. 1–31. Pergamon Press.
7. Orfila, M. P. (1814). *Traité des poisons tirés des regens mineral, végétal et animal, ou toxicologie generale.* Crochard, Paris.
8. Keys, T. E. (1945). *The history of surgical anesthesia.* Schuman's, New York.
9. Ruprecht, J., van Lieburg, M. J., Lee, J. A., and Erdmann, W. (eds) (1985). *Anaesthesia. Essays on its history.* Springer-Verlag, Berlin.
10. Atkinson, R. S. and Boulton, T. B. (eds) (1989). *The history of anaesthesia. International congress and symposium series, 134.* Royal Society of Medicine Services, London; Parthenon, Carnforth.
11. Squire, W. (1888). On the introduction of ether inhalation as an anaesthetic in London. *Lancet* ii, 1220–1.
12. Simpson, J. Y. (1847). *On chloroform.* Edinburgh. Reprinted in Thoms, H. (1935). *Classical contributions to obstetrics and gynecology*, pp. 29–34. Thomas, Springfield, Illinois.
13. Dewhurst, J. (1980). *Royal confinements.* Weidenfeld and Nicolson, London.
14. Balard, J. M. (1844). Mémoire sur l'alcool amylique. *Comptes Rendus de l'Académie des Sciences* 19, 634–41.
15. Guthrie, F. (1859). Contributions to the knowledge of the amyl group. I. Nitrite of amyl and its derivatives. *Journal of the Chemical Society* 11, 245–52.
16. Richardson, B. W. (1864). In *Report of the thirty fourth meeting of the British Association for the advancement of science*, pp. 120–9. Murray, London.
17. Brunton, T. L. (1867). On the use of nitrate of amyl in angina pectoris. *Lancet* ii, 97–8.
18. Sobrero, A. (1847). Sur plusieurs composes detonants produits avec l'acide nitrique et le sucre, la dextrine, la lactine, la mannite et la glycerine. *Comptes Rendus de l'Académie des Sciences* 24, 247–8.
19. Field, A. G. (1858). On the toxical and medicinal properties of nitrate of oxyde of glycyl. *Medical Times and Gazette* n.s. 16, 291.
20. Field, A. G. (1858). [Correspondence in the] *Medical Times and Gazette* n.s 16, 385.
21. Murrell, W. (1879). Nitro-glycerine as a remedy for angina pectoris. *Lancet* i, 80–1; 113–5; 151–2; 225–7.
22. Locock, C. (1857). [Report of discussion of a meeting] *Lancet* i, 528.
23. Balme, R. H. (1976). Early medicinal use of bromides. *Journal of the Royal College of Physicians, London* 10, 205–8.
24. Clouston, T. S. (1868). Experiments to determine the precise effect of bromide of potassium in epilepsy. *Journal of Mental Science* 14, 305–21.

25. Goodman, L. and Gilman, A. (1941). *The pharmacological basis of therapeutics*, pp. 155–65. Macmillan, New York.
26. Henry, T. A. (1913). *The plant alkaloids*. Churchill, London.
27. Turner, R. B. and Woodward, R. B. (1953). The chemistry of the cinchona alkaloids. In *The alkaloids. Chemistry and physiology*, eds R. H. F. Manske and H. L. Holmes, vol. 3, pp. 1–63. Academic Press, New York.
28. Larder, D. F. (1967). Alexander Crum Brown and his doctoral thesis of 1861. *Ambix* 14, 112–32.
29. Crum Brown, A. and Fraser, T. R. (1869). On the connection between chemical constitution and physiological action. Part 1—On the physiological action of the salts of the ammonium bases derived from strychnia, brucia, thebaia, codeia, morphia and nicotia. Part 2—On the physiological action of the ammonium bases derived from atropia and conia. *Transactions of the Royal Society of Edinburgh* 25, 151–203; 693–739.
30. Bynum, W. F. (1970). Chemical structure and pharmacological action. A chapter in the history of 19th century molecular pharmacology. *Bulletin of the History of Medicine* 44, 518–38.
31. Ure, A. (1841). On hippuric acid and its tests. *Provincial Medical and Surgical Journal, London* 2, 317–8.
32. Keller, W. (1842). Ueber Verwandlung der Benzoesäure im Hippursäure. *Annalen der Chemie und Pharmacie* 43, 108–11.
33. Peters, R. A. (1952). Lethal synthesis. *Proceedings of the Royal Society, Ser. B*, 139, 143–70.
34. Peters, R. A. (1954). Biochemical light upon an ancient poison. *Endeavour* 51, 147–54.
35. Butler, T. C. (1970). The introduction of chloral hydrate into medical practice. *Bulletin of the History of Medicine* 44, 168–72.
36. *A history of chemical achievement 1827–1937*. (1937) Darmstadt, E. Merck. [Pamphlet, no author stated. Copy seen in the library of the Royal Pharmaceutical Society of Great Britain.]
37. Weatherall, M. (1989). Research in the pharmaceutical industry: 19th century. *Pharmaceutical Journal* 242, 543–5.
38. Bäumler, E. (ed.) (1968). *A century of chemistry*. Econ Verlag, Düsseldorf.
39. Gershon-Cohen, J. (1964). A short history of medical thermometry. *Annals of the New York Academy of Sciences* 121, 4–11.
40. Brock, Lord (1972). The development of clinical thermometry. *Guy's Hospital Reports* 121, 307–14.
41. Filhene, W. (1884). Ueber das Antipyrin, ein neues Antipyreticum. *Zeitschrift für klinische Medizin* 7, 641–2.
42. Filhene, W. (1884). Ueber da Pyramidon, ein Antipyrinderivat. *Berliner klinische Wochenschrift* 33, 1061–3.
43. Erhardt, G. (1968). One hundred years of research. In *A century of chemistry*, ed. E. Bäumler, pp. 275–8. Econ Verlag, Düsseldorf.
44. Gerhard, C. (1853). Untersuchungen über die wasserfreien organischen Säuren. *Annalen der Chemie und Pharmacie* n.s. 11, 149–79.
45. Dreser, M. (1899). Pharmakologisches über Aspirin. *Archiv für die gesamte experimentelle Physiologie* 76, 306–18.
46. Wohlgemuth, J. (1899). Ueber Aspirin (Acetylsalicylsäure). *Therapeutische Monatschrifte* 13, 276–8.

47. Fischer, E. and von Mering, J. (1903). Ueber eine neue Klasse von Schlafmitteln. *Therapie die Gegenwart* n.s. 5, 97–101.
48. Mahoney, T. (1959). *The merchants of life.* Harper, New York.
49. Liebenau, J. (1985). Scientific ambitions; the pharmaceutical industry 1900–1920. *Pharmacy in History* 27, 3–11.
50. Boussel, P., Bonnemain, H., and Bové, F. (1986). *History of pharmacy and pharmaceutical industry.* Asklepios Press, Paris.
51. Liebenau, J. (1987). *Medical science and medical industry. The formation of the American pharmaceutical industry.* The Macmillan Press, Basingstoke.
52. Vallery-Radot, R. (1901). *The life of Pasteur*, (trans. R. L. Devonshire, 1937). Constable, London.
53. Dubos, R. J. (1950). *Louis Pasteur Freelance of Science.* Charles Scribner's Sons, New York; Gollancz (1951), London.
54. Vallery-Radot, note 52, pp. 38–42; 61–70.
55. Alsace and Lorraine were restored to France in 1919 and Strassburg again became Strasbourg.
56. Roll-Hansen, N. (1979). Experimental method and spontaneous generation: the controversy between Pasteur and Pouchèt, 1859–1864. *Journal of the History of Medicine* 34, 273–92.
57. Sakula, A. (1979). Robert Koch (1843–1910): founder of the science of bacteriology and discoverer of the tubercle bacillus. *British Journal of Diseases of the Chest* 73, 389–94.
58. Pasteur, L., Chamberland, C. and Roux, E. (1881). Compte rendu sommaire des expériences faites à Pouilly-le-Fort, près Melun, sur la vaccination charbonneuse. *Comptes Rendus de l'Académie des Sciences* 92, 1378–83.
59. Guthrie, D. J. (1949). *Lord Lister, his Life and Doctrine.* Livingstone, Edinburgh.
60. Lister, J. (1870). On the effects of the antiseptic system of treatment upon the salubrity of a surgical hospital. *Lancet* i, 4–6; 40–2.
61. Fraser-Moodie, W. (1971). Struggle against infection. *Proceedings of the Royal Society of Medicine* 64, 87–94.
62. Selwyn, S. (1979). Pioneer work on the 'penicillin phenomenon' 1870–1876. *Journal of antimicrobial Chemotherapy* 5, 249–55.
63. Guthrie, note 59, p. 95.
64. Truax, R. (1947). *Joseph Lister father of modern surgery*, p. 189. Harrap, London.
65. Ackerknecht, E. H. (1953). *Rudolph Virchow. Doctor, Statesman, Anthropologist.* University of Wisconsin Press, Madison.
66. Paton, W. D. M. (1979). The evolution of therapeutics: Osler's therapeutic nihilism and the changing pharmacopoeia. *Journal of the Royal College of Physicians, London* 13, 74–83.
67. Galaty, D. H. (1974). The philosophical basis of mid-nineteenth century German reductionism. *Journal of the History of Medicine* 29, 295–316.
68. The International Medical Congress of 1881, held in London, was fully reported in the *Lancet* and other medical journals at the time. See also Sakula, A. (1982). Baroness Burdett-Coutts' garden party: the international medical congress, London, 1881. *Medical History* 26, 183–90, and Paget note 69, pp. 306–14).
69. Paget, S. (ed.) (1901). *Memoirs and letters of Sir James Paget*, p. 308. Longmans, London.
70. Report (1881). *Lancet* ii, 210.

71. French, R. D. (1975). *Antivivisection and medical science in Victorian society.* Princeton University Press.
72. Rupke, N. A. (ed.) (1987). *Vivisection in historical perspective.* Croom Helm, London.
73. Report (1881). *Lancet* ii, 291; 299.
74. Huxley, T. H. (1881). The connexion of the biological sciences with medicine. *Lancet* ii, 272–6.

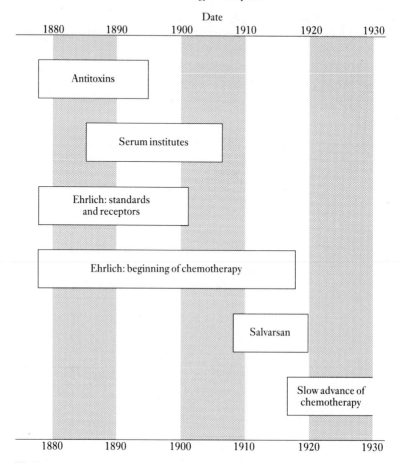

Chronology of Chapter 3

Date

1880 1890 1900 1910 1920 1930

Antitoxins

Serum institutes

Ehrlich: standards
and receptors

Ehrlich: beginning of chemotherapy

Salvarsan

Slow advance of
chemotherapy

1880 1890 1900 1910 1920 1930

NB The blocks show periods discussed in the text, and do not mean that there was no activity
about the subject at other times.

3

Immunotherapy and chemotherapy

Defence against germs

The discoveries of Pasteur and Koch opened the way to the new science of bacteriology and to a revolution in the prevention and treatment of infectious diseases.[1] First, more accurate diagnosis became possible. In the pre-bacteriological era, cholera was inevitably confused with enteric fever, dysentery, and other infections of the bowel; pulmonary tuberculosis with other lung infections, and so on. The findings of a post-mortem examination might establish a diagnosis, but that did not help the patient. By making bacteriological examinations, it became possible to make a more accurate diagnosis, and to choose treatment and predict the outcome accordingly. Consequently, every progressive hospital established a bacteriological laboratory of some sort. Diagnosis had a more objective basis than the authority of the most experienced clinician, and a new breed of laboratory-oriented doctor began to evolve.

Cultures of dangerous organisms could be manipulated with the aim of making them less virulent while keeping their power to confer some immunity. A Spanish bacteriologist, Jaime Ferran (1852–1929), who had worked under Pasteur, made a vaccine against cholera from cultures grown by Koch's method. It caused too much reaction when it was injected into patients, and was abandoned, but another pupil of Pasteur, Waldemar Haffkine (1860–1930) discovered that the microbes could be killed by heat and still promote immunity. He produced a vaccine[2] which saved many lives in India.[3] Regrettably, there was vigorous controversy with Ferran about priorities.[4]

To prepare good vaccines, one needed to know how bodily defences operated against microbes. It was then easier to discover how a vaccine stimulated immunity, and learn ways of making safer and more potent vaccines. Two schools of thought emerged. In Paris, one of Pasteur's pupils, the Russia zoologist Elie Metchnikoff (1845–1916), studied free-living cells, such as amoebae and mammalian blood cells, which surround particles, including microbes, and digest them. He called the process phagocytosis, and recognized that it played an important role in the body's defences. Might a vaccine stimulate phagocytes to attack a particular kind of germ? This was the beginning of what was later called cellular immunology. Metchnikoff established the importance of phagocytosis in defence against invading organisms, but other mechanisms turned out to be more immediately exploitable.[5]

The microbe which caused diphtheria was isolated by Theodor Klebs (1834–1913) in Zurich. A year later it was grown successfully in cultures by Koch's assistant Friedrich Loeffler (1852–1915). Two of Pasteur's younger colleagues and disciples, Emile Roux (1853–1933) and Alexandre Yersin (1863–1943) showed that a highly toxic material could be isolated from the cultures and that injection of this toxin into animals imitated some features of the disease, notably weakening of the heartbeat or heart failure. This discovery explained why diphtheria germs, which did not spread from the throat, nevertheless caused ill-effects in distant organs. Evidently the germs produced the toxin, which was set free into the patient's blood stream, and caused damage in any sensitive tissue.[6]

The discovery had profound practical consequences. Emil Behring[7] (1854–1917) and Shibasaburo Kitasato (1852–1931), in Koch's laboratory, injected less-than-lethal doses of toxin into animals and found that after a few days blood from these animals began to be able to neutralize fresh samples of toxin.[8] They called whatever it was which had appeared in the blood 'antitoxin' and showed that animals which had antitoxin in their serum were resistant to further infection with the diphtheria bacillus. Blood serum taken from such animals could be used to treat infection in otherwise unprotected animals. So they had found a mechanism which did not depend directly on cells like Metchnikoff's phagocytes, but on some protective substance, which could perhaps be purified, isolated, and used to cure patients. The problems, especially of producing enough material to treat even a single child, instead of a few mice, were overcome remarkably rapidly. According to accounts now difficult to verify[9] a preparation of antitoxic serum was injected into a little girl dying of diphtheria in a clinic in Berlin on Christmas Eve 1891. She recovered, the first of many who were dramatically rescued from tragedy.

This sequence of events gave much support to those who were primarily attending to the 'humoral' aspects of immunity, i.e. to materials carried in the body fluids which protected against infection. However, the practical application of antitoxins was limited to organisms which released powerful toxins. Such organisms were responsible for few major diseases. Beside diphtheria, the most familiar was tetanus or lockjaw, for which antitoxic sera soon began to be developed. Also, antitoxins were valuable only for treating established disease. The antitoxin had no permanent effect in protecting people who had not contracted the disease. The more ambitious objective, of preventing a disease as vaccination protected against smallpox, had yet to be achieved. Meanwhile many practical problems had to be solved.

Producing the remedies

The production of enough antitoxin to meet the needs of a whole nation or of the whole world was a very different problem from making enough in a laboratory for a few patients at a time. Production of vaccines and antitoxins in bulk was undertaken in different ways in different countries. Pasteur himself envisaged the founding of a model establishment in Paris, supported by donations and international subscriptions, but he waited until the success of his treatment for rabies was undoubted. At that time the mortality from dog bites in Paris ranged between 16 and 40 per 100 persons bitten. When 350 patients had been treated with the new vaccine and only one had died, Pasteur was able to affirm to the Academy of Sciences that the prophylaxis of rabies after a bite had been achieved. The Academy appointed a commission to make plans and raise funds for an 'Institut Pasteur', which was built and opened in 1888. It had three sections, one a great dispensary for the treatment of rabies, one for research on infectious and contagious diseases, and one a teaching centre. In the research area were laboratories under the direction of Metchnikoff for the study of problems of immunity.[10] When the value of diphtheria antitoxin was known, the institute was in a position to create facilities for its production. Also, separate Pasteur Institutes were established in French colonies, sometimes with the dual purposes of research and production, sometimes primarily as a centre for treatment.

In Germany, the chemical industry had strong academic connections and was prepared to manufacture antitoxin on a commercial scale and basis. The work was undertaken by Meister, Lucius and Bruning at their Hoechst factory. This was a remarkable undertaking for a chemical business, as it involved a substantial foray into the care and management of the animals in which antitoxin was raised, but it was achieved successfully.,[11] The quality of the product was subject to external control. Koch held the office of Professor of Hygiene and Bacteriology at the Friedrich Wilhelm University at Berlin, and also served as advisor to the Imperial Health Office. His laboratories were cramped and the German government provided a new research institute in Berlin, the Koch Institute for Infectious Diseases, which was opened in 1891 with Koch as the first director. Unlike the Pasteur Institute, these laboratories did not function for purposes of production and treatment, but were responsible for such experiments as were needed to make sure that the commercially manufactured antitoxin was of an acceptable standard.

England had no chemical industry on the German scale. Several businesses, some of long standing and some recently established, provided drugs, predominantly made from plants, and were developing mechanical means of production and formulation. Any move towards biological products led to difficulties with the strong national feelings about experiments on animals and the requirements of the 1876 Act which regulated such experiments (see Chapter 2). Injecting horses with toxin and drawing blood to obtain antitoxic serum did not qualify as an experiment, but attempts to estimate the strength of the sera by experiments in mice did require a licence. Considerable difficulties resulted, because commercial operators were not considered, in Victorian eyes, sufficiently respectable to hold such licences.[12] The first antitoxin made in England came from the Brown Animal Sanatory [sic] Institution, an institute set up by a charitable bequest in 1851 and associated with the University of London, which provided laboratory facilities and accommodation for large animals at a time when such essential facilities were very rare.[13,14]

Production of antitoxin was attempted by some firms, with the help of independent bacteriologists who were eligible to hold the necessary licences. Not surprisingly, the arrangements brought little success. The material produced was not potent, and also doctors were timid about dosage, so that clinical failures were common. In parts of England and Scotland results were so bad that the treatment was discarded and deaths from diphtheria continued unabated. The often militant journal, the *Lancet*, set up a special commission to investigate, and subsequently published figures which showed that British production was far behind the quality of some German manufacturers.[15]

A less commercial approach to the problem was to found an institute of preventive medicine, on the lines of the Pasteur Institute in Paris. A small band of distinguished men, among whom Lister was prominent, aimed to raise funds to build and staff laboratories, both for experimental work and for the production of remedies when they were discovered. The dual status presented difficulties. Endowments were sought to pursue research, while the production laboratory was expected to augment the income for research by profits on its sales, and at the same time to be open to students of any country to study its methods. The founders had also to fight official timidity about and public antipathy to vivisection.[16] They succeeded, and in July 1891 incorporated their institute. Its quarters were initially in Bloomsbury, but it moved to new buildings opened in 1898 close to Chelsea Bridge. By this time it had been renamed the Jenner Institute of Preventive Medicine, but the new title caused trouble with a small commercial undertaking with the same name. In 1903, the institute was finally renamed after one of its founders, by then Lord Lister. Like the Pasteur Institute, its laboratories were destined to make many contributions to human welfare, before the conflict between economics and innovation finally led to closure of the research laboratories in 1975.[17]

In the United States, the most advanced municipal authorities and the best equipped pharmaceutical businesses vied with each other to produce antitoxin.[18] In New York, an Institute named after Pasteur set out to make not only antitoxins but also glandular extracts containing hormones (see Chapter 5), which involved similar problems in making available biological materials for human treatment.[19] A much more modest activity, concerned with applied bacteriology though not with production and named the Hygienic Laboratory, was established about the same time within the Staten Island Marine Hospital, New York. This hospital catered for sick and disabled seamen and, by virtue of its position at the main port of entry into the United States, became a front line of defence against cholera brought to America by immigrants. The laboratory flourished as part of the United States Public Health Services and later, removed to Bethesda in Maryland, grew into the great National Institutes of Health, the foremost medical research laboratories in the world.[20]

Control of the quality of materials used in making vaccines, and of the vaccines themselves, also required laboratories. Koch and his colleagues were leaders in this field, in experience, staff, and resources. The outstanding advance in such methods was initiated by Paul Ehrlich (1854–1915) in the newly-created State Institute for the Investigation and Testing of Sera in Berlin. Ehrlich's range of work went far beyond the testing of sera.[21]

Standardizing the materials: the genius of Paul Ehrlich

Paul Ehrlich (1854–1915), the son of an innkeeper in Silesia,[22] studied medicine at Breslau, where he met Koch, and at the newly-founded University of Strassburg. He was attracted by work in laboratories, and at Leipzig wrote a dissertation on the theory and practice of staining material for examination under the microscope. The problem looks fairly straightforward and not very exciting. Why do particular cells, or parts of cells, combine with particular dyestuffs? But the question can be put more generally. Why does any substance of known chemical identity react with some or other unknown material in the cells? Then the question includes *the* fundamental problem of drug action, and in studying it Ehrlich was laying the foundations for his life's work.

For a time he worked in hospitals and their laboratories, finding, partly by chance, a method of staining the recently discovered tubercle bacillus and himself contracting the disease.[23] In 1890, after a period of convalescence in Egypt, he joined Koch at the Institute for Infectious Diseases in Berlin. Here Ehrlich began to work with Behring and performed many of the basic studies which were essential for Behring's success with diphtheria antitoxin. The collaboration began happily, but considerable ill-feeling developed later. There is some reason to think that Behring took advantage of Ehrlich's willing help and support.[24]

Ehrlich's quantitative work on the new antitoxins led to his appointment in 1896 as Director of the State Institute for Serum Research and Serum Testing at Steglitz, a suburb of Berlin. The institute existed in order 'to preserve standard preparations and to devise new and more accurate methods of determining the value of the serum, and to study the complex relations which govern the neutralization of toxin and antitoxin'. This description defines Ehrlich's work at this period so precisely that one wonders whether Ehrlich wrote it himself. He made great progress in devising new and more accurate methods and established principles of fundamental importance to what is now called biological standardization.[25]

The difficulty lies in measuring accurately minute quantities of substances which can be recognized *only* by their biological activity. Even now, nearly a century later, the exact chemical structure of diphtheria or tetanus toxins, and of the antitoxins produced in response to these diseases are just beginning to be worked out in detail. The molecules contain many thousands of atoms and exist in a milieu of many other similar molecules which have no activity in relation to diphtheria. Modern methods for measuring how much of the important material is present among the mass of impurities depend on physical and chemical techniques unknown in the 1890s.

There were no precedents for solving the problem. It had not previously been necessary to measure the amounts of potent substances present in complex materials so accurately. As long as medicines, crude and ill-defined, had obvious consequences when they were administered to patients, adjustment of the dose rested with the doctor. If he got a poor response, he could increase the dose; if the results of treatment were unexpectedly vigorous, he could perhaps take suitable measures and give a smaller dose next time. Even vaccines, which produced no visible consequences, did not present difficulties. Vaccines against smallpox were usually administered by scarifying the skin through a droplet of the active material. The proportion absorbed varied greatly, but as long as enough got in, a bit more did no harm. But manufacture of the new antitoxins raised serious problems of accurate dosage. Antitoxins are made by immunizing sheep or horses with the corresponding toxins. An unimmunized animal has no defence against the toxin and is easily killed by a sample of toxin which is too strong. So it is essential to have toxin of consistent potency, and to devise schedules for immunizing animals and producing a serum rich in antitoxin which can be used for treating patients. No schedule is worth having if one batch of toxin is two or three times stronger or weaker than its predecessor.

The toxin, produced by growing bacterial cultures, is a crude mixture of substances, quite beyond the chemical skills of the 1890s to purify, let alone analyse. At that time the only way of discovering how much toxin a culture produced was by showing that it actually did harm to or killed some animal; an inescapable experiment if life-saving antitoxin was to be prepared. The smaller the animal, the less was needed of the scanty supplies of the dangerous toxin, so mice were an obvious choice. Batches of toxin were assayed by determining the lethal dose in mice, and the potency of serum containing antitoxin was measured by seeing how many doses of toxin a given amount of serum would neutralize.

All these problems were related to each other. If pure toxin had been available, it would have been easier to devise dosage schedules. If one had had a reliable assay method, it would have been easier to purify toxin. The problems were interwoven, in a morass of uncertainties. Ehrlich's solution to the problem is described in a paper which is a landmark in the advance of medical and pharmaceutical science. In it Ehrlich states a principle superficially fairly simple, but often not properly grasped even by many experienced workers. The essential point is that the potency of a toxin, is a property of the toxin itself and not of the animal or other means by which it is tested. A unit of activity must be a unit of material, not a unit defined as the dose which will kill a mouse. Such a quantity depends on the size and health of the mouse as much as on the toxin, and inevitably varies. An assay may involve the response of an animal or a group of animals, but the reliability of the assay always depends on matching, directly or otherwise, the unknown material with a standard of defined potency.[25]

This principle, like Louis's insistence on the importance of clinical trials (see Chapter 2), has a curious history of being repeatedly forgotten or ignored. In the early stages of isolating a toxin no standard exists and the dose which achieves some particular response is the only feasible rough measure. But once the labour of establishing a stable reference standard has been performed, the gain in precision by making comparative assays is substantial. In time the principle came to be applied to the standardizing of such drugs as digitalis, which like all herbal materials varies greatly in potency, and to hormones and vitamins.[26] Its history is an important phase in the advance of the accurate and effective use of drugs and owes much to Ehrlich's clear understanding.

The basic principle of biological standardization was only one of the outstanding contributions made by Ehrlich to knowledge of infectious diseases and their treatment. Ehrlich studied extensively the processes by which animals became immune to foreign substances, and saw them, correctly, as chemical reactions between very complex materials. Knowledge of the composition of living cells, whether of micro-organisms or the hosts which they infected, was still rudimentary. Cells were supposed to be filled with an ill-defined material, perhaps a kind of giant molecule, called protoplasm. Ehrlich visualized protoplasm as having 'side-chains' with specific functions in nourishing and protecting cells. The side-chains enabled cells to receive and fix materials from outside, and so Ehrlich also used the term 'receptor' for the unidentified cellular constituents he was envisaging.[27] The state of chemical knowledge in 1900 gave no basis for exploring the ideas in detail,[28] but they have subsequently been amplified in many ways (chapter 6). The concept of receptors, part of the cell structure which reacts with specific foreign substances, has become a fundamental part of the science of drugs and very important for new discoveries (see Chapter 12).

Ehrlich's receptor theories

Ehrlich liked a chemical approach to biological problems. So he began to consider whether simpler substances than antitoxins might combine with the cells of micro-organisms and inactivate or destroy them. In his early days he had stained microbes with dyes, which meant to him that the dyes had combined with receptors in the microbes and quite possibly interfered with their normal activity. Was it possible to do this to organisms in a living patient?

The parasite which caused malaria was discovered in 1880 by the French physician Alphonse Laveran (1845–1922) during his military service in Algeria. Ehrlich observed that the parasites specifically took up the dye named methylene blue, and in 1891 he showed that the dye had a modest therapeutic effect in two patients with malaria.[29] The result was confirmed,[30] and later was an important basis for developing new drugs. Patients with malaria were rare in Berlin and quinine had become readily available, because extensive plantations had been established in the Dutch East Indies, so the discovery was not immediately pursued. But the idea of substances being fixed, visibly when dyestuffs were involved, with lethal consequences to the microbes, remained as a guiding principle throughout Ehrlich's life and became fundamental to later searches for synthetic chemicals which would kill microbes selectively. Ehrlich was fond of vivid phrases and coined the term 'chemotherapy' to distinguish this sort of action from the immunotherapy of antitoxins and vaccines.[31,32]

The laboratories provided for Ehrlich as Director of the Serum Institute at Steglitz were primitive and inadequate. They were replaced in 1898 by a larger state institute built for the purpose at Frankfurt-am-Main. With the support of a local chemical manufacturing business, which later became part of the great I.G. Farbenindustrie, Ehrlich had more opportunity for investigating new synthetic compounds for the treatment of infectious diseases. Malaria was not a good subject for study, because at that time, no means were known of producing it in experimental animals. Sleeping sickness, or trypanosomiasis, was more suitable, because it had recently been found that rodents could be infected with a particular species of trypanosome. The infection could be maintained by repeated passage from rat to rat or mouse to mouse, and provided a laboratory model on which Ehrlich could investigate the effect of dyes and any other promising substances.[33]

In 1904 Ehrlich and a visiting Japanese collaborator, Kiyoshi Shiga, showed that a particular dye, which became known as trypan red, cured trypanosomal infections in mice. The discovery was not important therapeutically, because the dye did not work against trypanosomiasis in man or cattle, but it provided a very useful laboratory situation for investigating how drugs might work against parasites. The thread-like trypanosomes could be seen, under the microscope, to vibrate vigorously in the presence of the drug, but experiments showed that they were unable to reproduce, and so the infection died out in treated mice. However, after repeated exposure to the dye, some strains of trypanosome became resistant and managed to multiply successfully. This was the first observation of a phenomenon which became immensely important as chemotherapy extended to wider fields. But how did the organisms manage to become resistant? Ehrlich tried to fit all these facts into his side-chain theory which received many modifications in consequence.[31–32]

At about this time, an organic compound containing arsenic, used for various skin affections and for anaemia, and named atoxyl because it was less toxic than inorganic substances containing arsenic, was shown to cure various kinds of trypanosome infection of animals.[34] Ehrlich's old chief, Robert Koch, was in East Africa investigating trypanosomiasis and showed that atoxyl was an effective remedy in the field.[35] Erhlich investigated the chemical properties of atoxyl, which had been synthesized in 1863 and was supposed to have a particular chemical structure. He showed that it reacted in a way which would have been impossible if the supposed structure had been correct, and suggested an alternative, which, if correct, would have great possibilities for developing compounds related to the synthetic dyes which were therapeutically active. Ehrlich had no pro-fessional training in chemistry and became involved in a memorable dispute with the chemists who were working with him. Two of them resigned on the spot rather than accept that he might be correct, but the one who remained established that atoxyl did in fact have the structure which Ehrlich proposed, and could be used for synthesizing homologues and a variety of derivatives. Later, atoxyl was found not to be as harmless as was hoped, because it caused blindness in a small proportion of patients, and its use was abandoned. But this was a further reason for seeking better drugs, and many new organic arsenicals were synthesized.

Ehrlich's facilities were greatly increased by financial support from Franziska Speyer, the widow of a wealthy Frankfurt business man. Her gift provided for the building of new laboratories adjacent to the serum institute. The laboratories, opened in 1906, were planned and used for the study of chemotherapy and named after her husband the Georg Speyer Haus. Soon afterwards, Ehrlich's earlier work on immunology received the highest scientific recognition by the award in 1908 of a Nobel prize. But Ehrlich now had little to do with the advance of immunology. His interest was concentrated on small molecules of precisely known composition, and the rest of his career was devoted to chemotherapy.

Although trypanosomes were convenient for laboratory experiments, they were much less important than the bacteria which caused diseases, such as pneumonia and typhoid fever, and the infection of wounds. Attempts to find effective drugs were baulked because the agents which killed the bacteria were too poisonous to be tolerated. Derivatives of phenol (carbolic acid), which Lister had used and discarded, and of quinine, which was tolerable but not itself antibacterial to a sufficient extent, were tested, but met with little success.[36]

About this time the organism which causes syphilis, a spirochaete called *Treponema pallidum*, was recognized. Special problems made it difficult to find: it does not take up any of the conventional stains, so special methods are necessary to visualize it, and it does not grow in ordinary culture media. Discovery of the bacterium,[37] soon followed by the invention of a reliable diagnostic test for syphilitic infection (the Wassermann reaction),[38] prompted Ehrlich to add the spirochaete to the orgnisms which he sought to attack. As there was no convenient laboratory model of the disease, he arranged for trials of two of his new arsenicals to be conducted in Java on apes infected with the spirochaete, and also in patients suffering from a disease called relapsing fever or yaws, caused by a closely related spirochaete.

The discovery that rabbits could be infected successfully with syphilis made progress easier and faster. A Japanese scientist, Sahachiro Hata, who came from the laboratory in Tokyo directed by Behring's former colleague Kitasato, brought the technique to Ehrlich and together they showed that a compound which had been made 3 years earlier and at that time rejected, was an effective agent against the spirochaete. This was Ehrlich's compound 606, later known as Salvarsan and as arsphenamine. It was soon to achieve world-wide fame as the 'magic bullet' against a widespread, untamed, and slowly destructive disease.[39]

The magic bullet

Once he was satisfied that compound 606 was capable of arresting syphilitic lesions in animals, Ehrlich arranged for it to be made on a larger scale and for samples to be sent to colleagues whom he knew and trusted to test the material in patients. Many of the trials took place in Germany, but Ehrlich had a wide reputation and an extensive circle of friends abroad, so the new drug was sent to them too. Some material came to England, to the leader of British medical bacteriologists, Almroth Wright (1861–1947), who had recently established his almost personal institute at St Mary's Hospital for the development of vaccines, and to William Bulloch (1868–1941), bacteriologist at the London Hospital in Whitechapel. Both deputed investigation of the material to younger colleagues. At St Mary's the work was carried out and published by Alexander Fleming (1881–1955) and Leonard Colebrook (1883–1967), one of whom is famous for his later work on penicillin and the other who was particularly involved in introducing the sulphonamides or 'sulfa-drugs' to England.[40] At the London Hospital the work was done by James McIntosh (1882–1948) and (later Sir) Paul Fildes (1882–1971),[41] whose fundamental contributions to bacteriology and antibacterial chemo-therapy we shall discuss later on (see Chapter 8). One may wonder how much the early involvement of Colebrook, Fleming, Fildes, and McIntosh influenced the later development of their respective interests and careers. The two short reports show quite different approaches to the same objective. Fleming and Colebrook describe apparatus for the as yet unfamiliar procedure of intravenous injection of the new drug. McIntosh and Fildes are concerned with clinical details and the effect of the drug on the Wassermann reaction.

Early clinical trials of 606 were satisfactory. Ehrlich announced publicly the discovery of the new remedy at the Congress for Internal Medicine held at Wiesbaden in April 1910. Doctors, patients, and the public in many parts of the world were greatly excited by the prospect of a new remedy, and there were many requests and demands for the new drug. It was impossible to meet even a few of these requests, and essential to confine the use of 606 at first to those who could be trusted to understand its difficulties. 606 was an unstable substance which deteriorated rapidly on exposure to air. Successive batches varied in chemical purity and in toxicity. To preserve the drug, it was necessary to supply it as a dry powder, which had to be made into a neutral solution in water immediately before use. To achieve a therapeutic effect it was necessary to inject the drug into tissues or, preferably, veins. Few doctors in 1910 knew about the requirements which would make water suitable for injection, and many were content with tap water, neither distilled nor sterilized. Intravenous injection was often regarded as a surgical procedure to be conducted by cutting the skin over a vein and trying a cannula into it.

Ehrlich, conscious of such difficulties and accustomed to detailed control of all that went on in his own laboratories, personally supervised the manufacture of every batch of 606 made at the Georg Speyer Haus, attempted to restrict supplies only to those clinicians he trusted, and sought information about the results of every treatment.[42] He had a good collaboration with the chemical industry, and obtained eager support. The Hoechst chemical works set up a pilot production plant, which relieved some of the pressure, and by the end of the year a larger production plant began to operate.[43] The name Salvarsan was chosen: the substance became commercially available, and the pressures were eased but not abolished. For Ehrlich, the discovery was not only a great success but also involved unattractive labours. He said,

Laboratory work is child's play in comparison: either a thing will go or it will not, and that is the end of it. But if you have to depend upon hundreds of collaborators, and each of them believes that he can do better than any other, life really can be made rather difficult and bitter.[44]

With all the hopes roused by a cure for a dreadful disease and all the difficulties in its use, Salvarsan was exposed to every attitude from indiscriminate praise to savage abuse. Much had to be learnt about how to use it best, about the right dose, the right frequency of dosing, and the right route of dosing. Supplies intended for such essential studies were diverted by fashionable physicians to the treatment of remunerative patients. Busy doctors passed material to enthusiastic juniors for investigation, just as Wright and Bullough had done. In the public domain,[45] the customs of Western society were strictly to refrain from the discussion of sexual matters, particularly when men and women were both present. Syphilis was an unmentionable topic until the excitement of a possible cure made some breaches in the conspiracy of silence. Nevertheless, it was shameful to have syphilis, and patients had no wish for the treatment they were receiving to be a reason for their disease to become known. Indeed, some physicians conducted the treatment without revealing the diagnosis to the patient. In such circumstances clinical trials were very difficult to conduct, nor at that time had adequate methods of evaluation been devised. There seems to be no doubt that after treatment with Salvarsan the Wassermann reaction of patients with primary or secondary syphilis became negative, and little doubt that after persistent treatment, possibly with several courses of weekly injections, each course lasting for 3 months, the disease could be eradicated. But many patients did not stay the course, and the disease was far from eliminated by the new therapy.

The last years of Ehrlich's life were filled with a multitude of problems raised by Salvarsan. A more manageable compound, numbered 914 (Neo-Salvarsan, neo-arsphenamine), was produced, which was more soluble and did not have to be neutralized immediately before use. But it too involved more developmental work. The burdens of success prevented Ehrlich's return to basic studies, and there was no final summing up of his theories of chemotherapy. Nor was the time ripe. Too little was known about the biochemistry of micro-organisms and about the structure and reactions of macro-molecules, and it was too easy to develop controversy about chemical and physical modes of action, a distinction which has faded in significance as knowledge of atomic and molecular behaviour has increased.[46]

Progress and decline of arsenical drugs

Ehrlich left no successor to continue with the development of his organic arsenicals. They were needed in every country in the world. Manufacture in Germany was insufficient, and made more difficult by the war of 1914. As the war provided grounds for over-ruling patent rights, pharmaceutical firms in other industrialized countries began to manufacture arsphenamine and neo-arsphenamine. Strict control of the quality of the product was essential and could not be ignored by governments. In the USA, the work was an added responsibility for the growing Public Health Service. Carl Voegtlin (1879–1960), chief of the Division of Pharmacology, and his colleagues not only tested compounds submitted by manufacturers, but also explored the mechanism of action of arsenicals, in more specific biochemical terms (see Chapter 8) than Ehrlich's very theoretical concepts of receptors,[47] and even developed a new arsenical, sulfarsphenamine, which had some practical advantages.[48]

The findings of Voegtlin and his colleagues suggested that arsphenamine probably behaved as a pro-drug (see Chapter 2), chemically altered by the body before it attacked the spirochaetes. If so, a simpler arsenical compound might be equally or more effective. The idea was pursued by Arthur Tatum (1894–1955), professor of pharmacology at the University of Wisconsin, and led to the introduction of *m*-amino-*p*-hydroxyphenyl-arsenoxide, a name which was contracted to mapharsen.[49] This compound had actually been tested by Ehrlich before he progressed to 606 and had been rejected because small doses were enough to kill rabbits quickly. However, as Tatum and Cooper showed,[50] by using sufficiently small doses and giving the drug more frequently than was acceptable with arsphenamines, it appeared to be both effective and safer in clinical use. It was more stable and certainly simpler to use. It was being used quite widely by the 1940s, when penicillin completely superseded arsenical compounds in the treatment of syphilis, and left them with a very limited range of uses.

Whether mapharsen was actually better in the long run will probably never be known. If Ehrlich's very proper concern with the safety of his compounds had not been marred by the limited value of his toxicity measurements, the whole saga of arsphenamine and neo-arsphenamine might have been avoided and the early days of chemotherapy would not have appeared quite so formidable to all the doctors who were not expert in the field. Moreover, the whole idea of chemotherapy might have appeared in a more attractive light. To most people, progress with vaccines and antitoxins seemed much more promising.

However, research in chemotherapy was not abandoned and notable successes were achieved in the specialized field of tropical medicine. Dysentery became easier to investigate after two quite distinct causes were seen: one was a large group of bacteria, and the other a much bigger organism named *Entamoeba histolytica*. The distinction between bacterial and amoebic dysentery resolved a long-standing argument about the merits of ipecacuanha. It became clear that emetine, the active principle which Pelletier had isolated from the plant (see Chapter 2), was a very effective drug against amoebic dysentery but useless against the bacterial variety,[51] and so emetine took its place alongside quinine as a potent chemotherapeutic alkaloid of plant origin. Attempts to find out how emetine worked were possible, because parasitic amoebae could be isolated and survived briefly in the laboratory. The results seemed very puzzling.[52] Drugs were expected at that time to kill microbes as soon as they met them, but emetine, in normal doses, did not kill the parasites very quickly. More detailed understanding of the reactions between parasites and drugs was essential before this could be understood. The problems of studying malaria in laboratories were overcome, and better antimalarial drugs than quinine began to appear (see Chapter 10). But a quarter of a century passed before chemotherapy achieved a major reappearance in the public eye (see Chapter 8). Meanwhile, many advances were being made in other fields of therapeutics and in more fundamental knowledge. Let us put chemotherapy aside and return to the 1880s to pursue quite different developments.

Notes

1. Parish, H. J. (1965). *A history of immunization*. Livingstone, Edinburgh. With less technical background, Parish, H. J. (1968). *Victory with vaccines.* Livingstone, Edinburgh.
2. Haffkine, W.-M. (1892). Inoculation de vaccins anticholériques à l'homme.

Comptes rendus hebdomadaires des Séances et Mémoires de la Société de Biologie 44, 740–1.

3. Powell, A. (1899). Further results of Haffkine's anti-cholera inoculation. *Journal of Tropical Medicine* 2, 115–6.

4. Bornside, G. H. (1982). Waldemar Haffkine's cholera vaccine and the Ferran-Haffkine priority dispute. *Journal of the History of Medicine* 37, 399–422.

5. Metchnikoff, E. (1908). On the present state of the question of immunity in infectious diseases. *Nobel Lectures Physiology or Medicine 1901–1921* 281–300. Published in 1967 for the Nobel Foundation by Elsevier, Amsterdam.

6. Roux, E. and Yersin, A. (1888). 'Contribution a l'étude de la diphtherie'. *Annales de l'Institute Pasteur* 2, 629–61.

7. MacNalty, A. S. (1954). Emil von Behring. Born 15 March 1854. *British Medical Journal* i, 659–63.

8. Behring, E. (1890). Untersuchungen über das Zustandkommen der Diphtherie—Immunitat bei Thieren. *Deutsche Medizinische Wochenschrift* 16, 1145–8.

9. Schadewalt, H. (1970). Behring, Emil von. *Dictionary of Scientific Biography* 1, 574–8.

10. Nield, T. (1987). Louis Pasteur and his institute. *New Scientist* 116, 40–1.

11. Bäumler, E. (ed.) (1968). *A century of chemistry*, Ch. 2, pp. 39; 278–80. Econ Verlag, Dusseldorf.

12. Tansey, E. M. (1989). Science, commerce and authority: the Wellcome Physiological Research Laboratories and the Home Office 1900–1901. *Medical History* 33, 1–41.

13. Wilson, G. (1979). The Brown Animal Sanatory Institution. *Journal of Hygiene*, 82, 155–76; 337–52; 501–21: 83, 171–97.

14. Anon. (1947). Sir Charles Sherrington's first use of diphtheria antitoxin made in England. *Notes and Records of the Royal Society of London* 5, 156–9.

15. Report (1896). Report of the Lancet special commission on the relative strengths of diphtheria antitoxic serum. *Lancet* ii, 182–95.

16. Chick, H., Hume, M., and Macfarlane, M. (1971). *War on disease. A history of the Lister Institute*. André Deutsch, London.

17. Anon. (1975). Lister Institute of Preventive Medicine. *Lancet* i, 54.

18. Borell, M. (1976). Brown-Séquard's organotherapy and its appearance in America at the end of the nineteenth century. *Bulletin of the History of Medicine* 50, 309–20.

19. Liebenau, J. (1987). Public health and the production and use of diphtheria antitoxin in Philadelphia. *Bulletin of the History of Medicine* 61, 216–36.

20. Koshland, D. E., Jr. (1987). National Institutes of Health: The Centennial Year. *Science* 237, 821.

21. The following references include English translations of the principal works of Paul Ehrlich:
Dale, H. H., Himmelweit, F. and Marquardt, M. (1956). *The collected papers of Paul Ehrlich. Vol. 1. Histology, biochemistry and pathology.* Pergamon Press, London.
Dale, H. H., Himmelweit, F. and Marquardt, M. (1957). *The collected papers of Paul Ehrlich. Vol. 2. Immunology and cancer research.* Pergamon Press, London.
Dale, H. H., Himmelweit, F. and Marquardt, M. (1960). *The collected papers of Paul Ehrlich. Vol. 3. Chemotherapy.* Pergamon Press, London.

22. Marquardt, M. (1949). *Paul Ehrlich.* Heinemann, London. (Martha Marquardt was Ehrlich's secretary for the last twelve years of his life and describes his personality and events in his laboratory).
23. Ibid, pp. 25–28.
24. Ibid, pp. 29–40.
25. Ehrlich, P. (1897). Die Wertbemessung des Diphtherieheilserums und deren theoretische Grundlagen. *Klinische Jahrbuch* 6, 299–326. English translation in Dale, Himmelweit, and Marquardt, (1957), note 21, pp. 107–25.
26. Burn, J. H. (1937). *Biological standardization.* Oxford University Press, Oxford.
27. Ehrlich, P. (1907). Experimental researches on specific therapy. I. On immunity with special reference to the relationship between distribution and action of antigens. *Harben Lectures for 1907 of the Royal Institute of Public Health*, Lewis, London. Reprinted in Dale, Himmelweit, and Marquardt (1960), note 21, pp. 106–17.
28. Rubin, L. P. (1980). Styles in scientific explanation: Paul Ehrlich and Svante Arrhenius on immunochemistry. *Journal of the History of Medicine* 35, 397–425.
29. Guttman, P. and Ehrlich, P. (1891). Ueber die Wirkung des Methylenblau bei Malaria. *Berlin Klinische Wochenschrift* 28, 953–6.
30. Parenski, S. and Blatteis, S. (1893). Ueber das Methylenblau (Merck) bei Malaria. *Therapeutische Monatschrift* 7, 16–19.
31. Dale, H. H. (1923). Chemotherapy. *Physiological Reviews* 3, 359–93.
32. Parascandola, J. (1981). The theoretical basis of Paul Ehrlich's chemotherapy. *Journal of the History of Medicine* 36, 19–43.
33. Laveran, A. and Mesnil, F. (1902). Récherches sur le traitement et la prévention du nagana. *Annals de l'Institut Pasteur* 16, 785–817.
34. Thomas, H. W. (1905). Some experiments in the treatment of trypanosomiasis. *British Medical Journal* i, 1140–3.
35. Koch, R. (1907). Bericht über die Tätigkeit der Deutschen expedition zur Erforschung der Schlafkrankheit bis zum 25 November 1906. *Deutsche Medizinische Wochenschrift* 33, 49–51.
36. Bechhold, H. and Ehrlich, P. (1906). Beziehungen zwischen chemischer Konstitution und Desinfektionswirkung. *Zeitschrift für Physiologische Chemie* 47, 173–99.
37. Schaudinn, F. and Hoffmann, E. (1905). Vorläufiger bericht über das Vorkommen von Spirochaeten in syphilitischen Krankheitsprodukten und bei papillonen. *Arbeiten aus dem Kaiserlichen Gesundheitsamte* 22, 527–34.
38. Wassermann, A., Neisser, A. and Bruck, C. (1906). Eine serodiagnostiche Reaktion bei Syphilis. *Deutsche medizinische Wochenschrift.* 32, 745–6.
39. Ehrlich, P. and Hata, S. (1910). *Die experimentelle Chemotherapie der Spirillosen.* Springer, Berlin. English translation of preface and closing notes in Dale, Himmelweit and Marquardt (1960), note 21, pp. 249–50 and 282–309.
40. Fleming, A. and Colebrook, L. (1911). On the use of salvarsan in the treatment of syphilis. *Lancet* i, 1631–4.
41. McIntosh, J. and Fildes, P. (1910). The theory and practice of the treatment of syphilis with Ehrlich's new specific '606'. *Lancet* ii, 1684–9.
42. Marquardt, M. (1949). *Paul Ehrlich*, pp. 185–206. Heinemann, London.
43. Liebenau, J. (1990). Paul Ehrlich as a commercial scientist and research administrator. *Medical History* 34, 65–78.
44. Note 42, p. 215.

45. Ward, P. S. (1981). The American reception of Salvarsan. *Journal of the History of Medicine* 36, 44–62.
46. Rubin, L. P. (1980). Styles in scientific explanation: Paul Ehrlich and Svante Arrhenius on immunochemistry. *Journal of the History of Medicine* 35, 397–425.
47. Voegtlin, C., Dyer, H. A. and Leonard, C. S. (1923). On the mechanism of the action of arsenic upon protoplasm. *United States Public Health Reports* 38, 1882–912.
48. Voegtlin, C., Johnson, J. M., and Dyer, H. (1922). Sulfarsphenamine. Its manufacture and its chemical and chemotherapeutic properties. *United States Public Health Reports* 37, 2783–98.
49. Swann, J. P., (1985). Arthur Tatum, Parke-Davis, and the discovery of mapharsen as an antisyphilitic agent. *Journal of the History of Medicine* 40, 167–87.
50. Tatum, A. L. and Cooper, G. A. (1932). Meta-amino-para-hydroxy-phenyl arsine oxide as an antisyphilitic agent. *Science* 75, 541–2.
51. Vedder, E. B. (1912). An experimental study of the action of ipecacuanha on amoebae. *Journal of Tropical Medicine and Hygiene* 15, 313–14.
52. Dobell, C. and Laidlaw, P. P. (1926). The action of ipecacuanha alkaloids on *Entamoeba histolytica* and some other entozoic amoebae in culture. *Parasitology* 18, 206–23.

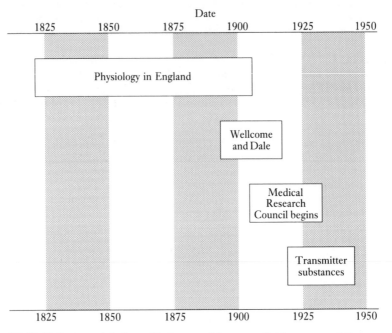

Chronology of Chapter 4

Date

NB The blocks show periods discussed in the text, and do not mean that there was no activity about the subject at other times.

4

The physiological basis of medicine

English physiology

At the International Medical Congress in 1881, T. H. Huxley claimed that 'the greatest steps which have ever been made towards the establishment of medicine on a scientific basis' included 'the elucidation of the action of medicaments by the methods and the data of experimental physiology'. He promoted the growth of physiology, especially at Cambridge University, where he had worked himself, and where his successor, (Sir) Michael Foster (1836–1907) developed an outstanding school of physiology. At this time England lagged far behind Germany in developing and applying scientific knowledge. England was leading the world in engineering, in the production of machinery and in the building of railways and roads, but men of higher intellect continued to study the past and seek to improve on it by reading, thinking, and writing rather than undertaking practical experiments. The study of nature, significantly called natural *history*, consisted of observing and classifying, often in meticulous detail. The observations could be used to support existing theological doctrine, as in Paley's *Evidences*, or provide the basis of a radically different interpretation, as in Darwin's *Origin of Species*. But these arguments alone did not convince everyone.

A few men saw the need for experimental evidence, which could be confirmed by repeating experiments and seeing whether what was predicted actually happened. For this they required laboratories and the opportunity to work with animals. A pattern existed in Germany, where physiology was being extended far beyond the bounds which Magendie and Bernard had reached. In England, despite much incomprehension, laboratories began to be built, both in existing universities and in new red brick establishments which had science faculties from their foundation. In London, University College, founded in 1826 in revolt against religious intolerance, and associated with its own hospital,[1] provided a climate particularly suited to medical science. In Cambridge, Foster was remarkably successful in developing a laboratory for research and for teaching, and in choosing men to work in it. By the end of the century, Cambridge had become *the* place to study medicine, and the students at that time included most of the future English medical scientists of the first half of the twentieth century.[2]

Physiology in Cambridge profited, partly, from the use of drugs as tools in investigating normal functions. A fundamental problem to be studied, and one which turned out to be very important indeed for the development of new drugs, was the mechanism by which one cell in the body influenced another; how, for instance one nerve cell might excite another into activity, or how an impulse reaching the end of a nerve could make a muscle cell contract. Physiologists debated how impulses might pass across the minute gap between cells. In 1877 the German physiologist Emil du Bois Reymond (1818–1896) had stated:

Of known natural processes that might pass on excitation, only two are, in my opinion, worth talking about: either there exists at the boundary of the contractile substance a stimulatory secretion in the form of a thin layer of ammonia, lactic acid, or some other powerful stimulatory substance; or the phenomenon is electrical in nature.[3]

These explanations became known as the chemical or humoral theory and the electrical theory of transmission. Conflict between them occupied physiologists and channelled the main stream of pharmacological research for the next 60 years. In Cambridge, J. N. Langley (1852–1925) extended Bernard's work on curare by detailed microscopic studies of the junction between nerve and muscle. Later he worked with the alkaloid nicotine, which in sufficiently high concentrations acts somewhat like curare. Langley described a special part of the muscle at the junction, which he named the motor end-plate, and proposed, correctly, that curare or nicotine combined with it. He gave reasons why 'chemical rearrangements set up in the muscle molecule by the combination of one of its radicles are different in the two cases'.[4] At this time, protoplasm was regarded as a giant molecule, but the concept of chemical combination between a drug and a receptor in this molecule was very advanced. However, as Langley himself noted, it had much in common with the theories which Ehrlich was putting forward about the combination of cells with antitoxins and with drugs. Ehrlich was passionately concerned with the advancement of therapy: Langley with the pursuit of physiology. Nevertheless, both were concerned with the actions of drugs on cells, and progressed by the same postulation of receptors.

Curare and nicotine were not the only drugs with actions closely related to the transmission of impulses from nerves to the muscles and other organs. Gradually it was seen that the vagus nerves, reaching from the lower end of the brain to the heart, lungs, and abdominal organs, were very important in the involuntary control of their activities: when the vagus was stimulated, the heart slowed and the gut contracted. These effects were also produced by the drug muscarine, from the poisonous mushroom *Amanita muscaria*, and by the drug eserine, which had been studied closely by Christison and by Fraser in Edinburgh (see Chapter 1). If the vagus was stimulated after eserine had been given, the response to stimulation was bigger, but this did not happen after muscarine. However, the effects of nerve stimulation, of eserine, and of muscarine, were all blocked by the well-known drug atropine, from deadly nightshade. If one accepted du Bois Reymond's idea of 'powerful stimulatory substances' secreted by nerve endings, one might have thought of something like muscarine as the stimulant, eserine as something which aided and abetted the stimulant, and atropine as something which got in its way. But the time was not right for establishing these ideas.

When another close match was observed between the effects of stimulating certain nerves and of administering an identified chemical substance (adrenaline) derived from animal tissues, a connection seemed more plausible. T. R. Elliott (1877–1961), in the Cambridge laboratory, wrote a short communication to the Physiological Society,[5] suggesting that the so-called 'sympathetic' nerves, part of the involuntary nervous system, released adrenaline to transmit their messages to the organs they supplied. However, he did not carry the idea forward into a subsequent, substantial, paper in which he described many physiological properties of adrenaline.[6] His colleague W. E. Dixon (1870–1931) attempted to show that muscarine or something like it was released when the vagus nerve to the heart was stimulated.[7] The experiments were unsuccessful and the idea that a powerful fungal poison could also be an important agent in animal tissues seemed ridiculous. There was general scepticism about the idea of identifiable substances acting as transmitters of nervous impulses, and most physiologists of the time continued to regard the process as electrical.

It remained for Elliott's colleague and friend Henry Hallett Dale (later Sir Henry Dale, O.M., F.R.S.) (1875–1968) to provide unequivocal proof 30 years later. Dale's work is important, not only for the great scientific advances made by his own research and that of his colleagues, but also for his contribution at a national level as a scientific director and counsellor.[8] We shall follow his career in both aspects.

Dale's mess of pottage

It is characteristic of Dale that his career developed along a thoroughly unorthodox route. He qualified in medicine at Cambridge and London, at a time when there was considerable social distinction between gentlemanly occupations, including university posts, and jobs in industry. However, not many academic posts existed for the numerous very able graduates who were emerging from the Cambridge school of physiology. Dale went to University College for a year or so, but thought that the academic future looked bleak. It was for this reason that he accepted a place in a laboratory, situated in a small park at Herne Hill in south London,[9,10] set up by the proprietor of a pharmaceutical business Burroughs, Wellcome and Company. Dale's friends believed that he was selling his birthright of academic freedom for a mess of industrial pottage. However, Dale wanted a salary on which he could afford to marry, and also wanted to work independently and, as he put it, 'make his own mistakes'.[11]

Henry Wellcome (later Sir Henry) (1853–1936) was among the first manufacturers of medicines to see the value of research laboratories under his direct control. He and Silas Burroughs (1846–1895), both American pharmacists who adopted British nationality, founded the firm of Burroughs, Wellcome and Company in 1880. Burroughs died at a relatively early age, and after 1895 the privately owned company was directed by Wellcome alone.[12] He used his business to finance many of his personal interests, including archaeological exploration in Egypt, medical research in tropical diseases, and studies in the history of medicine. In England he established chemical and physiological research laboratories, the latter being at some distance from the commercial and industrial units of the company. Their exact function was not clearly defined: they 'did a number of important services to the business of Burroughs Wellcome & Co., but they were also rather a pet project of the sole proprietor of that business'[11] and Wellcome appears to have prided himself on their independent academic standing. Certainly he selected staff who achieved high scientific distinction. Eight of the staff of the Wellcome Physiological Research Laboratory between its foundation and 1914 were later elected Fellows of the Royal Society,[13] and Wellcome's standing and contributions to science were such that he himself was elected F.R.S. in 1932.

Wellcome had good commercial reasons for investigating the medicines prepared from the parasitic mould, ergot, which grows on rye.[14,15] Ergot merits a brief digression. It has a long and romantic history, which includes strange epidemics of poisoning by bread made from infected rye. The victims had bizarre symptoms, which included burning pains, known as St Antony's fire, hallucinations, and, if eligible, miscarriages. Traditionally, the disease was relieved by making a pilgrimage to the shrine of St Antony. If it is correct that St Antony's shrines were located in districts where rye was not grown, the recovery of the pilgrims does not require a supernatural explanation. The ability of ergot to cause miscarriages was turned to medicinal use by accoucheurs and midwives because ergot accelerated the process of childbirth and prevented bleeding from the womb after delivery of the infant.[16] Its use to accelerate the progress of labour was a dangerous practice, and naturally the danger was greater if the potency of the material used, always somewhat variable, happened to be above average. Burroughs Wellcome included the liquid extract of ergot among the medicines which they sold, and Wellcome saw the obvious advantages of supplying a standardized preparation. He encouraged the men he had recruited to examine the possibilities. They included a chemist, George Barger (1878–1939), and Dale, whom Barger had already met in Cambridge.

An alkaloid had been isolated from ergot in 1875, but it did not account for the action on the uterus. Numerous other more or less purified fractions had been described, but none had the required properties. At Wellcome's request, Dale studied ergot, though without much pleasure, 'the pharmacy, pharmacology and therapeutics of that drug being then in a state of obvious confusion'.[11] Barger was already working on the problem and in 1907 isolated a new alkaloid, which was named ergotoxine.[17]

Ergotoxine provided Dale with his first major piece of independent research. It also gave him, as he put it, 'an immediate opportunity of making a mistake of my own—a really shocking "howler"'.[18] As a result of taking short cuts in testing adrenal extracts for the Burroughs Wellcome factory, he discovered that ergotoxine reversed the response of the blood pressure to adrenaline, i.e. that adrenaline caused a fall instead of a rise in blood pressure after ergotoxine had been administered. The observation was unexpected and puzzling, but started a very long trail of investigations which culminated after 60 years in the development of the beta-blocking drugs (see Chapter 12).

Wellcome evidently thought very highly of Dale. In 1906 the Directorship of the Physiological Research Laboratory fell vacant, and Dale was appointed to the post. For 8 years he undertook a wide range of administrative duties while continuing a very fruitful research career. The laboratories were deeply involved in the production of diphtheria and tetanus antitoxin. Improved methods of producing and assaying antitoxin were continually being sought. Dale was very interested in the technical side of this and extended his research interests, particularly to studying the problems which arose when patients or animals became sensitized to the proteins of horse serum and suffered severe allergic reactions after repeated doses of antitoxin. His interest in these problems took him to visit Ehrlich, for whom he developed a great respect and liking and whose collected works he edited many years later.[19]

Somehow he found the time to continue his work on ergot. The mould deteriorates in storage, and some of the products of decomposition had surprising physiological properties. One of the substances isolated and identified chemically was histamine, later recognized as a chemical mediator involved in inflammation, allergies, and some secretions. However, allergy was far from understood at this time. Dale commented long afterwards on his own failure to see the, admittedly most unexpected, connection between histamine from ergot and a chemical mediator of allergic responses.[20]

Barger and Dale also obtained a substance from ergot that had a powerful but very transient action on the heart and blood pressure, and on other organs. These actions resembled the effect of stimulating of the vagus nerve so closely as to suggest that the substance might have a physiological function. It was identified chemically as acetylcholine, some properties of which had already been described by Reid Hunt in Washington.[21] Again the origin from ergot[22] did not suggest that it had any physiological significance, but its imitation of the effects of the vagus was a reminder of Dixon's experiments with muscarine. As the idea of chemical mediators was still out of fashion, such thoughts remained in Dale's mind and did not appear in his published work.

Dale in his memoirs[11] is generous in his praise of Wellcome, for giving him opportunities and freedom to work as he thought fit. He also had the freedom to publish, though commercial considerations generated some conflicts of interest. The scientific productivity of the laboratories under his directorship was very substantial. Its immediate benefit to the commercial success of Burroughs Wellcome was perhaps principally in the antitoxin work. But Wellcome's interest and ambitions ran far beyond the immediate aim of profitability. The freedom afforded to Dale was of great service to the progress of medicine. Meanwhile events in the world outside the Wellcome laboratories created new scientific opportunities, and Dale was drawn into commanding positions in the development of British medical research.

A national institute for medical research

Public awareness, and appreciation, of the benefits of science was growing, but there were grave deficiencies in the national provision of facilities for research. For instance, tuberculosis was rampant, and means for its control were sought. The great authority Robert Koch stated publicly at a meeting in London that the bacillus which infected cattle was so different from the human bacillus that there was virtually no danger of its causing tuberculosis in man. As it turned out, he was wrong, but in the meantime a Royal Commission was appointed to inquire into the matter. The Commission is important to us, not so much for its views on tuberculosis as for the way it went about its business. It began, most unusually for such an august body, not 'by taking evidence, that is to say, by collecting the opinions of others,

... but by conducting experimental observations of [its] own', and it obtained money from public funds for the employment of staff and the cost of experiments.[23] One notes that the Chairman of the Commission was the shrewd creator of the Cambridge school of experimental physiology, Sir Michael Foster.

Tuberculosis was only one of many medical problems which afflicted the growing, mostly poor and urban, population. With the driving force and detailed interest of David Lloyd George (1863–1945), the first British Act of Parliament to provide a scheme of national insurance was passed in 1911. The Act included, in a rather roundabout way, provision of funds which could be used for research. It is not clear how this provision came to be included, but the Royal Commission on tuberculosis had provided a precedent for recognizing the value, and cost, of practical investigations. Presumably, too, it was hoped that the burden of illness and its cost to the community would be reduced in this way. Whatever the facts may have been, the Act was epoch-making not only in providing for national insurance but also for beginning the state funding of medical research.[24]

In order to use the funds, a Medical Research Committee, the forerunner of the present Medical Research Council, was appointed. It was chaired by a scientifically distinguished lawyer, and consisted of two parliamentary members and six eminent medical men each involved in a different aspect of research. The Committee decided that a central institute under its control was essential, and that it should be situated in London. Approaches were made to the Lister Institute, but the negotiations failed.[25,26] Instead a new institute was envisaged. A building was selected in Hampstead, where a hospital was being removed to new premises, and approaches made to certain individuals, including Dale, to become heads of the various proposed divisions. There was much hesitation on the part of those approached, because it was uncertain how the institute would develop. No director was proposed, but the post of permanent secretary to the Committee was created. This post was of enormous importance, because the secretary would be in the delicate position of ensuring the freedom essential for research workers to be innovative while demonstrating that political demands for accountability were met. The post was filled with great distinction by W.M. (later Sir Walter) Fletcher (1873–1933), until then a physiologist at Cambridge.[27,28] Fletcher's reputation stilled many doubts about the acceptability of the new posts. On the evening that Fletcher's appointment was announced, Dale signed his contract as head of the Department of Biochemistry and Pharmacology. He was accompanied in his move from Wellcome by Barger, who contributed the chemical skills to which Dale made no claim of expertise.[29]

These events were concluded during the summer of 1914. Within a month or so, the First World War started. Plans for the new institute were laid aside and temporary quarters were found for Dale's division in the Lister Institute, which had been partly depleted of its own staff by the war. Here he became increasingly occupied with a host of more pressing duties than the investigation of basic physiological mechanisms. Not least among these was a period of several months in which Fletcher was ill and Dale deputized as Secretary of the Medical Research Committee. After the war all the problems had to be solved of adapting a hospital in Hampstead to serve as a research institute, and of making its work fruitful. Much of the labour devolved upon Dale, but he was the youngest departmental head, and it was not until 1928 that he was appointed as the first Director of the Institute. Then at last he could bring matters fully under control and return to giving detailed attention to his own laboratory and the problems of chemical transmission of the nerve impulse.

Resolution of conflicting theories

In the meantime, important progress was made by the Austrian pharmacologist Otto Loewi (1873–1961), who had worked alongside Dale in University College in 1903 and, after the First World War, held the chair of pharmacology at Graz. He isolated and studied the beating heart of a frog to see whether a transmitter substance could be detected when its nerves were stimulated. For technical reasons it was convenient to use the vagus, which slows or, if over active, stops the heart. Loewi used a second heart to detect whether fluid from the first heart contained any substance which, like the vagus, would slow or stop its beat.[30] The experiment resembled Dixon's attempt in Cambridge 15 years earlier to find such a substance, but Loewi succeeded because, among other reasons, the lower temperature at which a frog heart beats gave more chance of survival of the transmitter. The choice was wise, because the transmitter turned out to be a very unstable substance and disappeared very quickly wherever it was detected.

For the first time there was good evidence that a chemical transmitter really existed. Loewi named it vagusstoff, but chemical identification of the minute amounts available was utterly impracticable with the methods in use at that time. One could guess that it might be acetylcholine, but acetylcholine had not been found in animal tissues: both acetylcholine and histamine were isolated from mammalian organs and identified chemically some years later.[31] Vagusstoff might just as well be any number of known or unknown substances. The one sure way it could be detected was by its action on the heart (but with a little ingenuity it was possible to show that it acted on other organs controlled by the vagus) and to build up a profile of its activity on different tissues, which could be used to see if particular substances imitated it and so could be identified with it.

As eserine augmented the action of the vagus on the heart, Loewi examined its effect on the transmitter, and found that it delayed the disappearance of the unstable substance. Furthermore, it became evident that a substance was present in the heart and in other tissues which inactivated the transmitter. This substance was later isolated and identified as an enzyme, or organic catalyst, i.e. a substance which accelerated a chemical reaction without itself being changed at the end of it. This enzyme was later named cholinesterase and was shown to be made inactive by eserine,[32] which explained why eserine prolonged the action of vagusstoff. The complex effects of eserine, which had been occupying pharmacologists since the time of Christison, were becoming understood and eserine became identified as a useful tool for recognizing whether an effect was attributable to the parasympathetic transmitter.

The next 10 years saw clear resolution of the alternatives propounded 50 years earlier by du Bois-Reymond. The case for regarding adrenaline as the sympathetic transmitter was strengthened by analogy, but a number of detailed discrepancies persisted. Almost at the same time as Loewi's investigations began to become known, Walter Bradford Cannon (1871–1945) in Harvard was carrying out experiments[33] which gave more information about a possible transmitter at sympathetic nerve endings. It was evidently related to but not identical to adrenaline, and he called it 'Sympathin'. This approach made some progress,[34] but eventually the transmitter was identified as noradrenaline, a compound differing slightly from adrenaline and having similar but not identical physiological actions.[35] This compound had in fact been studied by Dakin (see Chapter 5), by Loewi, and by Barger and Dale in or before 1910, but the significance of its exact spectrum of activity had been overlooked. As Dale observed with scrupulous self-criticism, he 'ought to have seen that nor-adrenaline might be the main transmitter'.[36] But the idea of chemical transmission at sympathetic endings, as distinct from the identity of the transmitter, ceased to be in much doubt by about 1930.

Chemical transmission of involuntary nerve impulses was accepted, but events at voluntary nerve endings, which supply the muscles of breathing, moving, and talking, occur very much faster. Transmissions occur in thousandths of a second and it was difficult to conceive that such fast transmission could be achieved by a chemical mediator, which would have to be released, diffused to its target, and be removed within this time. Electrophysiologists were quite sure that it could not, and for some years there was much controversy and polarization of ideas. However, by improving the techniques, especially for detecting minute amounts of transmitter by biological assay, chemical transmission was shown without doubt to take place at voluntary motor nerve endings. Much progress was made, particularly in physiological laboratories in Germany, until the ominous year of 1933 when Hitler came to power and a galaxy of brilliant scientists were obliged to flee from anti-semitism to England, America, or wherever refuge could be found. Some, already warmly esteemed colleagues of English scientists, came to London, Oxford, or Cambridge and provided a great concentration of skills in these fields. Dale's laboratory, 'famous all over the world, was regarded as the Mecca to which British and foreign scientists were anxious to come'.[37] With so much talent, success came quickly. By 1936, when Dale and Loewi shared a Nobel prize for discoveries on the humoral transmission of the nerve impulse, all but certain outstanding electrophysiologists were convinced.[8,38–40]

The practical applications which gave new drugs for the treatment of intractable diseases did not follow immediately. Once again a world war intervened. But the 20 years between the First and Second World Wars were crucial in understanding the chemical transmission of 'information', or messages, from cell to cell—and so provided a basis for choosing substances which would imitate, augment, or suppress the passage of messages crucial in the regulation of bodily functions.

Notes

1. Merrington, W. R. (1976). *University College Hospital and its medical school: a history.* Heinemann, London.
2. Geison, G. L. (1978). *Michael Foster and the Cambridge school of physiology.* Princeton University Press.
3. Du Bois Reymond's observations appear to have been forgotten by the early twentieth century, but were resurrected by H. H. Dale (1937–8). The William Henry Welch lectures. Acetylcholine as a chemical transmitter of the effects of nerve impulses. *Journal of the Mount Sinai Hospital* 4, 401–29.
4. Langley, J. N. (1906). On nerve endings and on special excitable substances in cells. *Proceedings of the Royal Society, ser. B* 78, 170–94.
5. Elliott, T. R. (1904). On the action of adrenalin. *Journal of Physiology* 31, xx–xxi.
6. Elliott, T. R. (1905). The action of adrenaline. *Journal of Physiology* 32, 401–67.
7. Dixon, W. E. (1907). On the mode of action of drugs. *Medical Magazine* 16, 454–7.
8. Feldberg, W. (1970). Henry Hallet Dale 1875–1968. *Biographical Memoirs of Fellows of the Royal Society* 16, 77–174.
9. Feldberg, note 8, p. 92f..
10. Tansey, E. M. (1989). Science commerce and authority: the Wellcome Physiological Research Laboratories and the Home Office 1900–1901, *Medical History* 33, 1–41.
11. Dale, H. H. (1953). *Adventures in physiology, with excursions into autopharmacology* p. xi. Pergamon Press, London.
12. Turner, H. (1980). *Henry Wellcome: the man, his collection and his legacy.* The Wellcome Trust and Heinemann, London.

13. Kellaway, C. H. (1948). The Wellcome Research Institution. *Proceedings of the Royal Society, ser. B* 135, 259–70.
14. Barger, G. (1931). *Ergot and ergotism.* Gurney and Jackson, London.
15. Berde, B. and Schild, H. O. (eds) (1978). Ergot alkaloids and related compounds. *Handbook of Experimental Pharmacology*, vol. 49. Springer-Verlag, Berlin.
16. Stearns, J. (1807). The introduction of ergot. *New York Medical Repository*, vol. xi. Reprinted in Thoms, H. (1935) *Classic contributions to obstetrics and gynaecology* pp. 21–5. Thomas, Springfield, Illinois.
17. Barger, G. and Carr, F. H. (1907). The alkaloids of ergot. *Journal of the Chemical Society* 91, 337–53.
18. Dale, H. H. (1953). *Adventures in physiology, with excursions into autopharmacology* pp. xi–xii. Pergamon Press, London. The studies recorded are in Dale, H. H. (1906). On some physiological actions of ergot. *Journal of Physiology* 34, 163–206.
19. Dale et al., ch. 3, note 21.
20. Dale, H. H. and Laidlaw, P. P. (1910). The physiological significance of β-iminazolylethylamine. *Journal of Physiology* 41, 318–44.
21. Hunt, R. and Taveau, R. de M. (1906). On the physiological action of certain cholin derivatives and new methods for detecting cholin. *British Medical Journal* ii, 1788–91.
22. Ewins, A. J. (1914). Acetylcholine, a new active principle of ergot. *Biochemical Journal* 8, 44–9.
23. Interim report of the Royal Commission on Tuberculosis. Quoted by Thomson, A. L. (1973). *Half a century of medical research. Vol. 1: Origins and policy of the Medical Research Council (UK).* Her Majesty's Stationery Office, London.
24. Ibid, p. 11.
25. Chick, H., Hume, M., and Macfarlane, M. (1971). *War on disease, a history of the Lister institute* pp. 120f. Andre Deutsch, London.
26. Thomson, A. L. (1973). *Half a century of medical research. Vol. 1: Origins and policy of the Medical Research Council (UK).* Her Majesty's Stationery Office, London.
27. Ibid, p. 31.
28. Fletcher, M. (1957). *The bright countenance. A personal biography of Walter Morley Fletcher.* Hodder and Stoughton, London.
29. Feldberg, W. (1970). Henry Hallett Dale 1875–1968. *Biographical memoirs of Fellows of the Royal Society* 16, 77–174 at p. 103.
30. Loewi, O. (1921). Über humorale übertragbarkheit der herznervenwirkung., *Archiv für die gesamte Physiologie* 189, 239–42; 193, 201–13.
31. Dale, H. H. and Dudley, H. W. (1929). The presence of histamine and acetylcholine in the spleen of the ox and the horse. *Journal of Physiology* 68, 97–123.
32. Loewi, O. and Navratil, E. (1926). Über das Schicksal des Vagusstoffs. *Archiv für die gesamte Physiologie* 214, 678–88.
33. Cannon, W. B. and Uridil, J. E. (1921). Studies on the conditions of activity in endocrine glands. VIII. Some effects on the denervated heart of stimulating the nerves of the liver. *American Journal of Physiology* 58, 353–4.
34. Cannon, W. B. and Rosenblueth, A. (1933). Studies on conditions of activity in endocrine organs. XXIX. Sympathin E and Sympathin I. *American Journal of Physiology* 104, 557–74.

35. Euler, U. S. von (1946). A specific sympathomimetic ergone in adrenergic nerve fibres (sympathin) and its relations to adrenaline and nor-adrenaline. *Acta Physiologica Scandinavica* 12, 73–97.
36. Dale, H. H. (1953). *Adventures in physiology with excursions into autopharmacology* p. 98. Pergamon Press, London.
37. Feldberg, W. (1970). Henry Hallett Dale 1875–1968, *Biographical Memoirs of Fellows of the Royal Society* 16, 77–174, at p. 124.
38. Dale, H. H. (1934). Chemical transmission of the effects of nerve impulses. *British Medical Journal* i, 835–41.
39. Bacq, Z. M. (1975). *Chemical transmission of nerve impulses. A historical sketch.* Pergamon Press, Oxford.
40. Feldberg, W. (1977). The early history of synaptic and neuromuscular transmission by acetylcholine: reminiscences of an eye witness. In *The pursuit of nature* pp. 65–83. Cambridge University Press, Cambridge.

Chronology of Chapter 5

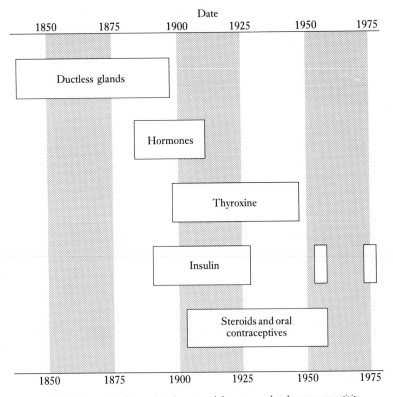

Date

| | 1850 | 1875 | 1900 | 1925 | 1950 | 1975 |

Ductless glands

Hormones

Thyroxine

Insulin

Steroids and oral
contraceptives

| | 1850 | 1875 | 1900 | 1925 | 1950 | 1975 |

NB The blocks show periods discussed in the text, and do not mean that there was no activity about the subject at other times.

5

Replacement therapy

Early ideas about ductless glands

The sequence of discoveries which led from Du Bois Reymond's idea of a messenger 'substance that might pass on excitation' (see Chapter 4) to the proof of its existence extended over 60 years, and has taken us far from its starting point. The theory of a messenger substance has wider possibilities, and variations of it came to other minds during the nineteenth century, when anatomists and physiologists were studying several small organs in various parts of the body. They included the pituitary at the base of the brain, the thyroid in the front of the neck, and the adrenals (also called suprarenals) at the back of the abdomen near the kidneys. Under the microscope these organs differed considerably but they all looked like glands, i.e. tissues which made some kind of juice. Unlike most glands, they had no ducts to carry their secretions away: the only escape would be by the blood stream or the lymph.

There were many speculations about their purpose. T. W. King (1809–1847), a lecturer in pathological anatomy at Guy's Hospital in London, wrote in 1836 about the thyroid gland,

We may one day be able to shew, that a particular material principle is slowly formed, and partially kept in reserve; and that this principle is also supplementary, when poured into the descending cava, to important subsequent functions in the course of the circulation.[1]

This was remarkably accurate as far as it went, but there was little or no concept of what the important functions might be, and even less of the identity of the principle.[2]

A clue about another ductless gland came from the post-mortem observations in 1855 of the physician Thomas Addison (1793–1860), also at Guy's Hospital, that the adrenal glands were diseased or destroyed in patients who had died of a wasting disease now known by his name.[3] A year later the French physiologist Édouard Brown-Séquard[4] (1817–1894) showed that the removal of the adrenal glands of animals had fatal consequences,[5] and his colleague D. Vulpian (1826–1887) described characteristic colour reactions of the central part of the adrenal gland when iodine or ferric chloride was applied to it, and showed that material with the same reactions was present in the blood which flowed away from the gland.[6] At the Medical Congress in London in 1881, the President of the Section of Medicine, Sir William Gull, (1816–1890) referred in his address[7] to 'the strange conditions of Addison's disease' and its relation 'to the suprarenal bodies, themselves probably but nerve centres, and related, at least by structure, to the system of the pituitary gland'. Facts and ideas were accumulating, but it was too early to sort them out.

Apparently Gull did not refer to some observations of his own on the failure of the thyroid gland, which he recognized in certain patients who suffered from apathy, thickening of the skin, and loss of hair. Cases continued to be recognized, and attention was drawn to the jelly-like swelling of the subcutaneous tissues, for which the name myxoedema was invented.[8] The Clinical Society of London set up a committee to review the subject of myxoedema. Its report appeared in 1888, but by that time a good deal of experimental evidence had accumulated which alone gave much value to the deliberations of the committee.

Progress in surgical techniques, and particularly in prevention of sepsis of operating wounds, allowed many new kinds of operation to be attempted. In severe cases of simple goitre, or swelling of the thyroid gland, the windpipe immediately behind the gland was liable to become obstructed, and could cause suffocation if the obstruction was not removed. The new surgery made this possible and operations were performed, particularly in Switzerland where goitre was frequent. Reports were published about the consequences of partial or total removal of the thyroid gland in patients and experimental animals. If too much thyroid was removed, patients, and animals, became myxoedematous, but dogs could be kept well if thyroid glands from another dog had previously been transplanted into the abdominal cavity.

In 1890 two surgeons in Lisbon tried introducing the halves of a sheep's thyroid gland under the skin of a myxoedematous woman and reported that her condition improved rapidly. George Murray (1865–1939), a physician at that time in Newcastle-upon-Tyne, saw in the Lisbon report that the patient began to improve on the day after the operation. This was too rapid to be due to a growth of blood vessels into the transplanted gland, so he inferred that the 'juice' of the gland had been absorbed into the patient's system, and that the 'juice' was all that was needed. On this basis he prepared a glycerine extract of sheep's thyroid, with carbolic acid (phenol) added as a preservative, and treated a myxoedematous woman aged 46 with subcutaneous injections of the extract. Her movements became easier and her mind clearer; the thickening of her eyelids diminished; her hair began to grow and menstruation returned.[9] This was the first successful treatment of a patient with a hormone deficiency.

Other doctors promptly showed that thyroid extract, or whole thyroid, was also efficacious by mouth, and that treatment by injection was unnecessary. So a means of treating 'Gull's disease' had been found, and further experience showed that it was generally reliable. An early failure of such treatment was later traced to the supply of thymus gland instead of thyroid by a butcher. Murray's original patient lived for 28 more years before she died of heart failure.[10]

Hormones

Murray's discovery was one of a family of observations, made at about the same time, which showed that various glands contained powerful and specifically active substances. Brown-Séquard's work on the adrenals was full of interesting observations but not crowned with success. He extended his studies to testicular extracts and published papers in 1889 on the therapeutic, rejuvenating, role of testicular extracts.[11,12] He was himself over 70 years of age by this time, and his experimental evidence was slender, but the idea of rejuvenation had great popular appeal. The manufacture of glandular extracts supposed to possess this property followed more swiftly than the scientific assessment of the evidence. Moreover, enthusiasts extended the idea considerably, so extracts of the heart were made and recommended for heart disease, extracts of grey matter from the brain for neurasthenia, of the kidney for uraemia and so on,[13] but the remedies proved worthless.

With hindsight, one can distinguish charlatans from speculators and speculators from serious and conscientious investigators. In 1890 it must have been more difficult. Discoveries of toxins and antitoxins were being made about the same time. Who could tell which of the novelties to trust? The effectiveness of Behring's diphtheria antitoxin when properly manufactured was confounded by the failure of ill-made material. The hopes and failures associated with a material called tuberculin, which Koch had produced in the hope of immunizing against tuberculosis, added to the confusion about the value of biological products. The Pasteur Institute in New York, which had been set up for the preventive treatment of hydrophobia (rabies) and for the study of contagious diseases, also undertook the preparation of extracts of organs for therapeutic purposes. It was obviously convenient to manufacture glandular extracts and serum preparations and perhaps other biologicals in one place and to dispense them from a single institution, but far from easy to establish control over such activities, to identify which products had reliable effects, and to ensure an adequate standard of quality in the products.[13]

The best progress was made by those who sought to analyse and purify the supposed active principles of the various biological materials. The observations made by Vulpian 40 years earlier of a reactive and potentially identifiable substance in the adrenal glands and in the veins which drained them provided a basis for research at University College, London. Professor Sharpey-Schafer (1850–1935) and his colleague George Oliver (1841–1915) discovered that simple extracts of the glands were very powerful indeed in raising the blood pressure.[14] The action was fleeting, but easy to detect, and obviously important in physiology. As sufferers from Addison's disease were apt to have a low blood pressure, it looked as if it might be important also for medical treatment.

As it happens, the active principle in Oliver and Schafer's extracts is composed of small molecules each containing only 22 atoms and well within the range of chemical knowledge and techniques at the time. Isolation of the active principle was attempted in several laboratories, most notably by John J. Abel (1857–1938) at Johns Hopkins University in Baltimore. Its chemical identity was established[15] and was confirmed by synthesis in several laboratories.[16,17]

Two commercial firms showed an early interest, Parke-Davis of Detroit in the USA and Meister Lucius und Bruning of Hoechst in Germany. Both acted in ways which gave concern to academic scientists. Parke-Davis registered the name 'adrenalin' as a trade mark in the United States, and Dale in England insisted that the name was public property. Dale and Wellcome were at variance on this matter, Wellcome wishing to respect trade rights. Dale won, as far as England was concerned. In America the trade name stood, and American scientists use the name 'epinephrine' for what is known in Europe as adrenaline.

The German firm was not involved in the dispute about names. They worked in secret and did not disclose their methods of making adrenaline until they were patented and the patents had been published. The English physiologist and chemist H. D. Dakin, at the Lister Institute, commented very firmly in his paper on 'The synthesis of a substance allied to adrenalin',

A considerable part of my work has, however, been anticipated by workers in the laboratory of Meister, Lucius & Bruning (D R-P., 157,300) and whilst I wish to disclaim any pretensions to priority, I take the opportunity of stating that my results were entirely independently arrived at, and that, owing to the method of publication adopted (Patent Specifications), it is only recently that I have become acquainted with the main portion of their work'.[18]

Both firms were left with the problem of maintaining commercial practices without offending academic scientists who insisted on freedom of knowledge and whose discoveries were the life blood of commercial innovation.

Initially it seemed obvious that adrenaline should provide a remedy for adrenal deficiency (Addison's disease), but matters did not work out so simply. Much had still to be learnt about the secretions of the adrenal gland and their relation to other bodily activities. In the meantime, the idea of substances being secreted by one organ and transferred in the blood stream to act on other organs became more familiar. E. H. Starling (1866–1927), who had succeeded Schafer as professor of physiology at University College London, showed their importance in controlling the activity of the gut during digestion of food,[19] and introduced the name 'hormone' for such chemical messengers.[20] The name 'endocrine', for the system of glands which secreted hormones into the blood stream, came soon afterwards.[21]

The thyroid problem

Investigation of the active principle in the thyroid progressed slowly. An important advance was made in 1895 when the thyroid was shown to be very rich in the element iodine.[22] It did not follow that the thyroid hormone must contain iodine, but it was worth attempting to isolate the iodine-material, essentially a chemical operation which needed little associated biological testing. Success was achieved by Edward Calvin Kendall (1886–1972), who had been deeply attracted by chemistry from his earliest years and, after brief and unsatisfactory periods with industry and clinical chemistry, spent most of his professional life at the Mayo Foundation in Rochester, Minnesota.[23]

His account of isolating the iodine material starts with an interesting comment on the state-of-the-art in 1915.

Despite the many years of investigation of the thyroid, some of the most fundamental problems concerning this ductless gland are still the subject of speculation and controversy, and the certain progress which follows quantitative knowledge has not yet materialized. Since quantitative values cannot be obtained by clinical observations alone, and since pathological studies of the glands cannot of themselves solve the function of the thyroid, it becomes apparent that nothing short of the actual isolation in pure crystalline form of the chemical substance or substances within the gland, which are responsible for its activity, can furnish the necessary knowledge with which to gain quantitative relationship. Having accomplished the isolation in pure crystalline form of the active agents within the gland, quantitative results may be obtained and ultimate solution of the thyroid problem is within the power of the investigator.[24]

Kendall noted five factors, such as temperature and the concentrations present of particular reagents, which influenced the outcome of his attempts to extract the iodine material. He commented:

These five factors finally became apparent after a consideration of all the results obtained over a course of 2 years investigation. It is remarkable that all these factors were unconsciously controlled during the first purification, especially as it took many months to find out that there were so many separate and distinct influences at work causing the destruction of the compound.[25]

One may wonder about the exact meaning of 'unconsciously controlled'. Would the investigation have progressed at all if the conditions of the first experiments had not, by chance and good sense, been favourable to a successful extraction? It was fortunate for such pioneering work that the iodine-material is fairly stable and is not destroyed by boiling. As usual, it is difficult to separate the influence of profound intuition, good luck, and sound judgement. The fruit of this work was the isolation of about 7 g of pure crystalline material, which contained 65 per cent of iodine, and which Kendall named 'thyroxin'. As with adrenalin, the terminal 'e' was added later when new conventions of chemical nomenclature were established. From the analytic data and various reactions given by the material, Kendall arrived at the empirical formula of $C_{11}H_{10}O_3NI_3$ and suggested a possible structural formula. The biological activity of the material was demonstrated at stages during its isolation by its effects on the heart rate of dogs and later, for the pure substance, in patients with a deficiency of thyroid function.[26]

It is difficult today to imagine such a major advance not attracting widespread attention and being repeated immediately in numerous laboratories, to confirm or refute, and to amplify the findings. But the number of active laboratories in 1919 was very small, and Kendall's undertaking, which involved the collection and processing of about 3 tons of thyroid gland, was not to be copied lightly. The pharmaceutical firm of E. R. Squibb took up Kendall's method to manufacture thyroxine for medicinal purposes. Therapeutic doses were a fraction of a milligram, but the product was immensely expensive and no more effective than the dried thyroid gland from which it was isolated.

No one was in a position to confirm the chemical structure, and Kendall's proposition raised doubts in the mind of an English chemist, C. R. (later Sir Charles) Harington (1897–1972), who was spending a post-doctoral year at the Rockefeller Institute in New York.[27] Encouraged by H. D. Dakin, who had emigrated there from the Lister Institute and was experienced in the isolation of physiologically active substances, Harington synthesized a compound with a structure which resembled the one which Kendall had proposed for thyroxin.[28] Harington's compound had no physiological activity, and the negative result increased Harington's doubts about Kendall's formula. Later he wrote:

... a careful study of Kendall's paper reveals the very slender nature of the evidence from which the formula is deduced. Indeed, it is justifiable to say that, from a chemical point of view, there is no evidence.[29]

To settle the matter, it was necessary to make a new analysis, for which a prohibitively expensive quantity of thyroxine was needed. Harington embarked on a new method of extraction, and achieved a yield of 27 mg of thyroxine per 100 g of fresh weight of thyroid, over twenty times what Kendall had obtained. The elementary analysis differed slightly from Kendall's and on the basis of a variety of chemical reactions a different structure was proposed. With the collaboration of George Barger, by this time professor of medical chemistry in Edinburgh, a compound with the projected structure was synthesized. Part of the synthetic material was administered to a patient by Barger's colleague D. Murray Lyon in the Edinburgh Royal Infirmary. The product of the laboratory induced a therapeutic response as satisfactory as had been achieved with material obtained from the thyroid glands of animals.[30] This was perhaps the first occasion on which a purely synthetic hormone had replaced the missing 'natural' product of the patient's own gland. What would Wöhler, who had long ago triumphantly prepared synthetic urea 'without the help of a kidney' or Berzelius, his sceptical teacher (see Chapter 2) have said to this further crumbling of the bastions of vitalism?

Again the scientific achievement did not lead to any advance in therapeutic practice. Even with the more efficient method of synthesis, pure thyroxine was very expensive and had no practical advantage over dried thyroid for the treatment of patients. Scientifically, however, pure thyroxine was now easier to obtain, and more accurate studies on the mode of action of the hormone became possible. Comparisons between the pure material and the properties of the dried gland showed some discrepancies, which were not fully resolved until many years later.[31] New technical methods then showed that thyroxine itself was altered in the body to a related compound containing three and not four iodine atoms in each molecule. The structure of this compound, tri-iodothyronine, was established by synthesis, and the potency of the new material shown to be several times that of thyroxine.[32,33] Much of the later work was done at the National Institute for Medical Research near London, where Harington had succeeded Dale as Director and continued to have an overview of work which had developed from his early researches.

The discovery of insulin

Several factors favoured the pursuit of the thyroid hormone, most of which could not have been anticipated before the research was done. The thyroid was an easily accessible gland. Its hormone was particularly stable. It contained much iodine, which served as a convenient marker during extraction and separation, and minimized the need for tedious and uncertain bioassays. Of course, the iodine might have been contained in some non-hormonal material, and if so its isolation would have led down a blind alley. Fortunately it was not so. The hormone was readily absorbed from the alimentary canal, so administration of thyroid gland by mouth was an effective therapy and was easily shown to be so. The disease caused by its deficiency was a sad condition, but it was uncommon and had not the tragic features of, for instance, sugar diabetes or diabetes mellitus in a young person, a condition which until the 1920s led inexorably to death within a few months of the onset of symptoms. These circumstances make investigations very much more difficult once success is in sight. The intense pressures to provide life-saving material compete with the painful but vital need to withhold it for investigations essential to establish methods of production.

The cause of diabetes mellitus began to be understood late in the nineteenth century, and its connection with the pancreas (sweetbread) was recognized at about the same time as the thyroid was associated with myxoedema. The Austrian physicians Mering and Minkowsky found that removal of the pancreas from dogs resulted in thirst, wasting, and death, and that the urine of the dogs contained large amounts of sugar.[34,35] Later experiments[36] showed that the pancreas was adding something to the blood as well as providing juices needed for the digestion of food. Under the microscope, particular groups of cells could be seen. They looked like small islands in the rest of the tissue and so were called, after their discoverer, the Islets of Langerhans.[37] In patients who had died of diabetes, they looked different.[38] At least four independent workers, in Berlin, Chicago, New York, and Bucharest, prepared extracts of the pancreas and treated diabetic dogs or diabetic patients with them, but for one reason or another, none established a reproducible therapeutic effect.[39] The nature of the internal secretion of the pancreas was unknown, but it seemed likely to be a protein. Any process of extraction was seriously complicated by the digestive juices. It was feared that these were likely to be released in any attempt to extract the hormone, and might digest and destroy it.

The problem of making a life-saving extract of pancreas was solved in Toronto in 1922-23. The research progressed uneasily and at times stormily. The four principals were Frederick Banting[40] (1891-1941), who having failed to establish himself in general practice, was tempted towards research (for which he had no training whatever) and was filled with an immense drive to solve the problem of diabetes; J. J. R. Macleod[41] (1876-1935), the professor of physiology and an expert in carbohydrate metabolism, who provided facilities for Banting's work and was in the difficult position of doubting whether they were being well used; Charles Best[42] (1899-1978), a very able medical student being trained in research in Macleod's department, and J. B. Collip[43] (1892-1965), a biochemist and visiting scientist on sabbatical leave from the University of Alberta. The summer was hot and the temperaments of all concerned differed widely, but work progressed in spite of much ill feeling.[39]

Banting sought to simplify the problem by tying the pancreatic duct in dogs some time before they were sacrificed to provide material from which it was hoped to extract a hormone. Cells which cannot get rid of their secretions wither, so this manoeuvre was directed towards killing the cells which made insulin-destroying enzymes. Macleod approved the idea and arranged that Best should work with Banting on the problem. The experiments with dogs did not go well, and many dogs died. Later, Macleod suggested that the islet substance might be sufficiently protected if the pancreas was chilled before extraction, and the elaborate duct ligation experiments were abandoned. By one means or another, extracts were obtained which kept diabetic dogs alive: but there is some doubt whether the dogs were truly diabetic or only partially deprived of their own secretion. Collip improved the methods and purified the extracts so that material could be administered to humans. There were many potential patients, and enough crude material was made to demonstrate that the inexorable decline of young diabetics could be arrested and reversed. Co-operation was sought with the Connaught Anti-Toxin Laboratories in Toronto so that larger amounts of insulin could be prepared and an agreement was made which covered immediate progress. At first the crude material, even after Collip's purification, caused fevers and extensive local inflammation on injection, but it kept patients alive. One or two patients, not many: there were no resources to make more than tiny amounts of the material and doses were needed daily by every patient treated,[39] and by a multitude more for whom no treatment could be provided.

The pharmaceutical firm of Eli Lilly of Indianapolis had in 1919 taken the then unusual step of appointing a director in charge of research. He was G. H. A. Clowes (1877–1957), an English born doctor with considerable experience gained in chemical laboratories in Germany, France, and in the chemical warfare department of the United States army. With great wisdom, the firm gave him exceptionally free terms of reference, so that he could seek benefit for the business where he thought fit, and could also continue research on his own.[44] Clowes attended the meeting of the American Physiological Society at Yale on 30 December 1921, heard Banting speak about his work[45] on insulin, and offered collaboration and guidance on patenting. Macleod was reluctant, despite advice from E. C. Kendall, but Clowes persisted. An agreement, dated 30 May 1922, was made between Toronto University and Eli Lilly which gave manufacturing rights to the firm and royalties to the university. Within a year or so, Eli Lilly, promptly followed by manufacturers in other countries, were making insulin in sufficient quantities to begin to meet the needs of diabetics and to convert a fatal disorder into one which, with proper management, was compatible with a normal life.

Personal conflicts broke out again in disputes about credit for the work. Banting and Macleod, who were the most irreconcilable pair, received a Nobel Prize in 1923. Banting, with great generosity, gave half his prize to Best, and Macleod followed suit to Collip. The whole process was neither a model of well-planned scientific research nor a happy example of human team-work. Nor did the Toronto workers by themselves solve a host of problems involved in the large scale collection of raw material, extraction and purification of insulin, and its supply in a state suitable for clinical use. Much of the credit for these tedious technical developments rested with the staff of Eli Lilly. Without effective development, the discovery of insulin would have remained intellectually exciting and without benefit to humanity.

Collip's work showed that insulin was certainly a protein, and so a substance many times more complex than adrenaline or thyroxine. There was at the time no more chance of determining its exact chemical composition and structure than there was of taking a bus ride to the top of Mount Everest. Nor was there any chemical means of measuring how much insulin was present in any solution or precipitate. Short of using patients or, just conceivably, human volunteers, the only way of making such measurements, essential for providing reliable clinical supplies of insulin, was by using laboratory animals. It was the same sort of problem that Ehrlich had handled in preparing toxins and antitoxins of known potency (see Chapter 3), and involved the same difficulties: the ethical questions of experiments on animals and the technical questions of how to perform the assays reliably and economically. In the face of an unequivocal life-saving measure, the arguments against vivisection did not act as a serious deterrent to devising reliable biological assays. Technically, it was desirable to have a stable, standard preparation of insulin, but none was known to exist. In spite of Ehrlich's work 25 years earlier, attempts were made, unwisely, to define a unit of activity in terms of the (highly variable) response of an animal or group of animals. The battle which Ehrlich had fought to define had to be fought again. Moreover, the problem was an international one. As soon as the news of the Toronto discovery was widespread and found to be reliable, every nation which had the facility embarked on the preparation of insulin and was inevitably involved in problems of standardization.

In Britain, the control and supply of biological standards, so far mainly for antitoxins and for organic arsenical substances, was carried out by a division of the National Institute for Medical Research, and the chief adviser on the subject was Henry Dale. At his instigation, H. W. Dudley, at the laboratories in Hampstead, made an essential advance by finding how to prepare insulin as a dry stable powder. An International Physiological Congress, held in Edinburgh in 1923, was a convenient opportunity for holding a conference of those concerned with the standardization of insulin in many countries. Dale attended, taking a sample of Dudley's preparation in his pocket. Later he described the meeting:

We had practically a whole day's discussion, largely futile, on the rival claims for a unit defined as the dose of insulin which would throw 2 out of 3 rabbits into convulsions, as put forward by MacLeod, who tried to dominate the situation on behalf of the Toronto team, and Krogh's insistence that the proper base of the unit would be to determine the dose which would produce the hypoglycaemic convulsions in 6 out of 12 mice. I listened to this wrangle with growing impatience during most of the day, and intervened when I thought that they were getting tired of the controversy without any prospect of a solution. My intervention took the form of insisting that it was complete nonsense to try to define any unit of any remedy in absolute terms of reactions in limited numbers of animals; and that, from the international point of view, the only sensible thing was to obtain the remedy in perfectly stable form, and define the unit in terms of an absolute quantity of such a standard sample, internationally accepted, leaving the methods of its determination to be the subject of indefinite possibilities of experimental improvement. Macleod replied that he had no doubt that such a policy would be ideal, but that he had no reason to believe that the preparation of a sample of insulin, in a dry and indefinitely stable form, was a practical possibility. At that point I was glad to be able to take from my waistcoat pocket a small tube of the preparation, which was to become the first international standard, and roll it across the table to Macleod with the statement—'Well, there it is'.[46]

Scientific advances in standardization allowed more reliable manufacturing and more effective supervision of manufacturers by governmental agencies. The manufacture of insulin was undertaken by pharmaceutical companies in many countries. Every batch of insulin had to be compared with the international standard, directly or usually via a secondary standard. For many years the assays required experiments in animals: without such experiments the insulin available would have been so variable as to be either ineffective or lethal when injected into patients. Not until the 1960s was a method of assay available by purely chemical means. By this time the chemical composition of insulin was fully established,[47,48] though the method of assay did not depend on such knowledge. The success of therapy with insulin was beginning to raise its own problems. The number of diabetics alive and requiring insulin was approaching the world's capacity to manufacture the hormone, because supplies were limited by the number of animals from which fresh pancreas could be obtained. This situation remained problematical until the advent of new methods to make and secrete insulin involving the modification of cultured cells or bacteria by genetic recombination. The process can be conducted on a large scale, and insulin made in this way has been available in several countries for therapeutic use from 1982 onwards.[49,50]

Steroid hormones and the contraceptive pill

The treatment of Addison's disease or adrenal failure remained ineffective until some time after diabetes was controlled. It was known from the time of Brown-Séquard that removal of the adrenal glands from animals was soon followed by death. Adrenaline was discovered, but no success was achieved by attempts to keep animals alive with adrenaline after their adrenal glands had been removed. Adrenaline came from the middle part or medulla of the gland, and so the outer part or cortex of the gland was suspected of having a separate and essential function. Many methods of extraction and assay of material from the outer part of the gland were tried, at first without success, but by about 1930 life-prolonging extracts were being made from the adrenal cortex. As procedures improved, batches of reasonably pure material could be produced and analysed chemically.

The work then converged on hormones secreted by the sex glands. The effect of the adrenals on sexual function was well known, particularly because tumours of the adrenals in young children were associated with precocious puberty.[51] It had long been known that removal of the ovaries was followed by loss of sexual characteristics and functions. About 1900 Knauer, in Vienna, and others showed that the transplantation of the ovaries to a new position prevented the effects of removal.[52] They were therefore acting by releasing a hormone, but no satisfactory way of detecting any hormone was devised until 1923. The basis for such a method was found when simple smears taken from the lining of the vagina of rats were shown to reflect changes in the oestrous cycle, and could be induced by injections of sex hormones.[54] Two kinds of female sex hormone were discovered. The oestrogenic produced the bodily changes necessary for pregnancy to take place and the progestogenic maintained the conditions necessary for pregnancy to continue. As ovaries provided very limited quantities of tissue to work on, progress was slow until it was found that oestrogenic substances could be recovered from the urine. In the early 1930s a substantial number of compounds were isolated and purified from the urine of large animals and were investigated to determine their exact chemical structure.

Much help came from advances made in the late 1920s and early 1930s in chemical studies of an apparently unconnected material. Bile, secreted by the liver, is an essential agent in digestion. Its analysis was an important step towards understanding the mechanisms by which food is converted to materials which can be absorbed and used for energy and body building. A major constituent of bile, cholic acid, was isolated, and, after some uncertainties, was found to have a structure with four linked rings of carbon atoms. Shortly afterwards the same structure was identified in each of the sex hormones and in materials isolated from the adrenal cortex.

Moreover, some of the sex hormones were discovered in the adrenal cortex. Substances with the four ring structure received the general appellation 'steroid', and further variants were found to fulfil a remarkable variety of functions. The steroids included cholesterol, which was known to be present in blood and bile, and possibly associated with abnormally high blood pressure; vitamin D (see Chapter 7); and a group of steroids derived from coal tar which produced cancer when painted repeatedly on the skin (see Chapter 11).[55] A further group of substances, not strictly steroids but closely related, were the active principles of digitalis and other plants with powerful actions on the heart (see Chapter 1). The many biological properties of steroids raised fascinating questions about their mode of action and their evolutionary significance.

At the Mayo Clinic in 1929, P. S. Hench made some puzzling clinical observations which suggested that certain illnesses, notably rheumatoid arthritis, improved during pregnancy or in circumstances in which the adrenal cortex was more active. He encouraged E. C. Kendall, who had already isolated thyroxine, to undertake a study of adrenal cortical hormones. It became evident that the adrenal cortex, like the ovaries, secreted steroid hormones, and a very large number were identified, some of which were identical with the steroids produced by the testes and ovaries. The physiological implications were considerable, and took a long time to work out; research was not helped by the war. In 1950 Reichstein of Zurich and Basel and Kendall shared a Nobel prize for their achievements in steroid chemistry.

The quantities of material available from natural sources were minute, and allowed isolation of pure material only for rare experiments in therapy. The first synthetic steroid with actions like those of the adrenal cortical hormones was made in 1937,[56-59] but its efficacy as a substitute for the natural hormone was limited and practical procedures for making it on an industrial scale were nowhere near in sight. A study of structure–activity relationships led E. C. (later Sir Charles) Dodds (1899–1973), a distinguished biochemist and physician at the Middlesex Hospital in London, to study compounds which imitated only part of the steroid structure, and to discover the oestrogenic activity of a substance later named stilboestrol (diethylstilboestrol in the USA).[60,61] It was the first effective synthetic oestrogen, and was introduced into clinical practice in England in 1937.

A fresh stimulus came in 1941 with a rumour that in Germany adrenal cortical extracts were being administered to aircraft pilots. This enabled them to fly above 40,000 feet without suffering ill-effects from lack of oxygen. The rumour had no scientific basis, but it led to work by Kendall and his colleagues and by the pharmaceutical firm of Merck in New Jersey to produce synthetic cortical hormones.[62] By 1948, 9 g of cortisone had been synthesized, long after the military rumour had been scotched and the original need had vanished. By complex means, some of the 9 g came to the Mayo Clinic and allowed Hench to investigate its possibilities in patients with rheumatoid arthritis. The results were dramatic, affording immediate relief from the crippling pain of the diseased joints, and within the next few years many other therapeutic applications of the cortical steroids were discovered. Only with time were the dangers of such therapy apparent. Meanwhile the use of steroid substances as anti-inflammatory or immuno-suppressive drugs was widely adopted, and the stimulus to the pharma-ceutical industry was powerful. The means of production were improved rapidly and the range of compounds available was extended. The new synthetic compounds included some which had clear advantages over cortisone for various therapeutic purposes. However, as experience grew, it also became clear that there were so many hazards and disadvantages in the use of steroid drugs that they were neither panaceas nor agents to be used lightly.

The new steroid hormones and their synthetic relatives were valuable agents for investigating fertility and sterility. Gregory Pincus, of the Worcester Foundation for Experimental Biology in Massachusetts, who had long been studying the physiology of reproduction, began early in the 1950s to investigate the possibility of making a safe oral contraceptive based on hormonal adjustment of the normal menstrual cycle.[63] As oestrogenic ovarian hormones create conditions in which pregnancy can be established, and progestogens maintain these conditions, he speculated that giving progestational hormones before oestrogens might produce a state of temporary infertility, because a pregnant, progesterone-controlled uterus resists further implantation of fertilized ova, and the shedding of ova from the ovaries stops. Rabbits were treated with massive doses of progesterone, were mated, and were found to be infertile. Later, after the effects of the progesterone had disappeared, their fertility returned to normal, so the effects were only temporary.

At this stage, Pincus collaborated with John Rock, a gynaecologist from Harvard University, who was investigating the treatment of infertility in women by using sequences of oestrogens and progestogens. The collabora-tion was fruitful, and made more so because Pincus obtained numerous progestational steroids from various pharmaceutical firms and, on the basis of trials in rats and rabbits, selected a number which were particularly suitable for clinical testing. The results showed that Rock and Pincus had found effective ways of controlling ovulation, and so had a remedy for one kind of infertility as well as a potential oral contraceptive.[64]

Field studies began in Puerto Rico in 1955 under the direction of Pincus and his associates. The progestin used was soon found to give effective control of fertility, provided that it was taken reliably. Later, some of the progestins used in the trials were found not to be quite pure and to contain small amounts of oestrogen. These preparations had worked particularly well, and it became evident that the oestrogen enhanced the suppressive effect of the progestin so that smaller doses were effective. The range of possible pills was therefore substantial, as there were many synthetic progestins and oestrogens to choose from and each was likely to have a different optimal dose. The results at Puerto Rico prompted many studies elsewhere and experience was soon available almost worldwide.[65]

Regular use of the 'Pill', whatever its exact composition, involved changing the quantities of hormone in a woman's body throughout every successive menstrual cycle and might be continued for years on end. The implications of such physiological trespass aroused much concern about the best ways of assessing the long-term hazards. The human sexual cycle and span of life differed from those of other species enough to limit the value of animal experiments. The problem was neatly if ironically put in the words, 'No woman should be kept on the pill for 20 years until in fact a sufficient number of women have been kept on the pill for 20 years'.[66] In practice the advantages of the pill were so great that fears of ill-effects were only a slight deterrent to its use. Oral contraception became widespread in 1963, and in a short time more experience had been gained of this kind of medication than of any other. The first serious reports of ill-effects, clotting of blood inside blood vessels, appeared in 1968,[67] and even now, after a quarter of a century, the various known ill-effects are relatively infrequent and call only for caution in particular circumstances.

Drugs and hormones

Other hormones were discovered and endocrinology became a distinct science.[68] Many diseases have been recognized that are due to the overactivity of a gland or to the insensitivity of tissues to hormones. As basic studies in physiology and biochemistry are extended, the synthesis of each hormone *in vivo*, its storage and release, its mode of action on the tissues, and the ways in which it is inactivated all begin to be understood. Other uses have been found for the hormones themselves, in the control of fertility, immunity, and cancer (see Chapter 11). Simpler substances have been found with hormone-like actions, and drugs discovered which promote the release or block the synthesis or the actions of hormones. The details of each of these advances are often highly technical, and each one would need more space than is appropriate here.

Notes

1. King, T. W. (1836). Observations on the thyroid gland. *Guy's Hospital Reports* 1, 429–47, at p. 443.

2. Harington, C. R. (1933). *The thyroid gland. Its chemistry and physiology.* Oxford University Press, London.

3. Addison, T. (1855). *On the constitutional and local effects of disease of the supra-renal capsules.* Samuel Highley, London.

4. Olmsted, J. M. D. (1946). *Charles Edouard Brown-Séquard. A nineteenth century neurologist and endocrinologist.* Johns Hopkins, Baltimore.

5. Brown-Séquard, E. (1856). Recherches expérimentale sur la physiologie et la pathologie des capsules surrénales. *Comptes rendus hebdomadaires des Séances de l'Académie des Sciences* 43, 422–5; 542–6.

6. Vulpian, D. (1856). Note sur quelques réactions propres è la substance des capsules surrénales. *Comptes rendus hebdomadaires des Séances de l'Académie des Sciences* 43, 663–5.

7. Gull, W. (1881). Medicine. *Lancet* ii, 224–6.

8. Gull, W. (1874). On a cretinoid state supervening in adult life in women. *Transactions of the Clinical Society of London* 7, 180–5.

9. Murray, G. R. (1891). Note on the treatment of myxoedema by hypodermic injections of the thyroid gland of a sheep. *British Medical Journal* ii, 796–7.

10. Murray, G. R. (1920). The life-history of the first case of myxoedema treated by thyroid extract. *British Medical Journal* i, 359–60.

11. Brown-Séquard, C. E. (1889). Du role physiologique et therapeutique d'un suc extrait de testicules d'animaux d'apres nombres de faits observes chez l'homme. *Archives de physiologie normale et de pathologie* (5e ser.) 1, 739–46.

12. Brown-Séquard, C. E. (1889). Des effects produits chez l'homme par des injections sous-cutanée d'un liquide retiré des testicules frais de cobaye et de chien. *Comptes rendus de la Société de Biologie* 1, 415–19.

13. Borell, M. (1976). Brown-Séquard's organotherapy and its appearance in America at the end of the nineteenth century. *Bulletin of the History of Medicine* 50, 309–20.

14. Oliver, G. and Schafer, E. A. (1895). The physiological effects of extracts of the suprarenal capsules. *Journal of Physiology* 18, 230–76.

15. Abel, J. J. (1903). On the elementary composition of adrenalin. *American Journal of Physiology* 8, xxix–xxx.

16. Aldrich, T. B. (1901). A preliminary report on the active principle of the suprarenal gland. *American Journal of Physiology* 5, 457–61.

17. Takamine, J. (1901). Adrenalin the active principle of the suprarenal glands and its mode of preparation. *American Journal of Pharmacy* 73, 523–31.

18. Dakin, H. D. (1906). The synthesis of a substance allied to adrenalin. *Proceedings of the Royal Society*, ser. B 76, 491–7.

19. Starling, E. H. (1905). The chemical correlation of the functions of the body. *Lancet* ii, 339–41; 423–5; 501–3; 579–83.

20. Ibid. p. 340, col. 1. The derivation is from the Greek 'ορμαο [hormao], I set in motion, urge on.

21. In 1913, according to the Shorter Oxford English Dictionary, 3rd edn, revised 1959.

22. Baumann, E. (1895). Ueber das normale Vorkommen von Jod im Thierkorper. *Zeitschrift für Physiologische Chemie* 21, 319–30.

23. Kendall, E. C. (1971). *Cortisone.* Charles Scribner's Sons, New York. This work is an autobiography of Kendall's scientific career and covers much more than the title suggests.

24. Kendall, E. C. (1919). Isolation of the iodine compound which occurs in the thyroid. *Journal of Biological Chemistry* 39, 125–47, at p. 125.
25. Ibid. p. 132.
26. Kendall, E. C. and Osterberg, A. E. (1919). The chemical identification of thyroxin. *Journal of Biological Chemistry* 40, 265–334.
27. Himsworth, H. and Pitt-Rivers, R. (1972). Charles Robert Harington 1897–1972. *Biographical Memoirs of Fellows of the Royal Society* 18, 267–308.
28. Harington, C. R. (1925). Synthesis of 3,4,5-triiodophenylpyrrolidone carboxylic acid, a possible isomer of thyroxin. *Journal of Biological Chemistry* 64, 29–39.
29. Harington, C. R. (1926). Chemistry of thyroxine. I. Isolation of thyroxine from the thyroid gland. *Biochemical Journal* 20, 293–9.
30. Harington, C. R. and Barger, G. (1927). Chemistry of thyroxine. III. Constitution and synthesis of thyroxine, with a note on physiological test of synthetic thyroxine, by D. Murray Lyon. *Biochemical Journal* 21, 169–83.
31. Chalmers, J. R., Dickson, G. T., Elks, J. and Hems, B. A. (1949). The synthesis of thyroxine and related substances. Part V. A synthesis of L-thyroxine from L-tyrosine. *Journal of the Chemical Society* 3424–33.
32. Gross, J. and Leblond, C. P. (1951). Metabolites of thyroxine. *Proceedings of the Society for Experimental Biology* 76, 686–9.
33. Gross, J. and Pitt-Rivers, R. (1952). The identification of 3:5:3'-L-triiodothyronine in human plasma. *Lancet* i, 439–41.
34. Mering, J. von and Minkowski, O. (1889). Diabetes mellitus nach pankreas extirpation. *Archiv für experimentelle Pathologie und Pharmakologie* 26, 371–87.
35. Houssay, B. A. (1952). The discovery of pancreatic diabetes. The role of Oscar Minkowski. *Diabetes* i, 112–16.
36. Hedon, E. (1909). Sur la technique de l'extirpation du pancreas chez le chien, pour réaliser la diabète sucre. *Comptes rendus de la Société de Biologie* 66, 621–4. According to Bliss (note 39), much of Hedon's work was done 10–15 years earlier.
37. Sakula, A. (1988). Paul Langerhans (1847–1888): a centenary tribute. *Journal of the Royal Society of Medicine* 51, 414–15.
38. Opie, E. L. (1910). *Diseases of the pancreas*, 2nd edn. Lippincott, Philadelphia.
39. Bliss, M. (1982). *The discovery of insulin*, pp. 45–153. McClelland & Stewart, Toronto.
40. Best, C. H. (1942). Frederick Grant Banting 1891–1941. *Obituary Notices of Fellows of the Royal Society* 4, 21–6.
41. Cathcart, E. P. (1935). John James Rickard Macleod 1876–1935. *Obituary Notices of Fellows of the Royal Society* 1, 585–9.
42. Young, F. and Hales, C. N. (1982). Charles Herbert Best 27 February 1899–31 March 1978. *Biographical Memoirs of Fellows of the Royal Society* 28, 1–25.
43. Barr, M. L. and Rossiter, R. J. (1973). James Bertram Collip 1892–1965. *Biographical Memoirs of Fellows of the Royal Society* 19, 235–67.
44. Kahn, E. J. (1976). *All in a century. The first 100 years of Eli Lilly and Company*, Eli Lilly, Indianapolis.
45. Banting, F. G. and Best, C. H. (1922). The internal secretion of the pancreas. *Journal of Laboratory and Clinical Medicine* 7, 251–66.
46. Feldberg, W. (1970). Henry Hallett Dale 1875–1968. *Biographical Memoirs of Fellows of the Royal Society* 16, 77–174 at p.122.

47. Sanger, F. (1956). In *Currents in biochemical research*, ed. D. E. Green, pp. 434–59. *Interscience*, New York.
48. Farber, E. (1963). *Nobel prize winners in chemistry, 1901–1961*, p. 282. Abelard Schumann, London.
49. Johnson, I. S. (1982). Authenticity and purity of human insulin (recombinant DNA). *Diabetes Care* 5, suppl. 2, 4–12.
50. Johnson, I. S. (1983). Human insulin from recombinant DNA technology. *Science* 219, 632–7.
51. Bulloch, W. and Sequeira, J. H. (1905). On the relation of the suprarenal capsules to the sexual organs. *Transactions of the Pathological Society of London* 56, 189–208.
52. Knauer, E. (1900). Die ovarientransplantation. Experimentelle Studie. *Archiv für Gynaekologie* 60, 322–76.
53. Stockard, C. R. and Papanicolaou, G. N. (1917). The existence of a typical oestrous cycle in the guinea pig with a study of its histological and physiological changes. *American Journal of Anatomy* 22, 225–65.
54. Allen, E. and Doisy, E. A. (1923). An ovarian hormone. Preliminary report on its localization, extraction and partial purification and action in test animals. *Journal of the American Medical Association* 81, 819–21.
55. Rosenheim, O. and King, H. (1934). The chemistry of the sterols, bile acids and other cyclic constituents of natural fats and oils. *Annual Review of Biochemistry* 3, 87–110.
56. Steiger, M. and Reichstein, T. (1937). Desoxy-cortico-steron (21-Oxy-progesteron) aus Δ^5-3-Oxy-ätio-cholensäure. *Helvetica chimica Acta* 20, 1164–79.
57. Kendall, E. C. (1950). The development of cortisone as a therapeutic agent. *Nobel Lectures Physiology or Medicine, 1942–1962*, pp. 270–88. Published for the Nobel Foundation, 1964, Elsevier, Amsterdam.
58. Reichstein, T. (1950). Chemistry of the adrenal cortex hormones. *Nobel Lectures Physiology or Medicine, 1942–1962*, pp. 290–310. Published for the Nobel Foundation, 1964, Elsevier, Amsterdam.
59. Hench, P. S. (1950). The reversibility of certain rheumatic and non-rheumatic conditions by the use of cortisone or of the pituitary adrenocorticotrophic hormone. *Nobel Lectures Physiology or Medicine, 1942–1962*, pp. 311–41. Published for the Nobel Foundation, 1964, Elsevier, Amsterdam.
60. Dodds, E. C., Golberg, L., Lawson, W. and Robinson, R. (1939). Synthetic oestrogenic compounds related to stilbene and diphenylethane. Part I. *Proceedings of the Royal Society*, ser. B 127, 140–67.
61. Dodds, E. C. (1949). Synthetic oestrogens. *Journal of Pharmacy and Pharmacology* 1, 137–47.
62. Polley, H. F. and Slocumb, C. H. (1976). Behind the scenes with cortisone and ACTH. *Mayo Clinic Proceedings* 51, 471–7.
63. Pincus, G. (1965). *The control of fertility*. Academic Press, New York.
64. Rock, J., Garcia, C. R. and Pincus, G. (1957). Synthetic progestins in the normal human menstrual cycle. *Recent Progress in Hormone Research* 13, 323–39.
65. Pincus, G. (1959). Progestational agents and the control of fertility. *Vitamins and Hormones, New York* 17, 307–24.
66. The remark is attributed to Sir Alan Parkes, and is quoted by D. M. Potts, (1970). The pill pilloried. *Nature* 226, 187–8.

67. Vesey, M. P. and Doll, R. (1968). Investigation of relation between use of oral contraceptives and thromboembolic disease. *British Medical Journal* ii, 199–205.
68. The first journal devoted solely to endocrinology, the *Journal of Endocrinology*, was first published in 1939.

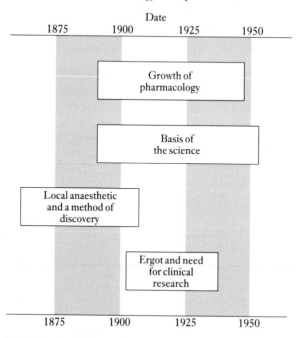

Chronology of Chapter 6

Date

1875   1900   1925   1950

Growth of
pharmacology

Basis of
the science

Local anaesthetic
and a method of
discovery

Ergot and need
for clinical
research

1875   1900   1925   1950

NB The blocks show periods discussed in the text, and do not mean that
there was no activity about the subject at other times.

6

The investigation of drugs

The formative forces

We have talked about the developments in bacteriology which gave us antitoxins, vaccines, and the first antimicrobial drugs, and in physiology and chemistry which gave us hormones. The discoverers studied some particular branch of science or medicine, and found remedies as their work developed: they were not primarily experts in drugs and had no special interest in medicines outside their own field. Two kinds of people did have reason to study medicines as a coherent subject. They were the medical students, who had to learn about all the medicines they might need in their practice as doctors, and the pharmacists and chemical manufacturers who made the medicines. For both these groups a unified science of medicines, the science called pharmacology, was essential. We have seen its academic beginnings in Dorpat, in Strassburg, and in Edinburgh, and its industrial roots, especially in Germany, and we need now to see how the science developed.

In Germany, the chemical industry was far ahead of any other country, and collaboration with universities, research institutes, and clinics was good. We have seen how well Hoechst supported Behring in the manufacture of diphtheria antitoxin, and took over the production of Salvarsan from Ehrlich. Much progress was made in producing pure drugs of known composition which superseded the less well-defined crude preparations—powdered leaves and roots and watery extracts of plants—and in synthesizing and testing new substances related to those of botanical origin. Local anaesthetics were among the earliest synthetic drugs to be developed in this way. Academic standards were high and pharmacology was strongly based, especially since the work of Buchheim's pupil Oswald Schmiedeberg on drugs such as digitalis and muscarine at Strassburg. In the 1880s the Institute in Strassburg acquired an international reputation, and students who came from many parts of the world returned to develop the subject in their own countries.

Among them was John J. Abel (1857–1938), a graduate of the University of

Michigan, who returned to fill the chair of materia medica and therapeutics at that University. He immediately converted the existing traditional course, largely botanical, into one on experimental pharmacology, and set up a research laboratory for his own investigations. In 1893 he was invited to become the first professor of pharmacology in the new medical school which was being founded at the Johns Hopkins University in Baltimore, and there he remained until he retired in 1932. He had great influence on his subject and is often referred to as the 'Father of American Pharmacology'. He recognized the necessity for a society in which pharmacologists could meet to advance their subject, and was the principal founder of the American Society for Pharmacology and Experimental Therapeutics. Abel believed firmly in the importance of chemistry for medicine. His own most important research was on hormones, and, as we have seen, he was one of the first to isolate adrenaline (see Chapter 5). Later he developed experimental methods which paved the way for artificial kidneys, and in 1925 insulin was crystallized[1,2] in his laboratory.

The American Society for Pharmacology was founded at a time when the USA was rapidly developing. The manufacture of drugs depended on small businesses, devoted to pharmaceuticals of plant origin. They had limited or no scientific basis and some of the products and some of the methods of salesmanship were far from creditable. When news of antitoxins reached America, few firms were competent to attempt to make them: much of the early materials were produced by public health authorities, especially in Philadelphia and in New York.[3]

One would have expected a Society for Pharmacology to provide a unique opportunity for collaboration between academic and industrial pharmacologists. However, the United States pharmaceutical industry was in such a condition that the Society, under Abel's presidency, explicitly ruled that membership was not open to workers in industry, and that any person who accepted such a post must resign his membership. Although the industry grew greatly both in size and responsibility, the provision persisted throughout the 1920s and '30s. By that time several eminent pharmacologists had accepted posts as consultants to various firms, and their collaboration often had scientifically fruitful results. These distinguished men usually belonged to the old and more highly valued sister society for physiology, and could hardly be rejected by the pharmacologists without much embarrassment.[4] In the end, what has been called the 'preposterous provision'[5] was rescinded in 1941.

Pharmacology grew throughout the world with much help from American wisdom and charity, including the Rockefeller Foundation. Early in the twentieth century the Foundation supported an enquiry into the basic medical sciences by sending Abraham Flexner (1866–1959) to study medical education in Europe.[6,7] His detached critical vision illuminated the gap between the old style of therapeutics and the faltering advances of modern science which were creating a sense of therapeutic despondency. In 1912, he wrote:[8]

The development of physiology and chemistry as interlocking sciences was bound to result in an inquiring therapeutic mood. For chemistry extricated the active principles of the crude drugs traditionally employed, and experimental physiology created the conditions for accurate observation of their effects. Conflict of therapeutic opinion was general and acrid enough to suggest a test as soon as it became feasible to correlate cause and effect. Tradition ascribed a certain efficacy to, let us say, camphor or sarsaparilla. The 'experience' of one man vindicated tradition; the 'experience' of another went directly against it. Where does truth lie? Organic chemistry and experimental physiology put the pharmacologist in position to determine.

In the last sentence Flexner was developing a mood of over-confidence in laboratory research which had later hazards, but there was no doubt that the research was needed. He went on:

The juncture was also otherwise auspicious: while miscellaneous dosing was still generally prevalent, intelligent practitioners had been infected with nihilistic doubts from two highly divergent sources: the disclosures of the autopsy table brought an overwhelming conviction of the futility of elixirs and extracts to combat, to terminate, or to repair organic changes so profound and destructive; homeopathy, by appearing to demonstrate that minimal are as efficacious as larger doses, hinted at the perhaps frequent impotence of both. . . . The new science of pharmacology represented from the start a distinctly more hopeful therapeutic attitude: instead of discarding, it undertook to probe; not content with testing traditional and empirical claims, it ventured the effort to ascertain the physiological effect of drugs hitherto unemployed. Finally, proposing to itself definite clinical and theoretic problems, it sought to create agents capable of coping with them. Its most recent outcome, Ehrlich's salvarsan, is a deliberate effort in constructive therapy.[9]

In England the facilities for pharmacological investigation were often dreadfully poor[10] and in most universities, teaching was deemed sufficient if left either to physiologists or to practising clinicians. The Victorian opposition to the development of laboratory sciences made for difficulties, and the curious English distinction between 'pure' and 'applied' science ensured opposition from 'pure' academics to the development of a science whose chief reason for existence was that drugs are useful. As drugs, such as nicotine and curare, were studied in the laboratories of influential physiologists what need was there for a separate pharmacology laboratory? Clinicians, often willing to pay lip-service to science, were very reluctant to let their clinical practices be exposed to questioning scientists, and heartily asserted that pharmacology was a subject to be taught with anatomy and physiology before students came to the wards of hospitals. (This attitude persisted, especially in the London medical schools, for another 40 years. For a long time I had the interesting task of teaching students about drugs for diseases of which they as yet knew nothing.)

Better advances were made away from the hospitals. With strong support from the physiologists and money from the Rockefeller Foundation, the first chair in pharmacology in England was established at University College, London, in 1905. Its incumbent was an Englishman, Arthur Cushny[11] (1866–1926), who had graduated in Aberdeen, studied under Schmiedeberg and replaced Abel in the chair at Michigan when Abel moved to Johns Hopkins in 1893. Cushny remained in London until 1920, when he succeeded Fraser in the chair of materia medica in Edinburgh, and restored the experimental approach which had been so vigorous in the earlier part of Fraser's reign. Cushny's textbook of pharmacology was the first English work to enshrine the new subject effectively for students. His research interests were wide but he was particularly interested in digitalis and in the way in which drugs acted on cells.

However, there was little growth of the subject in England. The British Pharmacological Society was not founded until 1931, and it did not begin to sponsor its own journal until 1946. In the meantime, British research into drugs was published either in physiological, chemical, or clinical periodicals or went to the American Society's Journal. The British Pharmacological Society remained much under the wing of physiology and was nervous of the influence of unscientific clinical practitioners. It was not until 1970 that a clinical section of the Society was approved.[12]

Not surprisingly, during the early part of the century, the British contribution to pharmacology lay mainly in establishing techniques and principles of great importance to the subject, and advances came more from institutions outside universities, including the National Institute for Medical Research, the laboratories supported by the (now Royal) Pharmaceutical Society, and the Wellcome Physiological Research Laboratories. During the 1930s British and American pharmacology was strengthened by the arrival of numerous victims of Nazi persecution, including men and women who rose to great scientific distinction and remained in their countries of adoption, to the great benefit of their new hosts. The corresponding loss to German pharmacology was immense, and it took many years for it to be restored to its one time standing.

A science in its infancy

Pharmacology was still an infant science in the 1920s and in no position to answer quite simple questions, such as why one drug is more potent than another. Medical students were confronted with lists of the doses of drugs, based on clinical experience. While this formed an important part of their knowledge it was quite devoid of any rational foundation and educationally most discouraging. Why, for instance, does a hundredth of a grain (0.6 mg) of atropine achieve its desired effects whereas the traditional dose of morphine is fifteen times greater? Of course, these drugs have different effects, though they are often given at the same time to prepare patients for anaesthetics and surgical operations. But even drugs with the same sort of property, for instance the relief of pain, have quite different doses. Aspirin has little effect in doses less than 5 grains (300 mg), which is thirty times the usual dose of morphine. To solve these questions, one has to go to the roots of the subject.

Through the nineteenth century, it became slowly obvious that drugs react with specific bodily components: which both Ehrlich and Langley called 'receptors'. In the twentieth century progress was made towards understanding the properties of 'receptors' or 'the receptor substance' for each drug. Cushny in Edinburgh experimented with drugs, such as atropine and adrenaline, which exist, like tartaric acid (see chapter 2), or like the left and right hands, in two chemical forms, identical except that they are mirror images of each other. One structure often acts many times more powerfully than the other, so Cushny concluded correctly that the receptor must be of a particular shape, which could be fitted by the left-handed but not the right-handed molecules.[13]

His successor in the Edinburgh chair, A. J. Clark (1885–1941) was also deeply interested in drug receptors, and wrote two classic monographs[14,15] about them. His approach was more algebraical than geometrical and, surprisingly, he made little reference to molecular shape and to Cushny's work. He discussed ways in which drugs might block the actions of other drugs by competing for the same receptors, and laid the foundation for advances,[16] after World War II, which gave us beta-blocking and other drugs (see Chapter 12).

Another root of pharmacology lies in the use of isolated tissues or whole animals to detect, identify, and measure tiny quantities of drugs. In the nineteenth century biological tests for organic poisons began to be used for legal purposes.[17] It was very important to have some means of detecting digitalis from the foxglove, and similar poisons, because chemical tests were too insensitive to pick up minute but none the less lethal amounts of poison, and would-be murderers were beginning to appreciate the possibilities. Biological tests were not only sensitive but also could be made very specific, for instance, distinguishing digitalis from other drugs which acted on the heart.[17]

Bioassay was also valuable for the quality control of manufactured products, initially antitoxins and later some drugs of plant origin, like digitalis and curare (see below), as well as hormones (see Chapter 5) and vitamins (see Chapter 7). Ehrlich's principles were repeatedly confirmed, and the subject was transformed 'from the plane of an insidious means of self-deception to that of a well-ordered and progressive science'.[18] It was necessary not only to estimate how much drug was present but also to know how accurate the estimate was. The need called for statistical treatment, which became quite elaborate and profited greatly from the work of statisticians concerned with agricultural field trials of factors affecting crop production.[19,20]

Again care for the roots of the subject improved its fruitfulness. Reliable means of bioassay allowed the first therapeutic use of curare which had been used in a crude form in physiological studies by Bernard, Langley, and others. The surgeon Sir Benjamain Brodie (1783–1862) is said to have discussed the possibility of using curare to control the convulsions of tetanus. But the materials available were too variable in potency to be reliable drugs for patients, and only when standardized material was available could trials be made to treat tetanus,[21] and certain nervous disorders,[22] to moderate the muscular contractions in electro-convulsive therapy[23] and, especially, to relax muscles of anaesthetized patients[24] so that surgery was easier and safer. Reliable assay also helped chemists who sought to isolate pure active principles, and, once this was achieved,[25] and methods established for their regular production, bioassayed crude material became obsolete.

Bioassay has become less important because modern precise physical and chemical techniques often provide sufficiently sensitive and specific methods of detecting and estimating drugs, and it is less important than it used to be in the control of manufactured products. But it continues to be an essential method in the investigation of substances with physiological activity and unknown chemical composition.

A pattern for inventing drugs

Most pharmacologists in the early twentieth century were investigating familiar medicines and sometimes seeking to improve on them. Especially when there were chemists eager to collaborate, or when chemists found pharmacologists willing to undertake tests for them, a pattern of research developed which was and still is a standard approach to finding new drugs. The discovery of local anaesthetics was one of the earliest journeys down this route.

Many pains have a well-defined local origin and local applications have been used to control them. Ointments and embrocations have ancient origins, but in the nineteenth century more subtle procedures were attempted. Simpson, who had introduced chloroform as a general anaesthetic, experimented with its use locally, but without success. Richardson tried various materials including amyl nitrite (see Chapter 2), and physical methods, including cold and electricity. A drug which achieved the desired effect was already known at this time, but there was a period when its potential was not recognized.

Nineteenth century explorers in South America discovered a plant, named *Erythroxylon coca*, which natives chewed to give themselves strength and prevent tiredness. Chewing the leaves made the mouth numb, but the value and dangers of this property seems to have been overlooked. Samples of the plant were brought to Europe and analysed. In Wöhler's laboratory, about 1858, an alkaloid was isolated and named cocaine. It was tasted by Wöhler himself, who confirmed that it made the tongue insensitive for a time. The drug was prepared on a moderate scale by the well-established firm of Merck at Darmstadt, notable pioneers in the isolation of pure alkaloids, and its effect on fatigue was studied by physiologists. Experiments were done in which cocaine was injected into the skin, which became insensitive to pinpricks,[26] but several more years passed before the value of this property was seen, and it was applied in medical practice.

In 1884 Sigmund Freud (1856–1939), then a young neurologist in Vienna, contemplated the use of cocaine, because of its euphoriant effect, as a substitute for morphine, and made some trials in collaboration with a colleague, Carl Koller. Koller was seeking an anaesthetic for operations on the eye. He tried morphine, and also chloral and bromide, and then, following Freud's interest, tried cocaine, with good results in experiments on animals, on himself, and on numerous patients.[27] Koller was working in a laboratory and the work naturally involved collaborating clinicians. Claims for priority were disputed viciously, but the objective benefits of cocaine as a local anaesthetic were indisputable and its use spread widely.[28]

Cocaine had other properties which made its use hazardous. Once enough research had been carried out to establish its chemical structure, the way was open to synthesize and test similar but simpler molecules and see whether they would act more safely than cocaine. It would have been rash to test new compounds directly on man without any knowledge of what they might do, but tests in animals were easy to devise and allowed many compounds to be examined. These were just the conditions which the growing fine chemical industry could exploit: a simple biological test and a theme on which chemists could invent variations for testing, gradually selecting features which gave an optimum combination of activity, stability, and any other desirable properties. So a series of compounds became available for clinical use. One of the simplest and earliest, Novocain or procaine,[29] is still widely used. Later we shall meet many families of synthetic drugs derived from a potent substance of known structure.[30]

The discovery of local anaesthetics was enormously valuable to surgery and dentistry, but did not add much to basic science. The mechanism by which nerves conduct impulses was established by research, particularly in the physiology laboratory in Cambridge, before and shortly after World War II.[31,32] Local anaesthetics, which stopped nerves from conducting impulses, had been known for 40 years or so, but they played little part in the physiological discoveries, nor did knowing how nerves work immediately give much help in understanding local anaesthetics.[33] It is unusual for a specific kind of drug not to be an aid to physiological discoveries.

The pattern of variation on a chemical theme, where the theme is a substance with some useful medicinal property, has been widely followed ever since. Like variations on a musical theme, the changes may be very substantial, or they may be so small as to seem insignificant. But the responses of living tissues are very selective, and there are many instances of minute changes radically altering the properties of drugs.[34] Much work, particularly in industrial laboratories, consists of operating suitable test systems, or screens, which are used to select the most active from a range of novel compounds. 'Synthesize and screen' can become an unproductive ritual, but it had led to many modest and some remarkably beneficial new agents. Even small improvements in the properties of drugs are not despised by patients, and 'molecular roulette', as it is sometimes called, does not always deserve the contempt it receives.

Scientists, doctors, and the ergot problem solved

A source of difficulty, especially in Britain, was the continued lack of understanding between scientists and clinical doctors. When contact was achieved, valuable results followed. The ergot problem, which had been of great interest to Wellcome and had provided Dale with his earliest independent research was not resolved until pharmacologists and obstetricians collaborated effectively. Dale's colleagues Barger and Carr isolated, in 1907, an alkaloid, ergotoxine, which was soluble in alcohol but not in water and which made the uterus contract, but midwives and doctors attending childbirths were not impressed by the new alkaloid and continued to use the old watery extract of ergot. Its status was officially recognized by its inclusion in the British Pharmacopoeia of 1914.

Dale himself had doubts about the action of ergotoxine. He observed in experimental animals that it acted in a clearly different way from the watery extract, but Barger could find no other alkaloid with the appropriate properties. Barger and Dale stated explicitly among the conclusions of their 1907 paper, '3. The action of the pharmacopoeial extracts appears too great to be accounted for by the small amounts of ergotoxine which they contain, and it seems likely that some other active principle is present in them'.[35] There, for a time, the matter rested.

In Switzerland, when drugs were in short supply because of World War

I, the dyestuffs firm of Sandoz turned to pharmaceutical manufacture. A research department was founded in 1917, under the leadership of Arthur Stoll (1887–1971). Ergot was among the first subjects studied. In 1920 Stoll isolated ergotamine, the first of a long series of ergot alkaloids which have originated from the laboratories of that firm. Evidence accrued that ergotamine, rather than ergotoxine, was 'the' active principle, and, after a brisk public exchange[36–39] of views between Stoll and Barger's successors at Wellcome, Dale wrote, 'These, however, are the curious and rather humiliating facts. Ergotoxine, described in England in 1906, has not to this day received a proper clinical trial; ergotamine, described in Switzerland in 1921, and identical in action with ergotoxine, according to laboratory tests, was promptly and adequately studied in the Swiss and German clinics'.[40] He had already referred to the 'presumption of laboratory pharmacologists' in assuming that their findings could be applied without further trial to a clinical setting and also to the difficulty of attempting clinical trials. 'More than one eminent gynaecologist was willing to carry out a test; but only by handing the extracts to a resident officer or a ward sister, with an instruction to administer them to alternate patients as a routine and to record impressions of their respective values'.[41] It was not the only occasion on which scientists had difficulties in persuading clinicians to undertake trials by reliable procedures.

However, as Dale was well aware, more effective use was being made in England of experimental methods in clinical research. In 1927 Harold Burn (1892–1981), who had worked with Dale at the Wellcome Laboratories and more extensively at the National Institute at Hampstead, collaborated with the gynaecologist Alec Bourne (1886–1974) in experiments in which a rubber bag was inserted in the uterus of women during labour and used, by means of transmitted pressure changes, to record uterine contractions. The method was not original: Bourne and Burn[42] refer to three accounts of its use between 1872 and 1893. It was not without some danger as it could introduce sepsis into the uterus, a condition that was very serious since neither sulphonamide nor penicillin had been discovered. All went well, but the experiments were abandoned as too risky.[43] Burn was at the time mainly concerned with problems of standardization of pituitary extracts. Ergot received only a minor mention, but Burn's words imply quite clearly that he expected the liquid extract of ergot to be ineffective, and describing ergotoxine or ergotamine as 'the specific alkaloid of ergot'.[44]

The same method of recording uterine contractions was used some years later by John Chassar Moir (1900–1977), obstetrician at University College Hospital, immediately after childbirth instead of during labour, when the risks were smaller. Moir's careful comparison of liquid extract of ergot and ergotoxine showed quite clearly that the liquid extract caused effective uterine contractions, and that those caused by ergotamine or ergotoxine were irregular and poor in comparison.[45] Harold Dudley (1887–1935) at the National Institute for Medical Research took up the problem and, supported by Moir's clinical tests, isolated a new alkaloid which matched the efficacy of the old liquid extract.[46] Moir's paper,[45] in 1932, aroused great interest elsewhere, and within a year the same alkaloid was isolated independently by Stoll in Switzerland[47] and in three other laboratories in America.[48–50] The ensuing claims for credit did least credit to those who most sought it and resulted in a multiplication of names for what was soon discovered to be the same substance. The alkaloid is still known in Britain by the name ergometrine, while the Council on Pharmacy and Chemistry of the American Medical Association established a name, ergonovine, which was better in accord with its own policies of nomenclature.[51] Ergometrine or ergonovine continues to be a useful drug. More refined modern analysis has shown that ergotoxine is a mixture of three closely related alkaloids and not, as was long believed, a single substance,[52] and advances in the chemistry of ergot alkaloids have led to other new drugs with remarkable properties.[53]

Confidence that laboratory research could provide more reliable answers than clinical observation received an uncomfortable but salutary jar from Moir's observations. The use of adequate methods of recording and measuring in clinical practice took a long time to come. The difficulties were great, not least in balancing responsibility to individual patients with responsibility for providing better therapy for all future patients. But such observations were vital, because humans do not always react to drugs in the same way as other species, and blind reliance on results from the laboratory can be misleading. Modern physical methods and improved instruments make such observations easier and impose less upon the comfort and safety of patients.

Limitations

So the infant science of pharmacology made small advances here and there, but it was still gravely handicapped by the general ignorance of bodily mechanisms. Another 30 or more years were to pass before drug receptors could be isolated or even partially isolated, and in the meantime new remedies for disease were being discovered in quite different ways.

The great therapeutic advance of the nineteenth century was the discovery of micro-organisms and acceptance of the germ theory of disease. The next great discovery was the recognition of diseases due, not to invading organisms, but to the lack of essential foodstuffs. It will be easier to understand what a big discovery and what a difficult discovery it was if we go back, initially to the seventeenth century, and follow a fresh pathway through history.

Notes

1. Parascandola, J. (1982). John J. Abel and the early development of pharmacology at the Johns Hopkins University. *Bulletin of the History of Medicine* 56, 1–18.
2. Parascandola, J. and Kenney, E. (1983). *Sources in the history of American pharmacology.* American Institute of the History of Pharmacy, Madison, Wisconsin.
3. Liebenau, J. (1987). *Medical science and medical industry. The formation of the American pharmaceutical industry.* Macmillan, Basingstoke.
4. Chen, K. K. (1969). *The American Society for Pharmacology and Therapeutics Inc. The first sixty years 1908—1969.* American Society for Pharmacology and Experimental Therapeutics, Bethesda, Maryland.
5. Parascandola, J. (1987). In title of address at symposium, Wellcome Institute of the History of Medicine.
6. Parker, F. (1961). Abraham Flexner (1866–1959) and medical education. *Journal of Medical Education* 36, 709–14.
7. Bonner, T. N. (1989). Abraham Flexner as critic of British and continental medical education. *Medical History* 33, 472–9.
8. Flexner, A. (1912). *Medical education in Europe.* Bulletin no. 6, Carnegie Foundation, New York.
9. Ibid. p. 87 f.
10. Anon. (1906). The Oxford Medical School. *British Medical Journal* i, 1479–91, at p. 1487.
11. Dale, H. H. (1926). Arthur Robertson Cushny (1866–1926). *Proceedings of the Royal Society*, ser. B, 100, xix–xxvii.
12. Bynum, W. F. (1981). *An early history of the British Pharmacological Society.* British Pharmacological Society.
13. Parascandola, J. (1975). Arthur Cushny, optical isomerism, and the mechanism of drug action. *Journal of the History of Biology* 8, 145–75.
14. Clark, A. J. (1933). *The mode of action of drugs on cells.* Edward Arnold, London.
15. Clark, A. J. (1937). General pharmacology. *Handbuch der experimentelle Pharmakologie*, vol. 4 (eds W. Heubner and J. Schüller). Springer, Berlin.
16. Ariens, E. J. (ed.) (1964). *Molecular pharmacology.* Academic Press, New York.
17. Fagge, C. H. and Stevenson, T. (1865). On the application of physiological tests for certain organic poisons, and especially digitalis. *Proceedings of the Royal Society* 14, 270–4.
18. Burn, J. H. (1930). The errors of biological assay. *Physiological Reviews* 10, 146–69.
19. Bliss, C. I. and Catell, McK. (1943). Biological assay. *Annual Review of Physiology* 5, 479–539.
20. Gaddum, J. H. (1953). Bioassays and mathematics. *Pharmacological Reviews* 5, 87–134.
21. Florey, H. W., Harding, H. E., and Fildes, P. (1934). The treatment of tetanus. *Lancet* ii, 1036–41.

22. West, R. (1932). Curare in Man. *Proceedings of the Royal Society of Medicine* 25, 1107–16.
23. Bennett, A. E. (1967). How 'Indian arrow poison' became a useful drug. *Anesthesiology* 28, 446–52.
24. Griffith, H. R. (1967). An anaesthetist's valediction. *Canadian Anaesthetists' Society Journal* 14, 373–81.
25. King, H. (1935). Curare alkaloids. I. Tubocurarine. *Journal of the Chemical Society* 1381–9.
26. Anrep, B. von (1880). Ueber die physiologische wirkung des Cocain. *Archiv für die gesamte Physiologie* 21, 38–77.
27. Koller, C. (1928). Historical notes on the beginning of local anesthesia. *Journal of the American Medical Association* 90, 1742–3.
28. Liljestrand, G. (1967). Carl Koller and the development of local anesthesia. *Acta physiologica Scandinavica*, suppl. 299.
29. Einhorn, M. (1899). Ueber die Chemie der localen Anaesthetica. *Munchen medizinische Wochenschrift* 46, 1218–20; 1254–6.
30. Hirschfelder, A. D. and Bieter, R. N. (1932). Local anesthetics. *Physiological Reviews* 12, 190–282.
31. Hodgkin, A. L. (1963). The ionic basis of nervous conduction. *Nobel lectures physiology or medicine 1963–1970*, pp. 32–48. Published in 1972 for the Nobel Foundation by Elsevier, Amsterdam.
32. Huxley, A. F. (1963). The quantitative analysis of excitation and conduction in nerve. *Nobel lectures physiology or medicine 1963–1970*, pp. 52–69. Published in 1972 for the Nobel Foundation by Elsevier, Amsterdam.
33. Shanes, A. M. (1958). Electrochemical aspects of physiological and pharmacological action in excitable cells. *Pharmacological Reviews* 10, 59–273.
34. Rose, F. L. (1965). *The changing pattern of drug research.* Sir Jesse Boot Foundation Lecture, University of Nottingham.
35. Barger, G. and Dale, H. H. (1907). Ergotoxine and some other constituents of ergot. *Biochemical Journal* 2, 240–99, at p. 296.
36. Smith, S. and Timmis, G. M. (1930). The alkaloids of ergot. Part I. *Journal of the Chemical Society* 180, 1390–5.
37. Editorial. (1930). The alkaloids of ergot. *Lancet* ii, 652.
38. Smith, S. and Timmis, G. M. (1930). The alkaloids of ergot. *Lancet* ii, 994; 1148–9.
39. Stoll, A. (1930). The alkaloids of ergot. *Lancet* ii, 873–4; 1148.
40. Dale, H. H. (1932). Note on communication by Moir, C. (1932). The action of ergot preparations on the puerperal uterus. *British Medical Journal* i, 1119.
41. Dale, H. H. (1930). The alkaloids of ergot. *Lancet* ii, 1149–50.
42. Bourne, A. and Burn, J. H. (1927). The dosage and action of pituitary extract and of ergot alkaloids on the uterus in labour, with a note on the action of adrenalin. *Journal of Obstetrics and Gynaecology of the British Empire* 34, 249–68.
43. Bülbring, E. and Walker, J. M. (1984). Joshua Harold Burn 1892–1981. *Biographical Memoirs of Fellows of the Royal Society*, 30, 45–89, at pp. 64–5.
44. Bourne, note 42 p. 267.
45. Moir, C. (1932). The action of ergot preparations on the puerperal uterus. *British Medical Journal* i, 1119–22. [cf. note 40]
46. Dudley, H. W. and Moir, C. (1935). The substance responsible for the traditional clinical effect of ergot. *British Medical Journal* i, 520–3.

47. Stoll, A. and Burckhardt, E. (1935). L'ergobasine, nouvel alcaloide de l'ergot de Siegle, soluble dans l'eau. *Comptes rendus de l'Académie des Sciences* 200, 1680–2.
48. Kharasch, M. S. and Legault, R. R. (1935). Ergotocin. *Science* 81, 388.
49. Thompson, M. R. (1935). The new active principle of ergot. *Science* 81, 636–9.
50. Kharasch, M. S., King, H., Stoll, A., and Thompson, M. R. (1936). The new ergot alkaloid. *Science* 83, 206–7. H. W. Dudley (note 46) died in 1935 and H. King, his successor, assumed his responsibilities for the British chemical work on ergometrine.
51. Council on Pharmacy and Chemistry of the American Medical Association (1936). The new ergot alkaloid ergonovine. *Journal of the American Medical Association* 106, 1008.
52. Stoll, A. and Hoffmann, A. (1943). Die Alkaloide der Ergotoxingruppe: Ergocristin, Ergokryptin und Ergocornin. *Helvetica Chimica Acta* 26, 1570–83.
53. Berde, B. and Schild, H. O. (eds) (1978). Ergot alkaloids and related compounds. *Handbook of Experimental Pharmacology*, Vol. 49. Springer-Verlag, Berlin.

Chronology of Chapter 7

Date

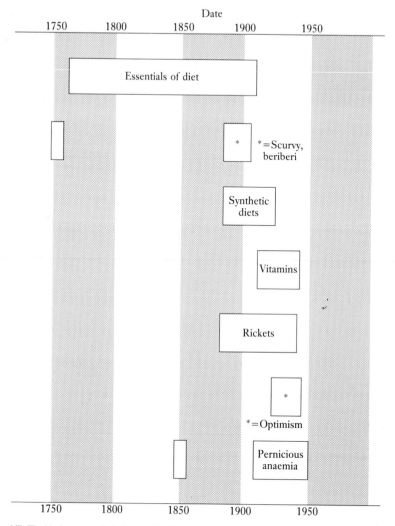

NB The blocks show periods discussed in the text, and do not mean that there was no activity about the subject at other times.

7

Deficiency diseases

Essential foods

Starvation was an obvious cause of debility and death and it was never thought of as a disease. It was not always clear that some diseases resulted from a lack of particular kinds of food. We can only guess how primitive man came to select particular materials to eat. People must have chosen foods because of their taste and smell, because they satisfied hunger, and because they had no ill-effects. Beliefs grew, and priests and others in authority made rules, sometimes prudent, such as prohibiting pork where meat from pigs infected with the parasite *Trichinia* was a serious cause of human illness,[1] and sometimes with no explicable basis, such as forbidding meat from animals that chew the cud but are not cloven footed.[2] The existence of too many rules and the absence of means for revising them stifled experiment, while ignorance of chemistry made it impossible to identify the important constituents of such complex materials as a loaf of bread or a joint of meat.

Physiologists of the late eighteenth century showed that food was broken down in the stomach and intestine into simpler materials and that this was a chemical and not simply a mechanical process. The great chemist Lavoisier and the mathematician Laplace showed in the 1780s that animal heat originated from chemical changes in food during its digestion and after absorption, though it was uncertain just how chemical changes produced heat. Although the idea that heat depended on the motion of particles had existed at least from the time of Francis Bacon (1561–1626), Humphry Davy in 1820 was still toying with the thought that heat might itself be a chemical substance produced by chemical reactions. Ideas became clearer when heat was recognized as a form of energy and not as a weightless element. In the early 1840s Joule demonstrated the mechanical equivalent of heat, and the law of the conservation of energy was stated by Mayer and Helmholtz. German physiologists then showed that the law held good for animal bodies, and began the accurate study of food as a source of energy.

At the time of these advances in physics and physiology, the chemically-minded physician Jonathan Pereira followed up his forward-looking treatise on medicines by one on food. Much of this latter work described very practical aspects of diet in health and sickness, disparaged fads, and was full of good sense, as applicable today as it was then. Practical applications were preceded by an account of the scientific foundations then available. Pereira claimed that his book contained 'a tolerably full account of the chemical elements of food,—a subject which has always appeared to the author of considerable importance, and to which the recent researches of Boussingault, Liebig, and Dumas have given additional interest. It is one, however, which preceding dietetical writers have altogether passed over, or only incidentally alluded to: and in no work, with which he is acquainted, has it been systematically treated'.[3] His claim was probably well founded, and his list of 13 elements of food is a landmark in the advance of dietetical knowledge. The list contains carbon, hydrogen, oxygen, nitrogen, phosphorus, sulphur, iron, chlorine, sodium, calcium, potassium, magnesium, and fluorine, which are undisputed today, but would be joined by iodine and zinc, and by some others unheard of in Pereira's time. He went on to describe 12 'alimentary principles' (the aqueous, the mucilaginous, the saccharine, the amylaceous, the ligneous, and so forth), in terms which would certainly not appear in a modern text. Water is all right, but advances of chemistry have regrouped the rest fairly thoroughly into carbohydrate, protein, fat, minerals, and accessory factors or vitamins. It is through these advances in chemistry, and through animal experiments (of which Pereira saw the necessity) that we must approach the discovery of vitamins.

Substances discarded by the body, such as carbon dioxide in breath and urea in urine, were found to be chemically simpler than other bodily constituents. The idea evolved that plants *made* complex materials, using energy obtained from sunlight, and that animals *broke* them *down*, releasing energy needed for bodily warmth and work in the process. Ure's discovery that animals could also perform synthetic operations, making more complex substances from simpler ones, began to undermine this apparent distinction between animals and vegetables. Gradually ideas of chemical changes or metabolism became widespread and were conveniently divided into anabolism, the building up of tissues, and catabolism, the breaking down of complex substances into simpler components.[4]

The complex materials which came from plant and animal tissues were analysed and found to fall into three broad classes, which we now call carbohydrate, fat, and protein. Carbohydrates and fats contained only carbon, hydrogen, and oxygen, and serve mainly as a source of energy. proteins also contain nitrogen and sometimes sulphur, and provide the nitrogenous constituents of the body. Egg white is a typical protein, and from its Latin name came the word albumin, still used for one class of protein. Other classes of protein were identified, but not until about 1900 did their composition begin to be understood. German biochemists established that proteins could be broken down under suitable conditions in the presence of water, first to peptides and then to simple small molecules, the amino acids, of which about a dozen were discovered during the nineteenth century.[5] They had a general structure:

$$NH_2 - R - COOH$$

where R stands for an organic radicle, such as CH_2 (in glycine) or $CH_3.CH$ (in alanine). Proteins could be envisaged as chains of amino acids linked together by union between the NH_2 group of one molecule and the COOH group of another in indefinitely large sequences.

$$NH_2 - R - CO - [OH \mid H] - NH - R' - CO - [OH \mid H] -$$

Amino acid 1   Amino acid 2

[Peptide link   [Link]
with loss of H_2O]

$$- NH - etc - COOH$$

Amino   etc
acid 3

However, these chemical details attracted less interest from late nineteenth century nutritionists than did the energy value of foods. Ways of measuring the amount of energy consumed by humans doing more, or less, work were devised, and the amount of energy released by a given weight of a given food was measured. So began the study of 'calories' and calculations about the quantities of particular foods which would meet the daily requirements of an animal or man. For a time the idea reigned supreme that an adequate diet consisted only of enough protein for body building and enough protein plus fat, plus carbohydrate, for energy needs. Common salt and some other mineral salts were also required, and some observers were concerned about palatability. Optimists believed that most of the facts about nutrition were known.

Given that this was all which needed to be known, it followed that it should be possible to prepare an adequate diet *entirely synthetically.* Attempts had already been made, not out of scientific curiousity but out of sheer necessity, in beleagured cities or in time of famine when every possible source of food was pursued by starving populations. During the siege of Paris in 1871, the great French chemist J. B. A. Dumas (1800–1884), one of Pasteur's teachers, tried to produce an artificial milk by emulsifying fat in a sweetened albuminous solution. It was given to starving infants, but it did them no good. The synthetic milk lacked something essential to life.[6]

So the preparation of a wholly synthetic diet did not have an encouraging start, but by 1900 about a dozen experiments had been made, mostly with mice or rats, to test artificial diets. None used *totally* synthetic materials, in the sense that they originated from pure elements, or pure compounds which were capable of being made from elements in a laboratory. The diets consisted of highly purified ingredients of normal foods which, as far as the skills of the time permitted, were completely identified. Some of the ingredients were not available ready made, let alone pure, and the work necessary to prepare even moderately pure materials for such diets in sufficient quantity must have been most discouraging. The experiments were not very successful, but an important clue came from the discovery that a very small amount of ordinary milk added to an otherwise completely identified diet made the survival and even the growth of young animals possible. The quantities of milk added were negligible in terms of the total fat, protein, and carbohydrate supplied by the rest of the diet, so something else was responsible and was waiting to be identified.

Scurvy and beri-beri

Discoveries about essential food constituents have rarely been made in everyday life. There are too many variables, and it is very difficult indeed to apply well-controlled experimental methods. Lind's work on scurvy (see Chapter 1) had the advantage of naval discipline and gave convincing results, so that lemon juice became established as part of the British navy's diet, but later events eroded this great advance. The navy, for good political and economic reasons, changed its source of citrus fruits from Mediterranean lemons to West Indian limes. Limes happen to have only about half as much of the material active against scurvy as lemons,[7] but this fact was, of course, unknown at the time.[8] Scurvy began to occur again. Long voyages without fresh provisions were becoming rare, but arctic and antarctic explorers suffered severely, and the good reputation of fresh fruits and vegetables began to decline. The germ theory of disease was becoming popular, and enthusiasts for seeking bacterial origins of disease offered new theories about the causes of scurvy. Bacterial infection did not appear to be a direct cause, but the notion gained hold that bacteria decomposed foods to produce poisonous substances which were called ptomaines. For a time scurvy was attributed to tainted meat. The British Antarctic expeditions of 1901 and 1910 were provisioned with no concern for the value of fresh fruit and vegetables. Their absence contributed to the fatal outcome for Scott and his companions.[9]

About the time at which scurvy was once again becoming an unsolved problem, another disease, mostly of Asiatic rice-eating peoples, and locally called beri-beri, attracted attention. The sufferers were weak, and had nerve pains and, sometimes, heart failure. The disease became very widespread from about 1860 onwards, when the European influence in south-east Asia was increasing. It appeared extensively in the Japanese navy, and was the subject of the most notable experimental study of nutrition after Lind.[10] Baron Takaki, the Director-General of the Japanese naval medical service, suspected protein deficiency as the cause, and largely eliminated the disease in the naval service by dietary changes, which included the replacement of polished rice by barley, and the addition of milk and meat. Not enough was known about the composition of foods for the exact implications of these changes to be understood. When Takaki added several sources of protein to the sailors' diet, he also added other substances which accounted for his valuable results.[11,12]

The Dutch government was much concerned about beri-beri in its East Indian colonies. A commission was sent in 1886 to survey the problem. In the fashion of the times, a bacterial cause was sought, but none was found. The commission went home, leaving a young doctor, Christiaan Eijkmann (1858–1930), with some laboratory facilities to investigate further. He tried to reproduce the disease in chickens which were fed on rice, and by good fortune, a spontaneous outbreak of polyneuritis occurred among his birds. The episode lasted from mid-July until late November, when it subsided as abruptly as it arose.[13] Eijkmann had no proper laboratory facilities and his birds were housed in a military hospital, although tended by civilians. As Eijkmann later discovered, the laboratory keeper had been feeding the birds on cooked, polished (i.e. de-husked) rice from the hospital kitchen. 'Then the cook was replaced and his successor refused to allow military rice to be taken for civilian chickens. Thus the chickens were fed on polished rice from 17th June to 27th November only. And the disease broke out on 10th July and cleared up during the last days of November'.[14]

Deliberately planned experiments confirmed that beri-beri, or something like it, could be produced in chickens, by feeding them on polished rice instead of the whole grain. Various explanations were possible. Eijkmann was inclined to think that rice contained a toxin which was neutralized by material lost when the outer husk was removed during polishing. Years passed before the true explanation was found, that the husks contained an essential nutrient now known as vitamin B_1.[15]

Whatever hypothesis might be entertained, it was necessary at this stage to see whether what applied to chickens applied also to humans. The civilian medical inspector for Java, by careful epidemiological study of prison diets (which sometimes gave prisoners the 'benefit' of the more palatable polished rice, and sometimes did not), established that beri-beri was some three hundred times commoner in those prisons where the prisoners were fed on polished rice. However, the authorities were uncooperative and scepticism was rife. No effective action was taken in spite of the clear cut evidence, and matters lapsed. Hardly any news of these adventures reached Europe until the twentieth century. When they did, and were amplified by laboratory experiments and analyses, it was appreciated that Eijkmann was the first scientist to reproduce experimentally in animals a human disease of dietary origin. Over 40 years later, in 1929, he shared the award of a Nobel prize for his work.

Synthetic diets

Eijkmann's partner in receiving the 1929 Nobel prize in medicine was Sir Frederick Gowland Hopkins (1861–1947), who spent much of his life seeking the exact chemical constituents of a diet which preserves life and health.[16-18] Hopkins began his scientific career as an articled pupil analyst in the business of Allen and Hanbury, an old established London manufacturer of pharmaceuticals.[19] He found the work unrewarding, and, after further study, he became an assistant to Dr (later Sir Thomas) Stevenson at Guy's Hospital, an expert in forensic medicine.[20] Hopkins obtained an external London degree in chemistry and then qualified in medicine, finding time also to pursue researches which appeared to be purely academic. He investigated the pigments of butterfly wings, and discovered that the opaque whiteness of the wings of some species was due to the presence of uric acid. He commented on the 'use of excretory substances in ornament, a phenomenon which may shock or please the aesthetic sense according to the point of view',[18] and went on to devise a method for estimating uric acid quantitatively. In 1898 he was invited by Sir Michael Foster to work in the physiological laboratory at Cambridge. Hopkins had never studied in Germany, then the fountainhead of all biochemical expertise, and was provided with very inadequate laboratory facilities at Cambridge, but he created a school of biochemistry which was a more than worthy complement to the distinguished school of physiology. He was elected F.R.S. in 1905 and became one of the founder members of the Medical Research Committee where his influence was of great importance to growth of the science of nutrition and to the advance of British biochemistry.[17,21]

Hopkins was one of the earliest scientists to crystallize a protein (egg albumin). The care with which he showed that repeated samples had the same composition went far towards establishing the very important fact that proteins are not mystical and undefined constituents of protoplasm but well-defined chemical compounds. Like others who were working in this field, he prepared diets consisting as far as possible of ingredients *all* of known chemical composition, and observed the progress of animals fed on the diet. He showed that the protein zein, from maize, lacked the amino acid tryptophane, and was inadequate as a source of protein to keep mice alive. When tryptophane was added to the food given to the mice, they remained alive, although not completely healthy.[22] This was clear evidence of the importance of a specific identifiable chemical constituent in the diet.

As addition of a little milk to artificial diets was necessary for young rats to gain weight, it was obviously desirable to investigate the separate constituents of milk. Hopkins took casein, the principle protein derived from milk, and showed that some samples were an excellent substitute for milk but that others failed. He went further and found that good samples lost their power if they were washed with water and alcohol, and that the washings restored the power to an ineffective diet. Plainly, some substance was present in minute amounts, quite insufficient to increase the calorie content of the diet, but essential for growth of the rats. As Hopkins said:[23]

By this time I had come to the conclusion that there must be something in normal foods which was not represented in a synthetic diet made up of pure protein, pure carbohydrate, fats and salts; and something the nature of which was unknown. Yet at first it seemed so unlikely! So much careful work upon nutrition had been carried on for half a century and more—how could fundamentals have been missed? But, after a time, one said to oneself, 'Why not?' The results of all the classical experiments had been *expressed* in terms of the known fundamental foodstuffs: but these had never been administered *pure*! If, moreover, the unknown, although clearly of great importance, must be present in very small amounts— again, why not? Almost infinitesimal amounts of material may have a profound effect upon the body, as pharmacology and the facts concerning immunity assure us. Why not then in nutritional phenomena? The animal depends ultimately upon the plant for the synthesis of materials which bulk largely in its food: there is no reason why it should not be adjusted so as to be in equal need of substances which the plant makes in small amount. Only if energy were the sole criterion of an animal's needs would this be impossible; but certainly it is not the sole criterion. . .[23]

Hopkins experiments between 1906 and 1912 provided the first observations under controlled laboratory conditions which showed the presence of what Hopkins named 'accessory food factors'.[24] The work was notable for the analytical control of materials, for the meticulous weighing and measurement of the quantities of food consumed, for the many days over which animals were studied (in contrast to calorimetry experiments, which were unlikely to last more than a few hours or a day or so), for consistent recording of the weight and growth of the rats, and for its independence from current dogma. Hopkins was fortunate also in his choice of butter as the source of dietary fat. Later, to economize, he changed to lard, with unforeseen and inconvenient consequences: several years had to pass before it became known that butter, too, contained other accessory food factors, the absence of which made experimental results very confusing when lard was used. But the significance of Hopkins's work was unmistakable. His publication was delayed by illness caused by overwork and the anxieties of a very inadequate income.[17] By the time his work appeared, fresh information was coming to light about scurvy and beri-beri which gave him good reason for putting forward his findings.

Accessory food factors and vitamins

While Hopkins was discovering the importance of individual constituents of diet, the Norwegian government was concerned about an illness affecting its sailors on long voyages. They became weak, dropsical, and short of breath, in a way so like beri-beri that the condition was named 'ship beri-beri'. But there was little neuritis, and sore gums and haemorrhages appeared, neither of which were usual in beri-beri. The symptoms suggested that scurvy was involved. An investigation was set up and reports on the condition were prepared in 1907 by Axel Holst (1860–1931) and his colleague T. Frölich. Once again, fashionable ideas about bacterial infections and tainted foods were offered, but did not bear critical examination. Holst[25,26] collected much information about the circumstances in which the disease occurred, and about diets which appeared to contribute to or delay its occurrence. He undertook experiments, first in pigeons and then in guinea pigs, fed on various diets. The choice of guinea pigs was fortunate because few species other than primates develop scurvy. Rats, for instance, do not, which was a great but unknown asset to most of the laboratory studies of the time. Holst's experiments showed that fresh fruits and vegetables contained material which protected the guinea pigs and which was easily destroyed by heating. Once the disease could be reproduced in guinea pigs, a way was open for recognizing and isolating the active principle.

This research did not, however, advance knowledge about beri-beri. The next progress in this field was made at the Lister Institute for Preventive Medicine in London. The foundation of this institute for research into the prevention of disease and for the manufacture of vaccines has already been described (see Chapter 3). The institute had many difficulties in its early days, but settled down under the direction of C. J. (later Sir Charles) Martin (1866–1955), a physician–scientist of wide experience. Like Hopkins, Martin saw the fundamental value of a chemical approach to the problems of life and death, and established in the institute one of the first biochemistry departments in Great Britain.[27]

Among Martin's friends and correspondents was a medical officer in the Federated Malay States, Leonard Braddon, who drew Martin's attention to the investigation of beri-beri and the work of Eijkmann.[28] Interest in the problems contributed to the development of a section of nutrition in the biochemistry department of the Institute. A guest worker from Warsaw, the chemist Casimir Funk[29] (1884–1967), and others began experiments, feeding pigeons on polished rice and treating the resulting polyneuritis with extracts of rice, bran, and yeast. Funk was successful in producing concentrates which prevented the disease and made progress in isolating pure material from them.[30]

Funk read and wrote widely and publicized the concept of diseases due to deficiency of organic substances present in trace amounts in normal diets.[31] He believed that the material he had isolated was an amine, and that other vital trace substances would be discovered and also be amines, so he coined the term 'vitamine'. Hopkins and others expressed dislike of the word, because many substances besides food factors were 'vital' and because the factors were not necessarily amines. As the ending '-in' was often used for substances of uncertain identity (e.g. insulin), 'vitamin' gradually displaced 'vitamine', but by this time the word had become too popular for the more exact phrase 'accessory food factor' to have any common use. There is no justification for the pronunciation 'Vittamin'.[16,32]

In the USA, experiments with rats, aiming to discover the minimum necessary ingredients of diets, were made both by T. B. Osborne (1859–1929) in Connecticut and by E. V. McCollum (1879–1967) in Wisconsin, and later at Johns Hopkins in Baltimore. McCollum and Davis found evidence of a factor present in butter and in egg-yolk, absent in lard and olive oil,[33] while Osborne and Mendel noted the occurrence of infectious eye diseases in animals inappropriately fed.[34] The terms 'fat-soluble A' and 'water-soluble B' were used to denote the essential substances and evolved into 'vitamin A' and 'vitamin B' as knowledge and agreement on vocabulary grew. But at this stage the chemical nature of the vitamins was unknown. Chemists faced the problems of identifying substances existing only in minute quantities and detectable only by prolonged experiments only too easy to mismanage.

It is hard to envisage the difficulties of long-term experiments on the diet of animals unless one has actually conducted them. Quite apart from the problems of determining the weight or volume of what animals have eaten and drunk (but not spilt) and preventing any eating of faeces (a valuable source of vitamins synthesized by intestinal bacteria), it is less foolproof than might be thought to prevent exchange of animals at times of cage cleaning, or illicit additions to diet by unauthorized visitors or ignorant assistants at unsupervised moments during week-ends, and so forth. It all sounds perfectly easy to control. But one mishap, occurring in a few seconds during an experiment of many weeks, can ruin half a year's work, whether it is detected or not.

Out of meticulous experiments in a growing number of laboratories, a clear picture began to emerge. Vitamin A prevented eye infections: we know now that it is necessary for the proper growth of the lining membranes, which are more easily infected when it is lacking, and also that it is the precursor of the pigment visual purple, necessary for the retina to respond to poor light. Vitamin B soon became known as the B-complex of several separate vitamins, starting with B_1 which prevented beri-beri. Scurvy was a separate problem: it did not occur in rats and mice, and was identified by testing the efficacy of purified extracts of fruits and vegetables in protecting guinea pigs from the disease. It became known as vitamin C, and in 1932 it was isolated and identified chemically in several laboratories.

Resolving the enigma of rickets

Like scurvy and beri-beri, rickets, the disease of malformed bones in infancy, was a battleground of conflicting ideas long before its cause was known. An association between sunlight and the formation of bone was recorded by the ancient Greek historian Herodotus (*c.* 480–*c.* 425 BC). He visited a battlefield where, many years earlier, the Persians had defeated the Egyptians. The bones of the dead still lay there, separately for the two nations. Herodotus noted that the skulls of the Persians were very thin, and those of the Egyptians were thick and tough. He was told that this was because Egyptians shaved their heads and exposed them to the sun, and that Persians wore felt skull caps to guard their heads from the sun.[35] The evidence was not very convincing, but experiments made more than 2000 years later showed that sunlight was good for making strong bones.

During those 2000 years rickets was described at various times, and knowledge increased about the formation and growth of bone. Rickets attracted renewed attention in England in the aftermath of the industrial revolution, because it became distressingly common in industrial towns, especially in the north of England and in Glasgow, where up to half the children between 1 and 2 years old were affected.[36] Dirt, overcrowding, confinement in squalid tenements, and lack of exercise were all seen as potential causes. Once again bacterial infection was a popular theory, but no particular bacteria were convincingly incriminated. Toxins in food were also popular. But nobody devised good experiments to identify any specific factor which needed attention in the diet or the environment.

Experiments in animals were possible only if rickets or something like it could be produced. It was in fact known to occur in animals kept in zoos. About 1880, (Sir) John Bland Sutton (1855–1936), a rising surgeon later of great distinction, studied the lions in the London zoo and found that the deformities of their cubs could be prevented by a traditional remedy, cod liver oil.[37] This evidence supported belief in its merits, but it also confirmed ideas that confinement and lack of exercise were culpable, because the lions, like the poor, were not spaciously housed. Properly controlled trials, such as Lind had performed a century or more earlier on the prevention of scurvy, were out of fashion. It would have been easy to treat 50 poor children with cod liver oil and 50 with olive oil through a year's growth, and make a convincing comparison: and how much suffering would have been prevented if an experiment had been done well and the results properly heeded! But the idea of controlled trials was dormant, and everything remained obscure.

When the Medical Research Committee was founded in 1911, rickets came second only to tuberculosis as a subject to be investigated. Several studies were supported, including one by Edward Mellanby (later Sir Edward, 1884–1955, and successor to Sir Walter Fletcher as Secretary of the Medical Research Council),[38] and one under the guidance of the Professor of Physiology in Glasgow, Noel Paton (1859–1928). Mellanby was one of the many notable students at Cambridge soon after 1900. He was well known to, and probably suggested by, Hopkins, a founder member of the Medical Research Committee. It seems clear and wholly probable that Hopkins thought rickets might be a deficiency disease. However, Mellanby was directed to approach the subject with quite different hypotheses in mind.[38,39]

There are several curious features about the support for Mellanby's research.[39] Instead of being given liberty to pursue the subject as he thought best, he was instructed to investigate a specific approach, viz., 'defects in the processes by which foods yield energy and the role of these defects as a causal factor'. This line of attack wasted a year or more unprofitably although Hopkins, a member of the instructing committee, had the right line in mind. One may guess that Hopkins, a quiet and modest man, did not press his point, and that some more forceful character succeeded in imposing his ill-founded opinion, as so often happens in committees. In later years the Medical Research Council was scrupulous in accepting or rejecting proposals without imposing any detailed direction. Did Mellanby's experience at this time make him persuade successive Councils of the wisdom of such a policy?[40]

Whatever the circumstances, Mellanby's first task was to reproduce the disease experimentally. This in turn required the choice of methods of demonstrating its presence and measuring its intensity. Working under the adverse conditions of the 1914–18 war, and using X-ray and microscopic examination of bones and estimation of their calcium content, Mellanby achieved the desired objective.[41] He produced rickets in puppies by changing what they were given to eat. A diet rich in cereals generated rickets, and the presence and kind of fat was important in determining whether the disease appeared. Cod liver oil was a particularly good preventive, and butter less so. Vegetable oils had no effect in delaying the onset of bony changes. Evidently a dietary factor was responsible, and its distribution accorded closely with the 'fat-soluble A' which had just been described by Davis and McCollum in America. But meat was also beneficial, and rickets had not been observed in animals deficient in fat-soluble A. Mellanby's careful work established rickets in puppies as a deficiency disease, but did not identify the missing substance, nor give any evidence about a connection with sunlight.

As always, it could be argued that puppies were different from humans, and many people remained unconvinced that the cause of rickets was being traced. The Great War came to an end, and in the famine in central Europe which followed, rickets became widespread. A research team from England, sponsored jointly by the Medical Research Committee and the Lister Institute of Preventive Medicine, went to Vienna, taking with them knowledge of wartime developments of nutritional science. There they obtained collaboration in trials of dietary supplements in children with rickets.[42,43] Shortage of food and supplements was so great that no ethical difficulty arose in maintaining some children on unsupplemented diets as a basis for comparison: there were not nearly enough supplements to go round. Results in the first winter of the study were most encouraging. The disease was checked in rickety children who received supplements of cod liver oil, and not in those who received other fats. But in the summer which followed, the controls did well too. This was confusing, but, fortunately, independent evidence came at this time from a number of studies in experimental animals and in children, and showed that the bony changes of rickets could be healed by exposure to ultraviolet radiation.[44] Experiments in the following winter and summer were designed to take this factor into account. The therapeutic value of cod liver oil was established unequivocally, but the way in which ultraviolet radiation contributed and the identity of the protective substance remained obscure for a little longer.

Mellanby himself was still impressed by the effects which he had observed when dogs were fed on cereals, and he investigated the possibility that they contained a toxic factor. It was an interesting line and led to the discovery of an important substance in some cereals, later identified and named phytic acid, which binds calcium and so restricts its absorption from food. Such a substance accelerates the appearance of rickets, but is not the main culprit. McCollum and his colleagues devised a diet which caused rickets in rats, and showed that the disease was prevented by material in butter clearly distinct from 'fat-soluble A'.[45] It soon became known as vitamin D. But the next big leap came from a diversity of observations on the effects of light.

After ultraviolet radiation was used successfully to cure children with rickets, the same treatment was applied to rats and found effective for them to thrive on a diet deficient in vitamin D. Then, no doubt disconcertingly at first, it was found that irradiation of the cages before the rats were placed in them worked equally well! As was later discovered, the rats ate sawdust in the cages, and it was the irradiated sawdust which protected them. Two independent groups showed that the irradiation of various foods gave them the power of preventing rickets. Separate ingredients of the foods were investigated and the property pinned down first, incorrectly, on cholesterol and later on a related substance, ergosterol, which was present in traces as an impurity in the cholesterol used. So the antirachitic factor had at last been identified, and was a product of the irradiation of ergosterol. The product, pure vitamin D, was isolated and crystallized in 1930 and its chemical nature was established. It was closely related to ergosterol and not very remote from cholesterol, from the sex hormones and the adrenal cortical hormones, and even from some carcinogenic substances which were then just becoming matters of interest.

So the knowledge was now available to prevent rickets and its ensuing deformities. The ideas which caused so much controversy each had some relevance—sunlight and smoke which obscured it, a substance in some foods which made it worse, a substance in other foods which made it better, poverty which led to unsatisfactory foods, and finally individual, personal lack of knowledge about nutrition, which remains the primary factor in generating the rickets that is still seen in Britain today.[46,47]

From scepticism to credulity

The idea of disease being due to a deficiency is now a commonplace, but at first it was difficult to adopt. The term 'growth factor' had come into use, and caused confusion. How could *lack* of a factor necessary for growth be the cause of a disease which occurred during growth? The Medical Research Committee in 1919 published a *Report on the present state of knowledge concerning accessory food factors (vitamines)*. It included the statement, 'Disease is so generally associated with positive agents—the parasite, the toxin, the *materies morbi*—that the thought of the pathologist turns naturally to such positive associations and seems to believe with difficulty in causation prefixed by a *minus* sign'.[48]

The difficulty in mental adjustment lingered for at least another 20 years, when it still appeared in the first edition of a very forward-looking textbook.[49] The 'theory' of vitamins was disturbing to contemporary thought and so was hotly contested. Hopkins, opening a discussion at the annual meeting of the British Medical Association in 1920 at Cambridge, remarked, 'In nearly every case that we are to consider, it is, I admit, still a hypothesis that the particular disease depends upon vitamine deficiency, but in respect to the broad aspects of nutrition as a whole, the importance of these factors is proven'.[50] There were, however, physicians ready to declare at the same meeting '. . . "vitamines", so far as their composition is concerned, seem to be a figment of the imagination', and, with reference to Mellanby and to Funk, '. . . All these observations are easily explained without invoking any recondite influence of "vitamines"'.[51] The sceptics were soon in a minority. Scientists who were satisfied by the experimental evidence used their imagination to invent ways of isolating, purifying and precisely identifying the accessory food factors so that no further doubt remained about their existence.

The search involved many laboratories. Vitamins occur in foodstuffs in very small concentrations, of the order of one part in a hundred thousand or less. At first, the only test for their presence was to feed suitable animals on specially prepared diets for many days at a time, and use the appearance or otherwise of some effect of deficiency as an indicator. Every step in purification of material known to contain the vitamin required further assays, to discover whether a more concentrated and less impure yield had actually been achieved, or whether the essential material had been lost altogether. In 1926 vitamin B_1 was isolated as a pure crystalline material, the first vitamin to reach this stage. Elementary analysis was possible, but methods of determining exact chemical structure then available required grammes rather than milligrammes of material. Heroic operations, like those of Kendall with thyroid glands 10 years earlier, were necessary to provide enough pure substance, but the structures of the principal vitamins known at the time were established by the early 1930s. Synthesis followed, and brought the crowning triumph when an entirely man-made substance was shown to be indistinguishable in its effects from organic material.[52]

Further investigations showed, perhaps inevitably, that the position was more complicated than the isolation of single vitamins suggested. Sometimes several closely related substances each acted in the same way, though they were not necessarily equally potent. The isolation and identification of vitamins came when detailed study of bodily chemistry was advancing rapidly, and vitamins took their place alongside enzymes as necessary parts of the processes for using food for energy and for growth. Hopkins had much to do with such developments, and did not shrink from trying to make them readily understandable. As an obituarist, most probably Sir Charles Sherrington, wrote:

I have a memory of an address given by him at a British Association meeting. In an uninviting gaunt lecture-room he decanted [sic] for nearly an hour on the cell as a theatre of chemical processes. Without any deliberate attempt at eloquence, and in a voice that as he became more interested fell to such a purely conversational level as to be a little difficult to hear, he conjured up for his audience—some 30 persons all told—a picture of the cell as a tiny sponge-work containing perhaps a thousand foci of different actions cooperatively confined within a unitary whole. An organized factory manifoldly hydrolysing, pulling to pieces, and contemporaneously constructing and reconstructing. And this unity bounded outwardly by a mosaic of countless chemical poles and leaking like a sieve. As I listened I felt I was being privileged for the time to see something of the microcosmic world in which my friend's scientific thoughts took shape and did his bidding. One mental factor which, it seemed to me, such thinking must demand, was a peculiar intensity of visual imagination, continuously checked in factual knowledge. The great organic chemist, conjuring with stereographic formulae, must have something of the same faculty.[53]

The 'organized factory manifoldly hydrolysing, pulling to pieces, and contemporaneously constructing and reconstructing' has become the central study on which all more modern advances in the discovery of drugs depend.

The availability, by the 1930s, of several vitamins and mixtures of vitamins for therapeutic use presented many problems. Suspicion gave way to credulity, so that new vitamins were postulated, some on the flimsiest of grounds. Some real or imagined substances thought to be vitamins remained in the uncertain realms of complex and sometimes conflicting experiments. Some belong more to the history of quackery than of drugs. Ill-founded beliefs grew about well-established vitamins, often in the direction of superstitious imagination and commercial exploitation. Small amounts of a vitamin fulfilled an essential function, but there was no reason to suppose that larger quantities had wider effects in promoting health. Indeed, evidence showed that excessive doses of several vitamins had toxic effects, though the margin between sufficiency and excess was usually wide. No satisfactory way of determining the minimum or optimum amount of vitamin to maintain health could be devised, and it was left to official bodies to set various standards at various times. 'Enough' depended on circumstances and could not be measured precisely. Popular belief was only too willing to accept magic in the name vitamin, and physicians have been happy to believe in 'subclinical deficiencies' to provide a therapeutic-ally simple diagnosis for unexplained symptoms. Indeed, as one very honest physician, a future Nobel prize winner in medicine, wrote:

the temptation to use them for every kind of ailment and to claim benefit for most of them proved irresistible. I recall taking time during the summer of 1938 to write what I thought was a conservative summary of the virtues of the then extra-popular Vitamin B_1, and I blush now to think of all the cures that I then ascribed to this magic compound. Anorexia, neuritis of all kinds, neurosis, heart failure, indigestion, loss of weight, even acute gout were relieved, more or less, if one took enough of it. For a brief period the new vitamin was everywhere and good for everything.[54]

Naturally, when such opinions were expressed by highly-qualified doctors, the commercial opportunities were not neglected by manufacturers.

Inevitably the consumption of vitamins increased. There were no obvious reasons for regulating their sale and use, apart from control of quality, so that zeal for self-medication and for the medication of children by parents was unrestrained. Nor were experimentally-minded doctors and scientists deterred from trying the effects of massive doses for this or that condition, and regarding the most favourable results with uncritical optimism. Gradually it became evident that vitamins, like any other substances, were dangerous in excess. Surplus water-soluble vitamins overflowed through the kidneys, but the fat-soluble A, D, and K were not easily got rid of. Tretinoin, a form of vitamin A, used for the treatment of acne,[55] has acquired a long list of ill effects, including the production of birth defects.[56] Peeling skin and other disorders, once familiar to arctic explorers who had eaten bear's liver, are also attributed to an excess of vitamin A. Too much vitamin D disturbs the delicate mechanisms which regulate the bodily distribution of calcium,[57] with complex consequences. The excesses necessary to achieve these effects are considerable, but are a reminder that potent substances are not harmless, whatever romantic reputation they may have acquired.

Subtler deficiencies

Euphoria about vitamins gave no more help to patients than any other misinformed optimism. A more analytical approach began to throw light on the exact role of vitamins in the working of the body, and to pinpoint the biochemical disturbances which resulted in loss of health. It became clear that vitamins, like drugs, were handled by the body in definite ways. Thus an ailment could be due to a deficiency in the metabolic process and might not simply be the result of the lack of a vitamin in the diet. In this way the cause of a baffling disease came to be understood as a failure of absorption of a vitamin rather than a consequence of its dietary absence.

This disease was a type of anaemia or bloodlessness. It was first described in the middle of the nineteenth century by Thomas Addison, the physician who also first recognized disease of the adrenal glands (see Chapter 5). The condition was named pernicious anaemia because of its remorseless progress to a fatal outcome. It differed from the commoner anaemias of northern Europe, in which lack of iron restricts production of the iron-containing, oxygen-carrying, blood pigment haemoglobin. Patients with pernicious anaemia had enough iron but they were short of blood cells, although the cells which persisted were unusually large and carried more than the usual quota of haemoglobin. As red blood cells are made in the bone marrow, it is interesting to recall an experiment made in the 1890s by Sir Thomas Fraser (see Chapter 1) in Edinburgh, who tried feeding bone marrow to a patient, and reported benefit.[58] After a flutter of enthusiasm in Britain, America, and France, nothing came of the treatment, and for 60 years the disease could be diagnosed, but only as a death sentence.

Early in the 1920s George H. Whipple (1878–1976), in California and then at the University of Rochester, was investigating the anaemia produced in dogs by repeated bleeding, and observed that meat, and especially liver, was valuable in accelerating their recovery. His report[59] stimulated George Minot (1885–1950), professor of medicine at Harvard, to suggest to a younger colleague, William P. Murphy, that one might try liver in pernicious anaemia, and Murphy, with dogged persistence, achieved cures of pernicious anaemia by persuading patients to eat very large amounts, a pound a day, of raw liver.[60,61]

The success was most welcome though the logic was open to criticism.[62] Whipple's experiments were about an anaemia quite different from the pernicious, and liver was good for his dogs because it contains a lot of iron. However, Minot and Murphy's achievement was as dramatically important as the isolation and provision of insulin for diabetics, and together they shared with Whipple the award of a Nobel prize. For patients, large amounts of raw liver were a most disagreeable diet, and chemical techniques were soon applied to concentrate the effective ingredient, and ultimately to isolate and identify it. A successful collaboration was established between the Harvard workers and the firm of Eli Lilly, and many of the problems related to profits and patenting, which had soured relations with the Toronto workers on insulin, were happily avoided.[63] But further research was difficult. There was no animal or other laboratory model for the disease, and every extract of liver had to be tested on patients with pernicious anaemia to see whether it contained active material or not. Patients suitable and willing to be subjects in assays, and co-operative clinicians, were not numerous. Several days were needed to be sure whether a response had occurred or not, and priority had to be given to assaying material for immediate clinical use, so advances in isolating an active principle were made terribly slowly. Meanwhile, pernicious anaemia was found to be more complicated than had so far appeared.

In Boston, about 1930, W. B. Castle showed that extracts of stomach added to certain foods, especially meat, relieved pernicious anaemia.[64] The use of stomach extracts was less satisfactory than the use of liver, so this discovery did not have great practical effect, but it was of considerable theoretical interest. At about the same time, Lucy Wills, at the Haffkine Institute in Bombay, during an investigation of maternal mortality, observed a condition which looked very like pernicious anaemia, but was not 'pernicious' (i.e. not usually fatal) and occurred only during pregnancy in women whose diet was very sparse.[65] Wills found that this particular anaemia was relieved by extracts of yeast, a rich source of water-soluble vitamins but of no particular benefit in pernicious anaemia.[66] She showed also that a similar anaemia could be produced in monkeys by feeding them on a diet like that taken by her patients. The anaemia in monkeys differed in various chemical details from the changes in typical pernicious anaemia, so it was unlikely that Wills's factor was relevant to pernicious anaemia. But no one could be sure.

Progress was made more quickly in solving the problems about the anaemia of pregnancy than about pernicious anaemia, mainly because the dietary deficiency could be imitated in monkeys and studied with more control and fewer complications than when progress depended on time-consuming assays in patients. Even so, many more investigations were needed before these tangled problems were elucidated. A new line of approach, which had its roots far from the problems of clinical medicine, began to shed light on the problem. We shall discuss the foods required by microbes, and the tremendous advances which came from studying them, in the next chapter. Here we need only note that a substance, discovered in spinach leaves and necessary for the growth of a particular bacillus, was found to be the cure for Wills's anaemia.[67] The substance was identified chemically by 1946 and named pteroylglutamic acid or, more conveniently, folic acid.[68] It relieved the anaemia of pregnancy, and so did a sufficient diet of green vegetables. It had complicated and important properties, especially in cells which were multiplying and reproducing, and it soon became a foundation for the discovery of anti-cancer drugs[69] (see Chapter 11). Hopes were raised that it would cure pernicious anaemia. But matters were not so simple. Folic acid did not cure pernicious anaemia, and sometimes made matters worse, so it was a most unsuitable treatment.

However, within a year the identity of the anti-pernicious anaemia principle was established independently by two sets of workers. Here too microbial approaches were invaluable. In the USA, microbes were discovered, which survived on the liver extracts effective in pernicious anaemia, and so could be used to guide the search for the active principle. This guidance was used with success by a group at the Merck laboratories.[70] In England, at the Glaxo laboratories, the approach based on clinical testing of purified materials bore fruit.[71] In both places a cobalt-containing pigment was isolated, which was named cyanocobalamin or vitamin B_{12}. Repeated injections restored the blood of patients with pernicious anaemia to normality and maintained them in good health. Massive oral doses were needed to produce a detectable response.[72] But with the pure vitamin it was easier to investigate the effect of stomach extracts which Castle had discovered 20 years earlier. Absorption of the vitamin from the gut was found to require the presence of a substance normally present in gastric juice, and this substance was found to be lacking in patients with pernicious anaemia. So the primary disorder in pernicious anaemia was eventually found to be in the stomach, but the illness of the patient resulted because an essential metabolite failed to reach the bone marrow where it was needed.

Since the discovery of folic acid and of vitamin B_{12} no more dietary ingredients of comparable importance have been discovered. One may well ask whether any are still waiting to be found. The answer depends on what exactly is meant by a vitamin. Now that much is known about the chemical processes in living organisms of all kinds, a number of substances are recognized as having key roles, as much in microbes as in higher animals and man, in keeping the 'organised factory manifoldly hydrolysing, pulling to pieces, and contemporaneously constructing and reconstructing',[53] which is the whole microbe or the cellular unit of more complex creatures. Most species make these key substances, often called 'essential metabolites', for themselves. Some take advantage of other species for some particular ingredients, and rely on what they eat to provide what they do not make for themselves. Perhaps, during evolution they lost the ability to do so. So we humans rely on a variety of edible plants and animals to provide some of our esential metabolites, and these are our vitamins. A committee of intelligent microbes would produce a very different list of essential metabolites. But probably they would not agree at all, because the requirements of different bacterial species are apt to be strikingly different from each other, and microbes are remarkably clever at adapting to new environments and new kinds of food.

In the 60 years between Eijkmann's rejection of microbes as a cause of beri-beri and their use to detect folic acid, much was learnt about the diseases caused by bacteria and about the behaviour of bacteria themselves. Let us return to the beginning of the twentieth century, after Pasteur and Koch had started mankind on the road towards the conquest of these diseases.

Notes

1. Zaman, V. (1983). Trichinosis. In *Oxford textbook of medicine*, eds D.J. Weatherall, J. G. G. Ledingham, and D. A. Warrell, p. 5.434–5. Oxford University Press, Oxford.
2. Leviticus, 11, 5–6.
3. Pereira, J. (1843). *A treatise on food and diet.* Longman, Brown, Green and Longmans, London.
4. McCollum, E. V. (1957). *A history of nutrition.* Houghton Miflin, Boston.
5. *Ibid.* p. 59.
6. Dumas, J. B. A. (1871). *Philosophical Magazine* 42, 129. Quoted from McCollum, note 4, pp. 202–3.

7. McCance, R. A. and Widdowson, E. M. (1960). The composition of foods. *Medical Research Council Special Report Series, 297*. Her Majesty's Stationery Office, London.
8. Henderson Smith, A. (1919). A historical enquiry into the efficacy of lime-juice for the prevention of scurvy. *Journal of the Royal Army Medical Corps* 32, 93–116; 188–208.
9. Carpenter, K. (1986). *The history of scurvy and vitamin C.* Cambridge University Press, Cambridge.
10. *Lancet* (1887). Health of the Imperial Japanese Navy. ii, 86.
11. Takaki, Baron (1906). The preservation of health among the personnel of the Japanese navy and army. *Lancet* i, †369–74; 1451–5; 1520–3.
12. Williams, R. R. (1961). *Towards the conquest of beriberi*, p. 16. Harvard University Press.
13. Eijkmann, C. (1897). Eine Beri Beri-ähnliche Krankheit der Hühner. *Virchow Archiv für Pathologische Anatomie* 148, 523–32.
14. Eijkmann, C. (1929). Antineuritis vitamin and beriberi. *Nobel lectures physiology or medicine 1922–1941*, pp. 199–207, at p. 203. Published in 1965 for the Nobel Foundation by Elsevier, Amsterdam.
15. McCollum, E. V. (1957). *A history of nutrition*, pp. 216 f, 244 f. Houghton Miflin, Boston.
16. Hopkins, F. G. (1929). The earlier history of vitamin research. *Nobel lectures physiology or medicine 1922–1941*, pp. 211–22, at p. 211. Published in 1965 for the Nobel Foundation by Elsevier, Amsterdam.
17. Dale, H. H. (1948). Frederick Gowland Hopkins 1861–1947. *Obituary Notices of Fellows of the Royal Society* no. 17, 6, 113–45.
18. Needham, J. and Baldwin, E. (eds) (1949). *Hopkins and biochemistry 1861–1947.* Heffer, Cambridge.
19. Chapman-Huston, D. and Cripps, E. C. (1954). *Through a city archway. The story of Allen and Hanburys 1715–1954.* Murray, London.
20. cf. Chapter 6, note 17.
21. Kohler, R. E. (1982). *From medical chemistry to biochemistry.* Cambridge University Press, Cambridge.
22. Hopkins, F. G. and Cole, S. W. (1903). A contribution to the chemistry of proteids. Part II. The constitution of tryptophane and the action of bacteria on it. *Journal of Physiology* 29, 451–66.
23. Hopkins, F. G. (1922). Chandler Medal Address. *Industrial and Engineering Chemistry* 14, 64. Quoted by Needham and Baldwin note 18.
24. Hopkins, F. G. (1912). Feeding experiments illustrating the importance of accessory factors in normal dietaries. *Journal of Physiology* 44, 425–60.
25. Holst, A. (1907). Experimental studies relating to 'ship beri-beri' and scurvy. I. Introduction. *Journal of Hygiene* 7, 619–33.
26. Holst, A. and Frölich, T. (1907). Experimental studies relating to 'ship beri-beri' and scurvy. II. On the etiology of scurvy. *Journal of Hygiene* 7, 634–71.
27. Chick, H. (1957). Charles James Martin 1866–1955. *Biographical Memoirs of Fellows of the Royal Society* 2, 173–208.
28. Chick, H., Hume, M., and Macfarlane, M. (1971). *War on disease. A history of the Lister Institute*, pp. 143–4. André Deutsch, London.
29. Griminger, P. (1972). Casimir Funk—A biographical sketch (1884–1967). *Journal of Nutrition* 102, 1105–14.

30. Funk, C. (1911). On the chemical nature of the substance which cures polyneuritis in birds induced by a diet of polished rice. *Journal of Physiology* 43, 395–400. For a discussion of the identity of Funk's 'pure' material, see Williams, (note 12) at p. 98.

31. Funk, C. (1912). The etiology of the deficiency diseases. *Journal of State Medicine* 20, 341–68.

32. Drummond, J. C. (1920). The nomenclature of the so-called accessory food factors (vitamins). *Biochemical Journal* 14, 660.

33. McCollum, E. V. and Davis, M. (1913). The necessity of certain lipins in the diet during growth. *Journal of Biological Chemistry* 15, 167–75.

34. Osborne, T. B. and Mendel, L. B. (1913). The relation of growth to the chemical constituents of the diet. *Journal of Biological Chemistry* 15, 311–26. The original observations of these workers and of McCollum and Davis (note 33) are spread over a large number of papers, especially in the *Journal of Biological Chemistry*. For a résumé see McCollum, note 4.

35. Herodotus. *The histories*, book 3.

36. Ferguson, M. (1917). A study of social and economic factors in the causation of rickets. Published by the Medical Research Committee, and later included in *Medical Research Council Special Report Series, 20*. His Majesty's Stationery Office, London.

37. Bland-Sutton, J. (1889). *Journal of Comparative Medicine and Surgery* 10, 1. Quoted from McCollum, note 4.

38. Dale, H. H. (1956). Edward Mellanby 1884–1955. *Biographical Memoirs of Fellows of the Royal Society* 1, 193–222.

39. Parascandola, J. and Ihde, A. J. (1977). Edward Mellanby and the antirachitic factor. *Bulletin of the History of Medicine* 51, 507–15.

40. Mellanby, E. (1938). The state and medical research. *Lancet* ii, 929–36.

41. Mellanby, E. (1919). An experimental investigation on rickets. *Lancet* i, 407–12.

42. Harden, A. (ed.) (1924). Report on the present state of knowledge of accessory food factors (vitamins). 2nd edn. *Medical Research Council Special Report Series, 38*. His Majesty's Stationery Office, London.

43. Chick, H., Hume, M. and Macfarlane, M. (1971). *War on disease. A history of the Lister Institute*. pp. 152–60. André Deutsch, London.

44. Huldschinsky, K. (1920). Die Behandlung der Rachitis durch Ultraviolett-bestrahlung. Dargestellt an 24 Fällen. *Zeitschrift für Orthopaedische Chirurgie* 39, 426–51.

45. McCollum, E. V., Simmonds, N., Becker, J. E. and Shipley, P. G. (1922). Studies on experimental rickets. XXI. An experimental demonstration of the existence of a vitamin which promotes calcium deposition. *Journal of Biological Chemistry* 54, 293–312.

46. Goel, K. M. *et al.* (1976). Florid and subclinical rickets among immigrant children in Glasgow. *Lancet* i, 1141–5.

47. Harris, R. J., Armstrong, D., Ali, R., and Loynes, A. (1983). Nutritional survey of Bangladeshi children aged under 5 years in the London borough of Tower Hamlets. *Archives of Disease in Childhood* 58, 428–32.

48. Harden, A. (ed.) (1924). Report on the present state of knowledge of accessory food factors (vitamins). 2nd edn. *Medical Research Council Special Report Series, 38*, p. 2. His Majesty's Stationery Office, London.

49. Goodman, L. and Gilman, A. (1941). The Pharmacological Basis of Therapeutics, p. 1243. Macmillan, New York.
50. Hopkins, F.G. (1920). The present position of vitamines in clinical medicine. *British Medical Journal* ii, 147–9.
51. Barr, J. (1920). Discussion on the present position of vitamines in clinical medicine. *British Medical Journal* ii, 150–1.
52. Sebrell, W. H. Jr. and Harris, R. S. (eds) (1967–1972). *The vitamins. Chemistry, physiology, pathology, methods.* Academic Press, New York.
53. Obituary notice (1947). *Lancet* i, 729.
54. Richards, D. W. (1964). A clinician's view of advances in therapeutics. In *Drugs in our society*, ed. P. Talalay, p. 29. The Johns Hopkins Press, Baltimore; Oxford University Press, London.
55. Peck, G. L. *et al.* (1979). Prolonged remissions of cystic and conglobate acne with 13-*cis*-retinoic acid. *New England Journal of Medicine* 300, 329–33.
56. Stern, R. S. (1989). When a uniquely effective drug is teratogenic. *New England Journal of Medicine* 320, 1007–9.
57. Fraser, D. R. (1981). Biochemical and clinical aspects of vitamin D function. *British Medical Bulletin* 37, 37–42.
58. Fraser, T. R. (1894). Bone marrow in the treatment of pernicious anaemia. *British Medical Journal* i, 1172–4.
59. Robscheit-Robbins, F. S. and Whipple, G. H. (1925). Blood regeneration in severe anemia. II. Favourable influence of liver, heart and skeletal muscle in diet. *American Journal of Physiology* 72, 408–18.
60. Minot, G. R. and Murphy, W. P. (1926). Treatment of pernicious anemia by a special diet. *Journal of the American Medical Association* 87, 470–6.
61. Thomas, L. (1983). *The youngest science*, p. 254–6. Viking Press, New York.
62. Gaddum, J. H. (1940). *Pharmacology*, p. 235. Oxford University Press, Oxford.
63. Swann, J. P. (1988). *Academic scientists and the pharmaceutical industry*, pp. 150–169. Johns Hopkins University Press, Baltimore.
64. Castle, W. B. (1934–35). The etiology of pernicious and related macrocytic anemias. *The Harvey Lectures* 30, 37–48.
65. Wills, L. (1931). Treatment of 'pernicious anaemia of pregnancy' and 'tropical anaemia', with special reference to yeast extract as curative agent. *British Medical Journal* i, 1059–65.
66. Wills, L., Clutterbuck, P. W., and Evans, B. D. F. (1937). A new factor in the production and cure of macrocytic anaemias and its relation to other haemopoietic principles curative in pernicious anaemia. *Biochemical Journal* 31, 2136–47.
67. Day, P. L. *et al.* (1945). Successful treatment of vitamin M deficiency in monkey with highly purified *Lactobacillus casei* factor. *Journal of Biological Chemistry* 157, 423–4.
68. Angier, R. B. *et al.* (1946). The structure and synthesis of the liver *L. casei* factor. *Journal of the American Chemical Society* 103, 667–9.
69. Jukes, T. H. and Stokstad, E. L. R. (1948). Pteroylglutamic acid and related compounds. *Physiological Reviews* 28, 51–106.
70. Rickes, E. L., Brink, N. G., Koniuszy, F. R., Wood, T. R., and Folkers, K. (1947). Crystalline vitamin B_{12}. *Science* 107, 396–7.
71. Fantes, K. H., Page, J. E., Parker, L. F. J., and Smith, E. L. (1949). Crystalline

anti-pernicious anaemia factor from liver. *Proceedings of the Royal Society, ser. B* 136, 592–613.

72. Berk, L. *et al.* (1948). Observations on the etiological relationship of achylia gastrica to pernicious anemia. X. Activity of vitamin B_{12} as food (extrinsic factor). *New England Journal of Medicine* 239, 911–13.

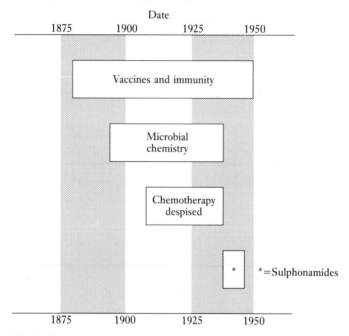

Chronology of Chapter 8

Date

1875   1900   1925   1950

Vaccines and immunity

Microbial
chemistry

Chemotherapy
despised

*   *=Sulphonamides

1875   1900   1925   1950

NB The blocks show periods discussed in the text, and do not mean that
there was no activity about the subject at other times.

8

War on germs

Biological remedies

It is difficult to imagine a time when only the natural processes of the body gave any defence against germs. Death in childhood from epidemic fevers was all too familiar. Promising careers were liable to be checked or ended by tuberculosis. Every childbirth brought a risk of death from infection of the womb and childbed fever. Pneumonia killed a quarter of its victims, often in the prime of life. The sexually imprudent might regard gonorrhoea as more trivial than some of its consequences deserved, but the progress to paralysis, insanity, and a lingering death from syphilis could no more be stopped than, at the time of writing, can the progress of AIDS (acquired immuno-deficiency syndrome).

Inroads against all these infections were made by antitoxins, by vaccines, and by chemotherapy with synthetic and antibiotic drugs. In the first half of the twentieth century, the human situation changed radically, and by 1950 few infections remained for which no effective remedy was known. Death from bacterial infection in a prosperous country became rare, and a false sense of security developed until new enemies—Lassa fever, legionnaire's disease, AIDS—appeared. Once again, we must go back to the beginning of laboratory medicine, when the germ theory of disease had become established doctrine, to see some of the many ways in which our knowledge of micro-organisms advanced, and how established methods were improved to meet new challenges.

The first useful treatments consisted of antisera and vaccines against the more amenable bacteria, and new synthetic drugs against protozoa, and against spirochaetes which caused syphilis and some less familiar conditions. One must distinguish clearly between the substances useful for treating disease and those useful for preventing it. Vaccines were preventive, antitoxins and synthetic drugs usually curative. Pasteur's rabies vaccine was exceptional as a treatment. It was useful only because infection, by the bite of an infected animal, was immediately obvious, and the long incubation period of the disease gave time for the vaccine to produce some immunity in the patient before the rabies virus spread. Usually vaccines were useful only for prophylaxis, and they were not called on for the treatment of sick patients.[1]

The value of antitoxins was limited to diphtheria and tetanus, and a few other organisms which released toxins into the patients' blood stream and so throughout the body. Most germs did not produce such toxins; they caused damage by less obvious mechanisms, and the ways in which immunity developed were far from clear. Substances which provoked immunity were extracted from microbes and called antigens, while the substances which animal bodies produced in response to an infection or a vaccine were isolated as far as was feasible and named antibodies. Both antigens and antibodies turned out to be proteins, substances with very large molecules and uncertain structures, and at that time not easy to separate and analyse more than crudely. It was soon apparent that many proteins, not only those that came from microbes, were capable of acting as an antigen. Likewise it was found that some of the reactions between antigen and antibody were not necessarily beneficial, and were responsible for allergies and for the much more dangerous condition called anaphylaxis.

In spite of such hazards, the theoretical possibilities of using antibodies for treatment were very great, and innumerable attempts were made to take advantage of them. Blood serum was the obvious source, either from animals which had been infected with the appropriate organism or from convalescent patients. Useful antisera were obtained against anthrax, and for a long time anti-pneumoccocal sera were used, with limited success, to treat lobar pneumonia. Difficulties arose because different types of pneumococcus existed, and the antiserum against one type was not always effective against another. Typing of an infecting organism was therefore necessary before obtaining serum to treat a patient. The antisera were most effective when given at the earliest possible moment in the disease, so good organization of supplies and testing was essential if any benefit was to be obtained.

All sera contained numerous proteins beside the required antibodies, and patients were liable to become sensitized to these irrelevant proteins if repeated use of the sera was necessary. Then the serious and sometimes fatal condition of serum sickness followed. Improved methods of purification reduced the problem, but such progress was slight until advances in chemistry after World War II permitted the separation of the proteins specifically concerned with immunity from other serum proteins, so that purer and safer agents became available.

The prevention of diseases by producing a vaccine containing microbial antigens was an obvious target. Jenner had made this possible for smallpox and Pasteur for anthrax in sheep and for rabies. The technical problems were formidable enough, but whenever they were solved, administrative and social attitudes were apt to delay progress. Vaccines are often unpopular. Whatever desire there may be for medicines in the face of illness, it goes strongly against deep-seated protective instincts to inject some potent substance into a healthy person, especially an infant. Vaccination against smallpox had its opponents as well as its protagonists, sufficient for a 'conscience clause' to be included in an act of parliament in 1898 to permit escape from earlier laws which enforced compulsory vaccination. Later campaigns came to rely on persuasion and propaganda to achieve their objectives, but it was difficult in the first place to arouse all the interested parties to recognize the benefits of a new vaccine.

Immunization could be applied more positively in the armed forces, where protection of troops from disease was of great military importance and where military discipline overrode personal objections. Typhoid fever was particularly important, both in the South African war at the beginning of the century and in the 1914–1918 war. In England, A. E. (later Sir Almroth) Wright[2] (1861–1947), professor of pathology in the British Army Medical Services from 1892 until 1902, developed a vaccine based on organisms killed by heat, probably at the suggestion of Haffkine of the Pasteur institute, who had developed a similar vaccine against cholera (see Chapter 3). Wright's proposals for vaccination of the troops destined for India and South Africa aroused much opposition from military men, so that Wright resigned from his post and moved to establish an inoculation department at St Mary's Hospital in London. More persuasive counsels were more effective and an extensive vaccination programme was carried out in the army. Statistical data about the efficacy of the vaccine, deemed adequate even by critical and experienced statisticians, appeared to show that it was undoubtedly beneficial.[3] The acquiescence of statisticians as critical as Major Greenwood (1880–1949) and G. Udny Yule (1871–1951) now seems surprising, because the evidence depended on comparison between volunteers who had accepted the treatment and others who refused. Greenwood and Yule gave careful reasons for dismissing the difference as unimportant, but a modern re-analysis of the facts, and recognition of the many differences in conduct and exposure to risk between men willing to volunteer and others, has cast doubts on the extent to which the vaccine was responsible for the recorded results.[4]

Apart from smallpox, immunization of civilians was rare, although typhoid and related fevers took a substantial toll. The first practical application of immunization on a large scale, apart from vaccination itself, was against diphtheria. There were difficulties in this kind of development. Once Behring had produced antitoxin for treatment, he and others attempted to use the bacterial toxin to produce immunity, giving at the same time enough antitoxin to protect the patient from the dangerous effects of the toxin. The procedure was used in horses, so that they could be given big doses of toxin to produce plenty of antitoxin for therapeutic use. In humans some success was achieved, but it was too easy to misjudge the right doses. Better results were obtained by modifying the toxin in some way. Partly by accident[5] the discovery was made that formalin, which reduced the toxicity of the toxin, did not interfere with its power of immunizing horses. After many years, modified toxins, or 'toxoids', suitable for human use, were developed, particularly by Alexander Glenny (1882–1965) at the Wellcome laboratories near London[6] and by G. Ramon at the Pasteur Institute in Paris,[7] and tried on a wide scale in the 1920s and 1930s. In the long run, the results were of the greatest value, but many obstacles had to be overcome.

At times, local advocacy achieved substantial results. In Hamilton, Ontario, a campaign from 1925 onwards to immunize all children eliminated deaths from diphtheria after 1930. In other places and other countries, years of delay preceded widespread immunization. In the United Kingdom, World War II and its problems made the prevention of disease more than usually urgent: also, the government of the time was commendably sensitive to the value of scientific advice. A national campaign against diphtheria came into action by 1942, directed mainly towards immunization in infancy. Naturally, a substantial part of the population was not covered, but the long-term effects were as striking as they had been years earlier in Ontario. In 1946 there were 11,986 reported cases of diphtheria with 472 deaths: in 1960, 49 cases with 5 deaths, 4 of which happened in the small minority of people who were still not immunized.[8] Plainly the virtual disappearance of the disease was not due to a decline in the virulence of the organism.

Although two of the earliest vaccines protected against the virus infections smallpox and rabies, very little progress was made in developing vaccines against other viruses. The study of viruses made little headway before the 1930s, when one of its first fruits was the development of a vaccine against the predominantly tropical, commonly fatal, disease, yellow fever. This vaccine was of great value in protecting troops in World War II. Research in this field is intimately linked to developments after the war, and will be considered later (see Chapter 10).

Microbial chemistry

The development of antitoxins and vaccines began with Pasteur's observations in 1880–1881 on fowl cholera and on anthrax in sheep. The practical applications were important and for many people took precedence over the more basic studies which had given the original impetus to Pasteur's concern with microbes. But the study of fermentation, i.e.

chemical changes which occurred only in the presence of micro-organisms, continued in various laboratories. The conversion of sugars to alcohol was perhaps the most striking. The further conversion of alcohol to vinegar was the subject which drew Pasteur towards the investigation of microbes. The souring of milk also depended on microbial activity. All these processes were challenges to pure chemistry: how could living organisms perform changes which did not happen in their absence?

But cells do not have to be intact to work. The German chemist Eduard Büchner (1860–1917) confirmed an earlier theory that the changes depended on enzymes which could be extracted from cells, and that the intact live cells were not necessary for fermentation to take place. Whether one regards this as a repudiation of Pasteur's discoveries or not is a matter of taste. The need for living cells to provide the means for fermentation was not discredited: the mechanism by which they did so now began to be analysed. Büchner made a cell–free filtrate of yeast which, despite the absence of living or even dead cells, was able to convert cane sugar to alcohol and carbon dioxide.[9] He attributed the process to a single enzyme, but it was soon shown that there were many intermediate steps in the fermentation and that a separate and distinct enzyme was required for each step. From these simple beginnings a very complex science of microbial biochemistry evolved, which in time provided vital tools for the discovery of drugs.[10,11]

The fermentation of sugars was important also to medical bacteriologists, who identified bacteria from patients by discovering what kinds of sugary culture media they would or would not grow on, and what kinds of fermentation went on when they did grow. I remember hours spent in 1940 as a medical student looking at test tubes containing bacteria growing in culture, media containing dyes which changed from pink to yellow if the acidity altered and which were full of bubbles if fermentation had occurred. Combinations of changes were specific for identifying bacteria, and so this was an important part of medical diagnosis. Like most people, I never wondered whether the microbes were capable of changing their dietary habits.

The orthodox view was established from about the beginning of the century, and sceptics who found evidence that microbes could change their habits did not prosper. Among them, Frederick William Twort (1877–1950), at the London Hospital in Whitechapel, showed that with care the microbe of typhoid fever could be persuaded to feed on milk sugar, which normal typhoid bacteria disdained and which was against all the rules for identifying this microbe. He did similar experiments with other bacteria, and published his results in a rather half-hearted way, writing that such micro-organisms were 'not to be regarded as distinct species but as varieties or hybrids of one or more species. If this be so' (he continued) 'one might expect them to be constantly varying, losing old characters and gaining new ones according to the conditions under which they are grown. . .'[12] This was about 1907, when Darwin's ideas of evolution had become firmly established, at least in scientific minds. Twort's discovery savoured of the inheritance of acquired characteristics. It conflicted with the new orthodoxy, established after much heated debate, and (what was probably worse) it undermined confidence in 'universal bacteriological experience'. So it was assumed to be wrong and was ignored, even though he was not the only person to make such an observation.[13]

Twort had not finished being unorthodox. A few years later, he described a means of culturing a bacillus recovered from a patient with leprosy. No one had previously grown a leprosy bacillus in artificial media. There is doubt whether it was actually the leprosy bacillus or a relative which Twort succeeded in growing: however, what is important is his idea that related organisms contained related substances, and that if one ground up dead organisms of a related species and added them to the medium, it might help a difficult species to grow. The idea of 'growth factors' aroused no enthusiasm at the time, but was taken up widely in the 1930s, and later became fundamentally important in the armoury of discoverers of drugs. This time Twort was only 20 years ahead of his colleagues.

Twort's third discovery[14] is mentioned later (see Chapter 9), but then too he failed to establish the importance of his work. His highly original early studies were published as preliminary observations and were never consolidated. He was appointed at the age of 32 as director of the Brown Institution (see Chapter 3) in South London, where he was isolated from his colleagues and accomplished nothing further, becoming embittered because his achievements were disregarded and he could not obtain further support.[15,16] (Eight years after Twort's death, in 1950, a Nobel prize was awarded to scientists who had established in a much more detailed and convincing way the capacity of microbes to create inheritable new enzymes as part of their adaptation to novel environments.)[17-19] Meanwhile mainstream microbial chemistry evolved during the 1920s in a growing number of laboratories, especially at the Rockefeller Institute for Medical Research in New York, under the direction of Simon Flexner[20] (1863–1946) and in Cambridge under Gowland Hopkins.

Hopkins was one of the earliest scientists to use bacteria as a means for chemical study, when he was investigating the amino acid tryptophane in 1903.[21] He continued to develop the subject of bacterial chemistry some 10 years later, when the availability of grants from the recently formed Medical Research Committee enabled him to recruit Harold Raistrick from Leeds, and a Cambridge student, Marjory Stephenson (1885–1948), when she returned to laboratory work in 1919 after serving with the Red Cross in Salonika. Raistrick left to apply microbiological methods in

Plate 1 Paracelsus (1493–1541), from a Dutch seventeenth century line engraving.

Plate 2 James Lind (1716–1794). Portrait by J. Wright after Sir G. Chalmers, 1783.

Plate 3 William Withering (1741–1799). Portrait by W. Bond after C. F. v. Breda.

Plate 4 Edward Jenner (1749–1823). Portrait by J. R. Smith, 1800.

MAGENDIE.

Plate 5 François Magendie
(1783–1855). Lithograph
Portrait by N. E.Maurin,
probably between 1842 and 1858.

Plate 6 Pierre-Joseph Pelletier
(1788–1842). Portrait by
Catherine Buisson, 1870, after
Eliza Dérivieères.

Plate 7 Joseph Caventou (1795–1877). Portrait by Catherine Buisson, 1870, after Eliza Dérivieères.

Plate 8 Claude Bernard (1813–1878). Undated photograph.

Plate 9 Friedrich Wöhler
(1800–1882). C. Cook,
after design by C. L'Allemand.
William Mackenzie, not dated.

Plate 10 Louis Pasteur (1822–1895) in his laboratory. From an engraving
published in *The Graphic*, November 1885.

Plate 11 Robert Koch (1843–1910) in his laboratory at Kimberly, South Africa, c. 1896–97.

Plate 12 Emil von Behring (1854–1917), c. 1909.

Plate 13 Joseph Lister, later Lord Lister (1827–1912), about the time of his experiments in antiseptic surgery, *c.* 1862.

Plate 14 Sir Thomas Fraser (1841–1920). Photographic portrait by A. Swan Watson, Edinburgh, not dated.

Plate 15 Paul Ehrlich (1854–1915), not dated.

Plate 16 H. H. Dale, later Sir Henry Dale (1875–1968), *c.*1918.

Plate 17 F. G. Banting, later Sir Frederick Banting (1891–1941).

Plate 18 C. H. Best (1899–1978).

Plate 19 Sir (Edward) Charles Dodds (1899–1973).

Plate 20 Sir Frederick Gowland Hopkins (1861–1947) in 1938.

Plate 21 Sir Edward Mellanby (1884–1955).

Plate 22 Discoverers of antibiotics. Back row: left to right, S. Waksman, H. Florey, J. Trefouel, E. Chain, A. Gratia. Front row: P. Fredericq, M. Welsch.

Plate 23 Sir Henry Dale (1875–1968) in 1959.

Plate 24 Dr G. H. Hitchings and Dr Gertrude Elion at the Wellcome Research Laboratories, New York, about 1948.

industry. Stephenson remained in Cambridge and established a research group which in time had great influence in the study of bacterial metabolism.[22]

As she herself described her work, 'Perhaps bacteria may tentatively be regarded as biochemical experimenters: owing to their relatively small size and rapid growth, variations must arise very much more frequently than in more differentiated forms of life, and they can in addition afford to occupy more precarious positions in natural economy than larger organisms with more exacting requirements'.[23] She founded the first major school of microbiological chemistry, attracted outstanding research workers, and after World War II was one of the first two women to be elected to the fellowship of the Royal Society. Her subject probably appeared to be highly academic to most people, but its practical consequences became apparent, not only in the search for antimicrobial drugs, but also in developing drugs against cancer (see Chapter 11), in wider fields of medical research, and in all industries which learnt to make use of the chemical skills of microbes.

Another unit which was developed soon afterwards had considerable influence on the development of vitamins and of drugs. Its originator was (Sir) Paul Fildes,[24] who had worked alongside Twort and later succeeded him as a lecturer at the London Hospital. There he worked with his colleague James McIntosh conducting some of the earliest English trials of Salvarsan and studying the chemotherapy of syphilis. Voegtlin's work on arsenical compounds and sulphydryl receptors (see Chapter 3) had been published, and, with this background and the influence of Twort's lines of thought, Fildes extended his work to wider aspects of the chemistry of bacteria.

By 1934 Fildes had established a research unit at the London Hospital, but, despite being elected F.R.S., was not appointed to the chair when it became vacant. Fildes moved to the Bland Sutton Institute of the Middlesex Hospital, where his old colleague McIntosh was Professor. In Fildes's unit,[25] various species of bacteria were grown in solutions containing different amounts of known nutrients, especially amino acids and vitamins of the B group. The concentration of each nutrient necessary for the microbes to grow and to reproduce was discovered and found to be specific and consistent. As well as throwing light on how microbes behaved chemically, this gave a novel way of measuring very small amounts of vitamins.

Methods like this were vastly quicker and needed much less material than those requiring the measurement of the rate of growth of small animals on deficient diets. So the fundamental research of Fildes's group had a considerable practical benefit. Such assays were valuable to manufacturers of vitamins, who needed to know how much vitamin their product contained. They were also invaluable to research workers seeking to identify and purify nutritional factors, whether they were necessary for man, for other animals, or for microbes. This aspect was soon exploited to good effect in relation to the chemotherapy of infectious disease, but only because more pragmatic approaches to the subject had, after a long period of doubt and hesitation, yielded marvellous results.

Chemotherapy despised

Ehrlich's success with Salvarsan (see Chapter 3) was evidence that microbes could be destroyed in patients by synthetic drugs, but Salvarsan and its more manageable successors had considerable limitations. A typical course of treatment consisted of an injection once a week for 10–12 weeks, repeated two or three times with intervals of a month or so to avoid toxic effects, which were sometimes serious or even fatal. Even the treatment of early syphilis took a year or more. To prevent relapses during the intervals between courses, other drugs were used. Preparations of mercury were traditional, but were largely superseded by salts of bismuth which appeared to be equally effective and less toxic.

The spirochaete which caused syphilis differed considerably from most bacteria. The limited success of Salvarsan neither gave much conviction that other microbes might be overcome by chemotherapy, nor encouraged a search for new agents. Without the knowledge of microbial chemistry which was slowly acquired in the 1920s and 1930s, there was little light with which to guide the search.[26] Ehrlich had already experimented with known antibacterial substances, such as phenol, and had investigated derivatives, which were good at killing bacteria, but unacceptably toxic.[27] He explored variants of quinine, which was tolerated by man in doses which cured malaria, but was not effective against bacterial fevers. Quinine-like compounds turned out to be more promising. A colleague of Ehrlich, Morgenroth, found that a relative of quinine, ethylhydroxy-cupreine, cured experimental pneumococcal infections in mice.[28] The compound was made available for clinical use under the name 'Optochin'. The benefits to patients were disputed[29] for a long time. With no adequate methods for estimating blood concentrations, it was difficult to know whether enough drug was being given. One ingenious group of investigators[30,31] assayed the antibacterial activity of the blood of patients being treated. The information was helpful in judging which dose to use, but did not advance the cure of patients. The use of Optochin was sometimes followed by damage to the optic nerve, impaired vision, and even blindness. The risks of optic damage were not regarded with the horror that such an adverse reaction would evoke today, because pneumococcal pneumonia had a mortality of 20–40 per cent and considerable risks were accepted as the price of survival. The standard treatment was with anti-pneumococcal serum, but this had its own difficulties and was far from reliably life-saving, if indeed it conferred any benefit at all. So the risk of using Optochin was sometimes taken, but its use declined. Related compounds continued to raise hopes, and are still not without interest.[32,33]

Some of the research workers whom Ehrlich had welcomed to his institute developed his ideas in their own laboratories. Carl Browning (1881–1972), a Scotsman who came to the Georg Speyer Haus in Frankfurt when it was newly opened, returned to Britain to continue the study of dyes as chemotherapeutic agents (see Chapter 3). An acridine dye had been found to be active against trypanosomes and named 'Trypa-flavin'. Browning brought the related yellow dyes acriflavine and proflavine from Ehrlich's laboratory, and showed that each was more toxic to some pathogenic bacteria than to host tissues.[34] By this time World War I was raging. Antiseptics were urgently needed for the treatment of wounds. Browning received no more supplies from Germany and no chemical support. He turned to Dale, during the period in which he was deputizing for Fletcher as secretary of the British Medical Research Committee (see Chapter 4). Dale arranged for more 'flavine' to be made and submitted to clinical trials. It was hailed popularly and prematurely as an enormous success, and, after a discreditable episode involving delays while a dyestuffs manufacturer pursued patent rights,[35] made available to the War Office and to the general public. In the following years, the flavine dyes proved useful, but they did not fulfil the exaggerated expectations which had been aroused. The acridine dyes have received much further study, particularly with regard to the relation of their precise chemical structure to their antibacterial action.[36]

The principal advances in chemotherapy between 1920 and 1932 were made in the laboratories of the Bayer Company in Elberfeld. Progress followed Ehrlich's method of choosing a suitable test system, usually a group of animals infected with a particular organism, and using it to test many substances, most of them newly synthesized. The most effective were taken for further detailed study. Such a method was readily applied to trypanosomes, which caused sleeping sickness and which had been used in experimental laboratory infections for the last 20 years. It became possible for malaria when the infection of small birds with a species of the malaria parasite was recognized[37] and adapted as a basis for laboratory tests. The method could be applied to any bacterial infection, but useful results from doing so were not apparent for some time. Much of the Bayer work was conducted in secrecy. Even the chemical identity of the first important compound which emerged, 'Bayer 205' or 'Germanin' for sleeping sickness, was not divulged initially.[38] Perhaps because of this secrecy and because neither malaria nor sleeping sickness were of any practical significance in northern Europe and north America, neither Germanin nor two important antimalarials[39,40] (Plasmaquin or pamaquin and Atebrin, alias mepacrine in England and quinacrine in America) raised doubts in the mind of those who were convinced that Ehrlich's hopes were as unsubstantial as dreams and that antibacterial chemotherapy was impossible.

Chemotherapy prevails: the defeat of streptococci

However, the staff of the Bayer laboratories was strengthened in 1927 by the appointment of Gerhard Domagk (1895–1964) as director of research in experimental pathology. In collaboration with the chemists Fritz Mietzsch and Joseph Klarer he developed work on antibacterial drugs. In accord with Ehrlich's dictum that substances had to be fixed by the cells on which they were to act (see Chapter 3), and with the self-evident fact that dyes were firmly fixed by the materials to which they were applied, dyestuffs were investigated thoroughly as a source of potential drugs. Among them was a red dye later named 'Prontosil'. It had been synthesized years earlier in the hope that it would be valuable for colouring leather. The possibiity that it would fix to other biological materials was good reason for testing it as a drug.[41] In December 1932 Domagk carried out an experiment on two groups of 12 and 14 mice infected with streptococci. One group was treated with the red dye and survived for at least a week. The others were not treated and died within four days. This was a most remarkable and encouraging result, but for 2 years nothing was published, while further information was being gained about the drug. Among the earliest patients to receive the new medicine was Domagk's own daughter, whom he treated, successfully, for a streptococcal infection.[42] By 1935 clinical trials were widespread, the drug was becoming well known, and Domagk reported on the mouse experiment[43–45] alongside reports by several German physicians of its use to cure streptococcal infections in patients.

Later in the same year, J. Tréfouel and his colleague at the Pasteur Institute in Paris showed that a much simpler compound than Prontosil was an equally effective drug.[46,47] This compound, chemically named p-aminobenzenesulphonamide or, more conveniently, sulphanilamide, was part of the whole molecule of Prontosil, and it was shown soon afterwards that Prontosil was broken down, after being absorbed into the body, to release sulphanilamide itself, so Prontosil was what would now be called a 'pro-drug' (see Chapter 2). This was a startling and, for some people, a disconcerting finding. Sulphanilamide was not a new substance. It had been synthesized and described in a publication in 1908, in the course of investigations on dyes,[48] and so could not be patented. If it reached the enormous market which was just opening for Bayer, it would divide the profit which the firm hoped to gain between as many manufacturers as chose to make the (much simpler) material. Indeed, that is exactly what did happen. It must have been a bitter blow for Bayer, but it may not have been unexpected. Scientifically, after discovering the action of the red dye, it would be obvious to test its component parts separately. Commercially, it would be an obvious step to prefer to market a patentable compound.

Why was sulphanilamide not recognized earlier? The simple answer is that there was no reason, in 1908, for its originator to suppose that this compound had any effect on bacteria. Nor was there any reason for Ehrlich, who had just published the results of a disappointing search for antibacterial compounds of low toxicity, the pay special attention to the substance, one among thousands of compounds being synthesized every year for all kinds of purposes. Some, expected to be antibacterial, were being tested with negative results. What would have been the point of searching more widely? Like aspirin in the nineteenth century, sulphanil-amide lay unappreciated in chemical research laboratories for many years. No doubt there are similar unrecognized potent drugs so placed today. To have missed a great success is a nightmare that haunts every medicinal chemist.

In fact sulphanilamide *was* tested as an antibacterial, in about 1921, by Michael Heidelberger at the Rockefeller Institute in New York, and its activity was not discovered. At that time it was expected that any anti-bacterial drug would be effective at once, if it was going to be effective at all. In Heidelberger's experiments sulphanilamide, mixed with bacteria, was injected into mice, and the mortality was no less than that caused by the bacteria alone.[49] As we now know, sulphanilamide takes hours to produce its effect and had no chance in the conditions which Heidelberger adopted. Similar expectations can be found in other publications of the 1920s. The British scientist Clifford Dobell (1886–1949), investigating the action of emetine against parasitic amoebae, noted that very strong concentrations (1 per cent or more) were needed to kill this parasite instantaneously but that very weak solutions (1 in 50,000 or less) would do so if allowed to act for a sufficient time. He described this situation as 'peculiar'.[50] Drugs which take hours or days before they act on microbes are now perfectly familiar, and the reasons why are usually well known. The difficulty in the 1920s was to get rid of existing notions which prevented discoveries from being made.

The fact that 'Prontosil' needed to be altered by the metabolism of the subject being treated, and that it was not active *in vitro*, is a quite different matter, but it nearly led to another failure to discover the drug. It is reputed that the chemists, Mietzsch and Klarer, were uneasy about doing initial tests in mice, and would have liked Domagk to screen compounds first on bacterial cultures in test tubes. If this had been done Prontosil would have been missed. However, Prontosil's lack of action in a test tube suggested that it underwent some modification after it was absorbed into the body, and so led the French workers to look for possible breakdown products such as sulphanilamide.

The next stage was to establish facts about the clinical value of the new drugs. In Britain the Medical Research Council organized trials in suitable clinical centres. The leading expert on streptococcal infections was Leonard Colebrook (1883–1967), who had recently moved from Almroth Wright's department to take charge of a new laboratory a few miles away at Queen Charlotte's Maternity Hospital. Colebrook was a devout disciple of Wright, but no stranger to the idea of chemotherapy, having made some of the first English studies of Salvarsan and investigated the possibilities of using arsenical compounds against streptococci early in the 1920s. Colebrook's studies in collaboration with the obstetrician Maeve Kenny did much to establish the reputation of Prontosil and, later, sulphanilamide in the English-speaking world as a powerful drug against puerperal sepsis, and incidentally to undermine Wright's conviction that chemotherapy was a useless subject to pursue.[51]

Within 5 years, the clinical efficacy of sulphanilamide against streptococcal infections was established beyond doubt. It had some unpleasant effects, but they were of minor importance compared with its therapeutic power. The mortality from the principal streptococcal diseases—puerperal fever, erysipelas (streptococcal cellulitis), and scarlet fever—tumbled. There was no longer any doubt that bacterial infections could be cured by a simple substance with no biological origin. Thousands of related compounds were synthesized and tested, and some were found to act against other pus-forming cocci as well as streptococci. Sulphapyridine from May and Baker ('M & B 693') was effective against pneumococci, which did not succumb to sulphanilamide, and elaborate investment in the production of anti-pneumococcal sera no longer had any purpose. More varieties appeared, and the collective name sulphonamide was chosen. Most outstanding of all, the mode of action of sulphanilamide was discovered (see Chapter 10), and the foundations of a rational search for fresh chemotherapeutic agents was established.[45,52]

From achievement to understanding: how sulphanilamide works

How did sulphanilamide stop some kinds of microbe from multiplying, but not others? How did it differ from all the familiar disinfectants like phenol and proflavine in not damaging human tissues to any serious effect? The problems were pursued in several laboratories, and evidence accumulated quickly from different sources. Sulphanilamide was a powerful and effective drug, but sometimes it did not work; for instance when the streptococci which it attacked were surrounded by dead tissues, blood clots, or pus. In the laboratory its effects were easily overcome by adding the same sort of materials to experimental cultures. Extracts of various micro-organisms, including the streptococci themselves, were also protective and so were extracts of yeast.[53-56]

Several lines of activity converged in 1938–1939 at the Middlesex Hospital in London. (Sir) Lionel Whitby (1895–1956) was conducting the first trials of a new sulphonamide, sulphapyridine, which was better than sulphanilamide because it attacked the coccus which caused pneumonia.[57] Fildes's group was joined by D. D. Woods (1912–1964), who came from Marjory Stephenson's laboratory.[58] The cross-fertilization of ideas between clinical users and microbiological chemists prospered, as Fildes and Woods applied the methods of both laboratories to finding out how sulphanilamide, the forerunner of sulphapyridine, worked.[25,59]

The properties of the active substance in yeast extracts suggested that it might be chemically related to sulphanilamide itself. The idea led to testing of a substance, p-aminobenzoic acid ('PABA'), which was found strongly to counteract the effects of sulphanilamide.[60] The substance was not at that time particularly familiar, but might well occur naturally. The most interesting thing about it was its very close chemical resemblance to sulphanilamide. It was soon found that PABA is very important to some species of microbe, which indeed cannot survive without it. These species are the ones which are poisoned by sulphanilamide. The studies were not simple: they depended on the kind of culture medium used, which might have more or less PABA in it, and on the number of organisms present (which determined the rate at which PABA was used up). But the broad picture was established and is very simple. The microbes which are sensitive to sulphanilamide need PABA as a building block for their growth. Sulphanilamide is a false building block; it is like PABA and can be put in its place, but it is the wrong shape to allow anything more to be built on it. So it jams the works, unless there is a lot of PABA available. When there is an excess of PABA, as in pus, the sulphanilamide has less chance of getting in and not enough is taken up to harm the microbes. If supplies of the essential PABA are scanty, the sulphanilamide wins, and brings the microbe's metabolism to a standstill, so that it becomes easy prey to the normal bodily mechanisms of immunity and phagocytosis, and the patient recovers.

The process is called competitive antagonism. The idea was not quite new even in 1940. The biochemist Leonor Michaelis (1875–1949) observed as early as 1914 that the splitting of some sugars by particular enzymes was reduced if a related sugar was also present, even though the related sugar was not itself being split.[61] J. H. Quastel, in Stephenson's laboratory, made similar observations on the breakdown of succinic acid.[62]

Karl Landsteiner (1868–1943), famous as the discoverer of human blood groups and the serological reactions by which they are recognized, noted that some such reactions, which take place between very large molecules, could be blocked by very small molecules.[63] All these showed that the type of mechanisms which Fildes and Woods proposed were quite likely to be correct. But this was the first time the practical importance of the problem took such competition far beyond being a laboratory curiosity.[64,65]

There was a second side to this story, so negative as to appear quite uninteresting, and hardly merit the title of a 'discovery'. PABA is not a necessary food or vitamin for man. Humans and other animals get along very well without eating it, though (as was discovered very soon after this time) it is a component of the vitamin folic acid, and is therefore part of an essential nutrient. Humans rely on folic acid made by plants or microbes, and are unaffected in this respect by sulphanilamide. The sulphonamide has other effects, some undesirable, on humans, but basically its chemo-therapeutic success depends on a step in the metabolic pathways specific to the organisms which it overcomes. This step is absent in humans and so *selective* poisoning of the microbes can take place.

Here was a basis for inventing any number of new drugs. Instead of starting with a drug and finding out how it worked, one could identify some substance which played an essential part in the cause of a disease (preferably one which did not have an important role in human physio-logy).[65] Then one synthesized related compounds and tested to find com-petitive blockers, from which one could choose the most suitable for medical purposes. It looked attractively simple. It presented immense possibilities, and the principle was sound. The practical difficulties became more obvious as time progressed. But fundamentally, it was an explicit working out, in chemical terms, of what pharmacologists had studied since the experiments of Claude Bernard on curare and carbon monoxide and of Fraser on physostigmine and atropine. What a great advance to have it formulated in chemical terms and in the simplest of cellular systems!

Woods did not stay for long in Fildes's laboratory. The exigencies of war required his service elsewhere, and he left the Middlesex Hospital less than a year after his arrival. It was a very short stay in which to achieve such a seminal discovery. But the time was ripe, and the identity and role of *p*-aminobenzoic acid could hardly have failed to come to light from the parallel researches of several other laboratories in Britain and the USA. Not only the effectiveness of chemotherapy but also a mechanism was now known. New chemotherapeutic drugs could be designed, and the principal was applied in laboratories all over the world. But at this moment an altogether different way of attacking microbes suddenly came into the limelight, and needs to be described in a chapter of its own.

Notes

1. Parish, H. J. (1965). A history of immunization, p. 141. Livingstone, Edinburgh.
2. Colebrook, L. (1954). Edward Almroth Wright 1861–1947. *Obituary Notices of Fellows of the Royal Society* 6, 297–314.
3. Greenwood, M. and Yule, G. U. (1915). The statistics of anti-typhoid and anti-cholera inoculations, and the interpretation of such statistics in general. *Proceedings of the Royal Society of Medicine* 8, Section of Epidemiology and State Medicine, 113–90.
4. Cockburn, W. C. (1955). The early history of typhoid vaccination. *Journal of the Royal Army Medical Corps* 101, 171–85.
5. Parish, H. J. (1965). *A history of immunisation*, pp. 151–3. Livingstone, Edinburgh.
6. Glenny, A. T. and Sudmersen, H. J. (1921). Notes on the production of immunity to diphtheria toxin. *Journal of Hygiene* 20, 176–220.
7. Ramon, G. (1924). Sur la toxine et sur l'anatoxine diphthériques. Pouvoir floculant et propriétés immunisantes. *Annales de l'Institut Pasteur* 38, 1–10.
8. Parish, H. J. (1965). *A history of immunisation.* pp. 157–8. Churchill Livingstone, Edinburgh.
9. Buchner, E. (1897). Alkoholische Gärung ohne Hefezellen. *Berichte des Deutsches Chemische Gesellschaft* 30, 117–24.
10. Bulloch, W. (1938). *The history of bacteriology.* Oxford University Press, Oxford.
11. Collard, P. (1976). *The development of microbiology.* Cambridge University Press, Cambridge.
12. Twort, F. W. (1907). The fermentation of glucosides by bacteria of the typhoid-coli group and the acquisition of new fermenting powers by *Bacillus dysenteriae* and other micro-organisms. Preliminary communication. *Proceedings of the Royal Society, ser. B* 79, 329–36, at p. 333.
13. Massini, R. (1907). Uber einen in biologischer Beziehung interessanten Kolistamm. (*Bacterium coli mutabile*). *Archiv für Hygiene* 61, 250–92.
14. Twort, F. W. (1915). An investigation on the nature of ultramicroscopic viruses. *Lancet* ii, 1241–3.
15. Fildes, P. G. (1951). Frederick William Twort 1877–1950. *Obituary Notices of Fellows of the Royal Society* 7, 505–17.
16. Wilson, G. (1979). The Brown Animal Sanatory Institution. *Journal of Hygiene* 82, 155–76. at p. 170; 83, 171–97, at pp. 185–92.
17. Beadle, G. W. (1958). Genes and chemical reactions in neurospora. *Nobel lectures physiology or medicine, 1942–1962*, pp. 587–99. Published in 1964 for the Nobel Foundation by Elsevier, Amsterdam.
18. Tatum, E. L. (1958). A case history in biological research. *Nobel lectures physiology or medicine, 1942–1962*, pp. 602–12. Published in 1964 for the Nobel Foundation by Elsevier, Amsterdam.
19. Lederberg, J. (1959). A view of genetics. *Nobel lectures physiology or medicine, 1942–1962*, pp. 615–36. Published in 1964 for the Nobel Foundation by Elsevier, Amsterdam.
20. Benison, S. (ed.) (1977). *Institute to university.* The Rockefeller University, New York. Simon Flexner was the older brother of Abraham Flexner.

21. Hopkins, F. G. and Cole, S. W. (1903). A contribution to the chemistry of proteids. Part II. The constitution of tryptophane and the action of bacteria on it. *Journal of Physiology* 29, 451–66.
22. Robertson, M. (1949). Marjory Stephenson 1885–1948. *Obituary Notices of Fellows of the Royal Society* 6, 563–77.
23. Stephenson, M. (1949). *Bacterial metabolism* 3rd edn., p. xi. Longmans Green, London.
24. Gladstone, G. P., Knight, B. C. J. G., and Wilson, G. (1973). Paul Gordon Fildes. *Biographical Memoirs of Fellows of the Royal Society* 19, 317–47.
25. Kohler, R. E. (1985). Bacterial physiology: the medical context. *Bulletin of the History of Medicine* 59, 54–74.
26. Hawking, F. (1963). History of Chemotherapy. In *Experimental Chemotherapy*, eds F. Hawking and R. J. Schnitzer. Academic Press, New York.
27. Bechhold, H. and Ehrlich, P. (1906). Beziehungen zwischen chemischer Konstitution und Desinfektionswirkung. *Zeitschrit für Physiologische Chemie* 47, 173–99.
28. Morgenroth, J. and Levy, R. (1911). Chemotherapie der Pneumokokken-infektion. *Berliner Klinische Wochenschrift* 48, 1560–1; 1979–83.
29. Wright, A. E., Morgan, W. P., Colebrook, L., and Dodgson, R. W. (1912). Observations on the pharmacotherapy of pneumococcus infections. *Lancet* ii, 1633–7; 1701–5.
30. Moore, H. F. and Chesney, A. M. (1917). A study of ethylhydrocuprein (optochin) in the treatment of lobar pneumonia. *Archives of Internal Medicine* 19, 611–82.
31. Moore, H. F. and Chesney, A. M. (1918). A further study of ethylhydrocuprein (optochin) in the treatment of acute lobar pneumonia. *Archives of Internal Medicine* 21, 659–81.
32. Maclachlan, W. W. G., Permar, H. H., Johnston, J. M., and Kenney, J. R. (1934). Some effects of quinin derivatives in experimental pneumoccus studies. *American Journal of Medical Science* 188, 699–705.
33. Smith, J. T. (1984). Awakening the slumbering potential of the 4-quinolone antibacterials. *Pharmaceutical Journal* 233, 299–305.
34. Oakley, C. L. (1973). Carl Hamilton Browning. *Biographical Memoirs of Fellows of the Royal Society* 19, 173–215.
35. Feldberg, W. (1970). Henry Hallett Dale 1875–1968. *Biographical Memoirs of Fellows of the Royal Society* 16, 77–174, at pp.109–11.
36. Albert, A. (1985). *Selective toxicity*. 7th Edn. Chapman and Hall, London.
37. Sergent, E. and Sergent, E. (1921). Etude experimentale du paludisme. Paludisme des oiseaux. *Bulletin de la société de pathologie exotique* 14, 72–8.
38. Dale, H. H. (1923). Chemotherapy. *Physiological Reviews* 3, 359–93.
39. Schulemann, W. (1932). Synthetic antimalarial preparations. *Proceedings of the Royal Society of Medicine* 25, 897–905.
40. Mauss, H. and Mietzsch, F. (1933). Atebrin, ein neues heilmittel gegen Malaria. *Klinische Wochenschrift* 12, 1276–8.
41. Colebrook, L. (1964). Gerhard Domagk 1895–1964. *Biographical Memoirs of Fellows of the Royal Society* 10, 39–50.
42. Biographical notes on G. J. P. Domagk, in *Nobel lectures physiology or medicine 1922–1941*, pp. 530–2. Published in 1965 for the Nobel Foundation by Elsevier, Amsterdam.

43. Domagk, G. (1935). Ein Beitrag zur Chemotherapie der bakteriellen Infektionen. *Deutsches medizinisches Wochenschrift* 61, 250–3.
44. Mietzsch, F. (1938). Zur Chemotherapie der bakteriellen infektionskrankheiten. *Berichte des Deutsches Chemische Gesellschaft* 71, 15–28.
45. Northey, E. H. (1948). *The sulfonamides and allied compounds.* Reinhold, New York.
46. Levaditi, C. and Vaisman, A. (1935). Action curative du chlorhydrate de 4'-sulfamido-2,4-diaminoazobenzene et de quelques derives similaire, dans la streptococcie experimentale. *Comptes rendus de la Société de Biologie* 119, 946–9.
47. Tréfouel, J., Tréfouel, J., Nitti, F., and Bovet, D. (1937). Activité du p-aminophenylsulfamide sur les infections streptococciques experimentales de la souris et du lapin. *Comptes rendus de la Société de Biologie* 120, 756–8.
48. Horlein, H. (1935). The chemotherapy of infectious diseases caused by protozoa and bacteria. *Proceedings of the Royal Society of Medicine* 29, 313–24.
49. Heidelberger, M. (1977). A 'pure' organic chemist's downward path. *Annual Review of Microbiology* 31, 1–12, at p. 8.
50. Dobell, C. and Laidlaw, P. P. (1926). The action of ipecacuanha alkaloids on *Entamoeba histolytica* and some other enterozoic amoebae in culture. *Parasitology* 18, 206–23.
51. Colebrook, L. and Kenny, M. (1936). Treatment of human puerperal infections, and of experimental infections in mice, with Prontosil. *Lancet* i, 1279–86.
52. Loudon, I. (1987). Puerperal fever, the streptococcus, and the sulphonamides, 1911–1945. *British Medical Journal* 295, 485–90.
53. Lockwood, J. S. (1938). Studies on the mechanism of the action of sulfanilamide. III. The effect of sulfanilamide in serum and blood on hemolytic streptococci in vitro. *Journal of Immunology* 35, 155–90.
54. Stamp, T. C. (1939). Bacteriostatic action of sulphanilamide in vitro. *Lancet* ii, 10–17.
55. MacLeod, C. M. (1940). The inhibition of the bacteriostatic action of sulfonamide drugs by substances of animal and bacterial origins. *Journal of Experimental Medicine* 72, 217–32.
56. Fleming, A. (1940). Observations on the bacteriostatic action of sulphanilamide and M & B 693 and on the influence thereon of bacteria and peptone. *Journal of Pathology and Bacteriology* 50, 69–81.
57. Whitby, L. E. H. (1938). Chemotherapy of pneumococcal and other infections with 2-(p-aminobenzenesulphonamido)pyridine. *Lancet* i, 1210–22.
58. Gale, E. F. and Fildes, P. (1965). Donald Devereux Woods 1912–1964. *Biographical Memoirs of Fellow of the Royal Society* 11, 203–19.
59. Woods, D. D. (1962). The biochemical mode of action of the sulphonamide drugs. *Journal of General Microbiology* 29, 687–702.
60. Woods, D. D. (1940). The relation of p-aminobenzoic acid to the mechanism of the action of sulphanilamide. *British Journal of Experimental Pathology* 21, 74–90.
61. Landsteiner, K. and van der Scheer, J. (1932). Serological reactions with simple chemical compounds (precipitin reactions). *Journal of Experimental Medicine* 56, 399–409.
62. Michaelis, L. and Pechstein, H. (1914). Ueber die verschiednartige Natur der Hemmungen der Invertasewirkung. *Biochemische Zeitschrift* 60, 79–90.

63. Quastel, J. H. and Wooldridge, W. M. (1928). Some properties of the dehydrogenating enzymes of bacteria. *Biochemical Journal* 22, 689–702.
64. Woolley, D. W. (1947). Recent advances in the study of biological competition between structurally related compounds. *Physiological Reviews* 27, 308–33.
65. Fildes, P. (1949). A discussion on antibiotic activity of growth factor analogues. *Proceedings of the Royal Society ser B*, 136, 145–7.

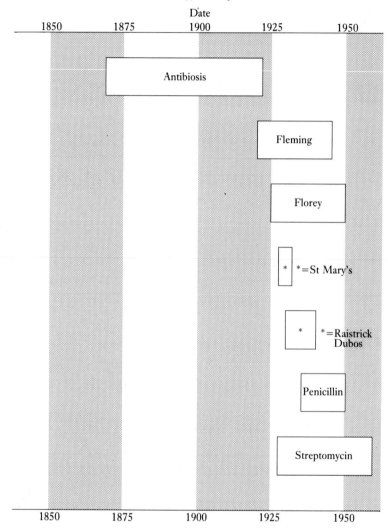

Chronology of Chapter 9

NB The blocks show periods discussed in the text, and do not mean that there was no activity about the subject at other times.

9

Antibiotics
and medicines

Microbes against microbes

It is widely believed that penicillin, a substance obtained from moulds of the family called *Penicillium*, was discovered by Sir Alexander Fleming at St. Mary's Hospital in London. However, the antibacterial properties of some *Penicillium* moulds were observed several times from the 1870s onwards, and Lord Lister treated a patient with a preparation of *Penicillium* before Fleming was born. When microbes were first cultivated and investigated, John Burdon Sanderson (1828–1905), later Regius Professor of Medicine in Oxford, observed that bacteria did not grow readily in the presence of certain *Penicillium* moulds.[1] Lister was then at the height of his struggle against infection in surgical operations and took up this observation in experiments with bacterial cultures. He found that growth did not occur in cultures infected with the mould *Penicillium glaucum*. The inference was obvious, and in a letter to his brother, a distinguished mycologist, Lister proposed to try using the mould for treating patients. No report of the treatment appears to have been published at the time, but evidence has come to light that he did indeed treat at least one patient, a young nurse with a persistent abscess resulting from injuries received in a street accident. She lived for many years, and preserved a record that she had been treated with *Penicillium*.[2] It is unlikely that the crude material did much good, and not surprising that no more was said about it at the time.

Simple application of a culture or scrapings of a mould to an exposed wound was an unsatisfactory business, and some sort of concentration of an active principle was needed. An Italian, Bartolomeo Gosio, in 1896, described the extraction of a crystalline substance from culture filtrates of a *Penicillium*: the substance was later shown to be a compound called mycophenolic acid, with interesting biological properties but no valuable antibacterial ones. Investigators in 1897 in Lyons, in 1908 in Vienna, repeatedly showed that various bacteria failed to grow or appeared to be destroyed in the vicinity of *Penicillium* moulds, but the practical application of this fact was elusive. The word 'antibiose' appeared in a paper by the French physician Jean-Paul Vuillemin in 1889. The allied word 'antibiotic' reached the English language when the botanist Marshall Hall used it 10 years later.[3,4]

There is nothing surprising about antibiosis. All living organisms are continually in competition with other species and other members of their own species in order to obtain their needs. The smallest organisms are no exception, and any mechanism by which some individuals can dominate others has value for survival. An organism able to check the growth of a competing species is fortunate and better equipped to survive, so we should expect that many living organisms are successful exponents of chemical warfare. They include various moulds and some bacteria, especially those of the species at first called *Bacillus pyocyaneus* and later *Pseudomonas pyocyanea*. In 1899, Rudolph Emmerich (1852–1914), professor of hygiene at the University of Munich, produced a preparation named Pyocyanase from appropriate cultures, which controlled some bacterial infections when it was injected into animals.[5] It was marketed for a number of years and was the subject of numerous clinical reports before its use declined. It may never have been genuinely effective, or it may have failed because the organism lost its antibiotic properties when it had been grown for years in artificial cultures, or because its commercial manufacture was inadequately supervised. It is said that the final preparations owed their antibacterial activity to the phenol which they contained as a preservative.[6]

Another enemy of bacteria was described in 1915 by F. W. Twort (see chapter 8). He observed an agent, invisible with the microscopes available at the time, which could be propagated like bacteria in cultures and which dissolved bacteria in cultures which were infected with it. This was his third notable discovery, and, like its predecessors, he did not pursue his observations, which were extended and given wider prominence by Felix Hubert d'Herelle (1873–1949).[7,8] He called it bacteriophage; later it was found to exist in many forms; or rather there were many bacteriophages, each associated with the bacteria of particular species. Indeed, the specificity of bacteriophages made them useful in identifying particular strains of bacteria, and the procedure called 'phage typing' was used for tracing the organism responsible for the spread of an infection in a community. The phage could be identified and acted as a kind of label, so that bacteria from a single source could be distinguished from others of different origin. Bacteriophages appeared to be harmless to man, and seemed to be promising therapeutic agents of revolutionary importance. But the results obtained by pioneer workers were not confirmed, when properly controlled experiments in infected animals failed to achieve any success. Much later, bacteriophages were found to transfer genetic information, from one bacterium to another, so they became immensely important tools for geneticists and have played a fundamental role in understanding the functions of deoxyribonucleic acid (see Chapter 11).

Fleming

By 1920 those searching for better defences against microbial infection had had many setbacks. Apart from quinine for malaria and arsenicals for syphilis and sleeping sickness, chemotherapy was unpromising, usually because the 'chemical' did as much damage to the patient as to the microbe. The obscure world of warfare between microbial species had suggested some promising substances, but was even less successful than chemotherapy. The pyocyanases and the penicillia did not appear to harm patients, but they were too difficult to extract and too unstable to be useful. Undoubtedly the natural defences of the human body held out the greatest promise of further success, and undoubtedly, in England if not in the world, Sir Almroth Wright[9,10] was the leading expert in devising ways of stimulating such defences. After leaving the British Army Medical Services, frustrated by the opposition of professional soldiers, he had created a research laboratory and clinic at St. Mary's Hospital where his theories were developed and his own authority was not to be questioned. He was elected a Fellow of the Royal Society, and had powerful support from politicians and writers. He and his ideas were portrayed on the London stage by G. B. Shaw in his play, *The Doctor's Dilemma*. Wright's combination of autocracy, determination to advance and understand medical science, love of logic and new words, contempt for women, and occasional humility make him one of the most fascinating characters in the history of therapeutic progress. he was supported by a band of younger doctors, of whom Leonard Colebrook[11,12] and Alexander Fleming[13-17] were outstanding characters in the search to cure infectious diseases. Each of them achieved success by discovering the sort of drugs which Wright dismissed as impossible or useless.

The collaboration between Colebrook and Fleming in the earliest trials of Salvarsan in this country, and Colebrook's work in establishing Prontosil and sulphanilamide have already been mentioned (Chapters 3 and 8). However Fleming has another place in this story. In 1921 he discovered a substance which he named lysozyme. The discovery was strictly in the tradition of Wright's philosophy of seeking natural defences against infection, though Wright may have grieved because it was made by chance[18] and was not the fruit of logical reasoning. Whatever its origin, this discovery had much importance in the future of penicillin. As will appear later, it probably contributed more than Fleming's own later observations on the mould *Penicillium*. It was, in fact, a double discovery, both of a new microbe and of an agent which destroyed it. Fleming's account, communicated to the Royal Society in 1922 by Wright,[19] began:

In this communication I wish to draw attention to a substance present in the tissues and secretions of the body, which is capable of rapidly dissolving certain bacteria. As this substance has properties akin to those of ferments I have called it a 'Lysozyme', and shall refer to it by this name throughout the communication.

The lysozyme was first noticed during some investigations made on a patient suffering from acute coryza. . . .

Fleming went on to describe how the nasal secretions of the patient (himself, in fact)[20] were cultured, and how a round microbe or coccus first grew and then was destroyed where it came close to the nasal secretion. He named the coccus *Micrococcus lysodeikticus* and used it as a test for identifying the destructive or lysing agent. ('Lysodeikticus' is derived from Greek and means 'an indicator of lysis'. The name was devised by Almroth Wright, who had a great love for such verbal inventions.)

In the first experiment nasal mucus from the patient, with coryza, was shaken up with five times its volume of normal salt solution, and the mixture was centrifuged. A drop of the clear supernatant fluid was placed on an agar plate, which had previously been thickly planted with *M. lysodeikticus*, and the plate was incubated at 37 °C. for 24 hours, when it showed a copious growth of the coccus, except in the region where the nasal mucus had been placed. Here there was complete inhibition of growth, and this inhibition extended for a distance of about 1 cm. beyond the limits of the mucus.

This striking result led to further investigations, and it was noticed that one drop of the diluted nasal mucus added to 1 c.c. of a thick suspension of the cocci caused their complete disappearance in a few minutes at 37 °C.

These two preliminary experiments clearly demonstrate the very powerful inhibitory and lytic action which the nasal mucus has upon the *M. lysodeikticus*. It will be shown later than this power is shared by most of the tissues and secretions of the human body, by the tissues of other animals, by vegetable tissues, and, to a very marked degree, by egg white.

At first sight this was a very promising substance indeed. It came from human tissues, and so could reasonably be expected to be harmless to them, unlike the synthetic antibacterial substances such as arsenicals and flavines which were the best known at the time. However, the promise did not hold. Lysozyme appeared to be effective only against more or less harmless micro-organisms, and indeed no organism was found more sensitive than the curious *M. lysodeikticus*. The microbes which caused dangerous diseases were unaffected by it. Indeed, it appeared to be a necessary property of any microbe capable of invading animal tissues that it was resistant to the action of lysozyme. So there was no place in therapeutics for lysozyme, but it had a more fundamental value: it showed that substances did exist which were lethal to some microbes and harmless to human tissues. What sort of a substance was it? Fleming had called it a ferment, or enzyme, but that did not take matters very far, and gave no hint of what it might ferment, or of what substance it attacked in susceptible microbes.

Fleming made no progress in purifying lysozyme.[10] The isolation of specific active substances from living cells is an intricate and laborious business, requiring the skills of a chemist more than those of a medical doctor. This was the time at which Harington was still grappling with the improved purification of thyroxine (see Chapter 5). Loewi and Dale were far from identifying acetyl choline convincingly (see Chapter 6), and isolation of the vitamins was straining the abilities of many chemical laboratories in the western world. The medical school of St Mary's Hospital, like the other London medical schools, was constitutionally part of the University of London, but was in fact academically isolated, and such research as might be undertaken had no facility for difficult chemical investigations. Fleming continued to experiment as a bacteriologist with his material, but he had no advanced training in chemistry, and his environment did not throw chemically expert colleagues in his way. Nor, had he sought to recruit a biochemist, was there any laboratory space available for his work. It was only in the 1930s, in Oxford, that lysozyme was isolated and crystallized, and its exact mode of action determined, as described later in this chapter.

History was destined to repeat itself. A few years later Fleming made another unpremeditated observation, this time on a mould of the genus *Penicillium*, and again failed, for lack of chemical support, to develop it.[21] Again the work was completed in Oxford, and was carried through by the leadership of a man who made no mistakes in seeking chemical collaboration or in letting matters drop for lack of it. His name was Howard Walter Florey.

Florey

H. W. Florey (later Lord Florey, P.R.S.)[22–24] (1898–1968) was the Australian born son of an Oxfordshire shoemaker who had emigrated to Adelaide in 1885. The son came to England for the first time in 1922 as a Rhodes scholar. At Oxford he worked under and received support particularly from Sir Charles Sherrington, whose immense skill and wisdom influenced him greatly. Florey found more interest in studies of the circulation than in Sherrington's own field, the nervous system, and presently moved to Cambridge with a studentship in pathology. A brief visit to the USA in 1926 was followed by a research fellowship at the London Hospital, where he established a long lasting friendship with Paul

Fildes (see Chapter 8). Unlike Fleming, Florey found the atmosphere of a London teaching hospital uncongenial and the conflicts between clinicians and research workers discouraging. Cambridge was easily accessible, and he returned there to begin his major research activities. He acquired a lasting scientific interest in mucus, possibly augmented by digestive problems of his own.[25]

Very little was known about mucus at the time. As a secretion of all the internal lining membranes of the body, it was important both for lubrication and protection against bacterial invaders. Lysozyme, recently described by Fleming, evidently had some bearing on the protective function. Florey, with a fellow Australian N. E. Goldsworthy, and with some help from Fleming in acquiring the technique of assaying lysozyme in samples of mucus, investigated the lysozyme content of samples of gastric secretion from various sources. The result was disappointing. No correlation was found between lysozyme content and resistance to bacterial infection.[26,27] In the course of this work, Florey and Goldsworthy made an incidental observation that some normal intestinal bacteria inhibited the growth of the test organism used in the lysozyme assays. It was another example of the often described, but at that time seldom remembered, phenomenon of bacterial competition, and may have contributed to Florey's later decision to pursue other examples of antibiosis. Meanwhile, his interest in lysozyme persisted. He sought collaboration from various colleagues, including Marjory Stephenson (see Chapter 8) in the nearby Department of Biochemistry, but made little progress before his departure from Cambridge to fill successively the chairs of Pathology at Sheffield in 1932 and Oxford in 1935.

It would be misleading to imply that mucus and lysozyme were Florey's only interest. At the time when he became a professor, his reputation probably depended largely on his studies of the capillary circulation and of the blood cells concerned with defence against bacterial invasion. He also studied spermicidal substances, in conjunction with H. M. Carleton, a histologist with whom he had made friends in Oxford. The work provided the foundation for the vaginal use of gels containing organic mercurial compounds as aids to contraception.[28] It was an important study in selective toxicity, because it required the discovery of substances which brought spermatozoa to a standstill without damaging the tissues of the vaginal wall. Later Florey collaborated with Paul Fildes in an experimental study of the use of curare to relieve the intractible muscular spasms which occur in fully developed infection with tetanus or lockjaw.[29] As curare was only just emerging from obscurity (see Chapter 6), this was a very prompt seizure of a new opportunity for therapy. Characteristically, the curare was obtained from the one source where the material obtained might be properly characterized, the laboratory of Henry Dale.

Perhaps the most important element in Florey's brief occupation of the Sheffield chair was that it brought him into close contact with Edward Mellanby (see Chapter 7). Mellanby was professor of pharmacology when Florey was appointed. They were colleagues for a year, during which time Mellanby no doubt formed a very definite appraisal of Florey's abilities. Mellanby left Sheffield to succeed Sir Walter Fletcher in the powerful position of Secretary of the Medical Research Council. Two years later, when the chair of pathology at Oxford became vacant by the untimely death of its incumbent, Mellanby was nominated as one of the electors responsible for finding a successor. It is said that, at the crucial meeting of the appointments board, Mellanby's arrival was seriously delayed by a late train. The board had decided on another candidate, an orthodox pathologist with predominantly clinical and descriptive interests and not an experimentalist. Mellanby arrived in time to be heard, and succeeded in changing the decision so that Florey was appointed.[30] Without the resources of the Oxford department in the following three years, Florey would probably not have been able to bring penicillin to clinical fruition. 'It is interesting to reflect that the course of the history of medicine—and, indeed, of the history of mankind—hung briefly on the timekeeping of the Great Western Railway'.[30]

Florey continued his interest in lysozyme and furthered it both by collaboration with the professor of organic chemistry, Sir Robert Robinson (1886–1975) and by recruiting an enthusiastic and skilful biochemist E. B. (later Sir Ernst) Chain (1906–1979).[31,32] Chain was the son of a Russo–German industrial chemist and himself a Ph.D. in chemistry of the Friedrich-Wilhelm University in Berlin. He had worked on the optical specificity of certain enzymes (one may remember that problems of optical activity first drew Pasteur from chemical to biological problems) but was also devoted to music and became the music critic of a Berlin evening paper. He nearly adopted music as a profession, but left Germany during the Nazi persecutions because of his Russian and Jewish ancestry. He found an opportunity to work in the biochemistry laboratory at Cambridge, and then came to Florey. He completed some work on snake venoms and, turned to lysozyme. E. A. H. Roberts, and later E. P. (later Sir Edward) Abraham, both students of Robinson's, succeeded in purifying and crystallizing lysozyme, while Chain collaborated in showing that it was indeed an enzyme, as Fleming had suggested, and that it decomposed an essential constituent of the cell wall of those organisms which were sensitive to its effect.[33,34] These discoveries would not be very difficult to make nowadays, but with no more than the techniques of the time, each was a great achievement. So the lysozyme story was completed. It did not produce any direct advance in therapeutics, but it remained as evidence that powerful antibacterial agents were not incompatible with living human tissues, and that chemical methods of investigation were essential for solving biological problems. In principle, the techniques were now established by which other antibacterial substances could be nailed down.

A new tool for bacteriologists

In 1929, while Florey was working in Cambridge, Fleming published a paper with the title 'On the antibacterial action of cultures of a *Penicillium* with special reference to their use in the isolation of *B. Influenzae*'.[21] It reported work done in the previous year, arising, like Fleming's earlier discovery of lysozme, from a chance observation. Later research[16] has shown that some of the statements in that paper do not agree with records made at the time in the laboratory notebooks of some of the people concerned. Also, it has proved very difficult to reproduce the events which led to Fleming's famous observation,[15] and it is clear that many versions of the story which were published 15 or more years later, after penicillin became famous, drew on faded but vivacious memories and are more or less imaginary. At the time, Fleming's observation of a contaminated culture plate aroused little interest in his colleagues. It looked like another lysoszyme, and, after 6 years with little progress, that subject may have become tedious. Fleming's colleagues most concerned were Ronald Hare (1899–1986), a St. Mary's graduate with a scholarship to do bacteriological research, F. Ridley, an ophthalmologist working on lysozyme, S. R. Craddock, another research scholar working under Fleming's direction on staphylococcal variants, and C. J. La Touche, a mycologist investigating the role of fungi in precipitating asthmatic attacks and the development of vaccines against them. Hare's account of the events has been published, and many obscure points in the history of the discovery have been investigated subsequently.[15–17] Professor Hare, as he later became, was an eminent bacteriologist and had worked in the laboratories where the discovery was made, so his account is preferable to some of those written by imaginative authors with no scientific training.

Fleming made his famous observation late in the summer of 1928, when he called in at his laboratory during his summer holiday. During his absence, Craddock was using the room, as the amount of space available in the department was very limited. Among the numerous culture plates awaiting inspection or disposal, Fleming noticed one unusual one, contaminated by a mould around which the staphylococci appeared to have been lysed (dissolved).[35] Fleming set up a culture of the mould, and preserved the plate by fixing it in formalin vapour. Then, having shown the plate to various not very excited colleagues, he resumed his holiday.

The first entry in Fleming's notebook about the mould is dated 'Oct.30.'28.' It describes an experiment, the first of many, to see what organisms were inhibited by the mould. This approach was followed for some months, so that an impressive list was produced of sensitive organisms, including many important pathogens. To simplify the experiments, Fleming set Ridley and Craddock to grow the mould and make extracts of the culture medium on which the mould had grown. Whatever inhibitory material was present appeared only after the mould had grown for several days, so there was an obvious advantage in having a stock of the active principle ready for bacteriological experiments. Ridley had taken a course in biochemistry, and so was probably the person best equipped to pursue the matter, which turned out to be very difficult. The mould juice was filtered, concentrated by evaporation to a 'sticky mass', and the proteins (derived from the original broth on which the mould was grown) were precipitated with alcohol, an orthodox procedure for their removal. Unexpectedly it was found that almost all the inhibitory substance passed into the alcohol. This was important, because it showed that it was not a protein, and so clearly differed from lysozyme. From this point on, progress was limited because of the great instability of the inhibitory substance. Craddock's notebook shows that he found the substance soluble in acetone, but this fact, like various other details of Ridley and Craddock's work, was not reported in Fleming's paper. Fleming was no chemist and appears not to have been interested in pursuing the chemical problem.[35]

He was, as a bacteriologist, interested in the microbiological aspects, and as an enthusiast for technical methods, in the use of *Penicillium* extracts to make selective culture media (i.e. media in which a required organisms will grow readily while unwanted contaminants fail either through lack of an essential nutrient or because something is present which stops them). This aspect was emphasized by the title of his paper on *Penicillium*. Requests for samples of his mould soon came from several laboratories, including the pathology laboratory at Oxford. Fleming sought help in identifying the mould from the mycologist La Touche, who worked in the same building and who thought that of the many species of *Penicillium*, it was most like *Penicillium rubrum*. This diagnosis was later upset, but it was of no great importance, because the mould which contaminated Fleming's plate was almost unique even within its species (*P. notatum*) in producing inhibitory material on an appreciable scale. The only other mould found to have equal activity was one kept in culture by La Touche, with which Fleming's mould appeared to be identical.[36] Hare had noticed that the windows of Fleming's laboratory were rarely opened and indeed were almost inaccessible across laboratory benches stacked with glassware. It must be admitted that the famous mould may well have strayed upstairs from the cultures on the floor below, and is perhaps to be included in the long list of profitable discoveries which arose from a lapse in maintaining the highest standards of laboratory practice.

The therapeutic possibilities of the inhibitory substance appear to have received little attention from Fleming. He showed that it was harmless to white blood cells, which are a principal defence against invading microbes, and he and Craddock administered enough to animals to discover that it was remarkably innocuous. Craddock had a chronically infected nasal antrum, and Fleming and he tried instilling some mould filtrate into it, with no obvious benefit. Another member of the laboratory staff received treatment for a conjunctivitis caused by pneumococci, with a successful result. Ridley, as an ophthalmologist, might well have followed up this observation with more studies of eye infections, but the opportunity seems not to have been pursued. A recently qualified medical student from St. Mary's, C. G. Paine, who had been taught by Fleming and who had gone to work in the Royal Infirmary at Sheffield, obtained a sample of the mould from his former teacher and carried out some clinical experiments with it. The notes of some of the patients treated have been rescued,[37,38] and appear to be the oldest surviving records of the clinical use of the material. When Florey came to the chair of pathology at Sheffield a year or two later, Paine mentioned the matter to him but aroused no obvious interest. However, Paine's observations were duly recognized when Florey and his colleagues published their substantial text on antibiotics in 1949.[39]

In his paper, in 1929, Fleming mentioned the possible use of penicillin as an antiseptic in surgical dressings, but the matter was taken no further. Fleming went on using mould extracts in selective media, and published once more on the subject before 1942. He had abundant other work with which to occupy himself. Craddock left to take up an appointment at the Wellcome Research Laboratories, and Ridley went to join the staff at Moorfields Eye Hospital.[17]

Trials and apathy in diverse places

In contrast to the unpremeditated way in which Fleming was taking up chance observations, various scientists were making methodical studies of microbes of all kinds. Notably, at the London School of Hygiene, a couple of miles from St Mary's Hospital, Harold Raistrick (1890–1971) was examining the biochemistry of moulds and other fungi. Like so many biochemists of the time, Raistrick had worked in Hopkins's laboratory, using chemical methods to study the living cells of micro-organisms. After 9 years in which he applied microbial biochemistry to industrial manu-facturing problems, he returned, in 1929, to academic work as professor of biochemistry at the London School of Hygiene, where he continued to identify the chemical constituents of fungi and discover their functions.[40]

Fleming's paper on an antibacterial substance from a *Penicillium* mould was a natural starting point for a branch of Raistrick's research. He obtained a sample of the mould from Fleming, and discovered that it had been incorrectly identified. After proper consultation and correction of the error, Raistrick set out to isolate active substances from Fleming's mould. He made an immediate advance by growing the mould on a medium consisting only of simple salts and glucose, instead of the broth which Fleming had used. This meant that any extraction procedure was much cleaner and more satisfactory: the mould might provide some tiresome debris, but at least the medium did not add the viscous remains of stale meat soup. Fleming's paper had given no details of Craddock's work, and indeed was misleading about it. Communication between Raistrick and Fleming appears to have been minimal. Unknowingly Raistrick followed the same path as the St. Mary's workers and came to the same conclusion: the instability of the antibacterial substance did not make further attempts worth while. Raistrick's prime interest was in the biochemistry of moulds, and he was not particularly aiming to isolate antibacterial agents from them.[41] A full-scale attack on the evanescent substance would have diverted many resources from other problems equally as or more important, and Raistrick had no reason for making such upheaval. He did not rate his findings on the antibacterial substance worthy of a full paper, and included them in a publication which dealt with several substances derived from *Penicillia*.[42]

Ten years later, when penicillin had been established as an effective drug, Raistrick gave fresh attention to the therapeutic possibilities of antibiotics. He achieved notable if modest successes by contributing to the discovery of Patulin,[43] which achieved a transient reputation for curing the common cold, and Griseofulvin,[44] which was developed commercially and is used for the treatment of fungal infections of the skin.

One or two other workers made attempts at isolating the antibacterial substance from *Penicillium* during the 1930s. One, Lewis Holt, actually worked in Fleming's laboratory, and took the purification a stage further than any of the previous workers,[45] but he did not publish his results and again what was achieved at St. Mary's had to be rediscovered by later workers.

A systematic search for the agents produced by microbes was also being made in New York, at the Rockefeller Institute for Medical Research, by René Dubos and his colleagues. Their rationale and objectives were somewhat different.[46] Dubos started from the assumption that all organic matter added to the soil eventually undergoes decomposition through the agency of micro-organisms. It followed that *somewhere* in the world there were soil micro-organisms with enzymes capable of attacking any substance which occurred in living tissues, and, given such enzymes, that one might use them, for instance, to destroy bacteria lethal to man. This was neither the basic approach of Raistrick to fungal metabolism, nor quite the same concept as that of antibiosis, but it led to a similar outcome, the possible discovery of a microbial agent which could destroy the microbes that caused disease.

The search prospered moderately. Dubos published two papers in 1939 describing the extraction from a soil bacillus of a substance, active against some pathogenic micro-organisms, which cured mice infected with pneumococci upon injection.[47] It was named tyrothricin and was later shown to contain two polypeptide components, which were named gramicidin and tyrocidine. Its potential use was limited because it was not absorbed from the gut and so was ineffective if given by mouth, and because tyrocidine, in concentrations liable to occur during therapeutic use, damaged or destroyed red blood cells. However, though not much of an advance on existing antiseptics, it is still included in some antiseptic lozenges for sore throats.

Tyrothricin can reasonably be regarded as the first antibiotic to be established as a therapeutic substance. Like the sulphonamides, it showed that bacteria could be attacked *in vivo* by single substances of external origin, and, like lysozyme, that such substances could be obtained from living organisms. But it took time to develop interest. Without the encouragement of interested colleagues, some scientists showed little initiative in applying their knowledge of antibiotic substances to practical ends. One notes S. A. Waksman (1888–1973),[48] who worked at the Agricultural Experimental Station of the State University of New Jersey. He pursued the antimicrobial properties of soil organisms, particularly actinomycetes, which had been known since the 1920s to produce inhibitory substances.[48] He had been studying soil fungi since 1916, and wrote a survey[49] of the known inhibitors in 1940, but his effective contribution to medicine came after the very modest success of tyrocidin and the very great success of penicillin.

Dubos had been a pupil of Waksman, and together they tried to organize a round table discussion at the meetings of the Society of American Bacteriologists in Saint Louis during Christmas 1940, on the subject of the production of antibacterial substances by micro-organisms. The subject attracted no interest, and they failed to secure enough participants for the proposed 2-hour discussion.[50]

English bacteriologists and microbiologists have little reason to claim that they were more perceptive than their American contemporaries, but there was one exception. In a laboratory in Oxford the purification of penicillin had progressed far enough to be reported in *The Lancet* in August 1940. This was 4 months before the St. Louis meeting mentioned above, so the publication evidently did not stimulate much excitement in the United States. It was, however, a preliminary report, made before clinical trials had taken place. To appreciate the progress that was being made, we must return to the career of Howard Florey, whom we left newly appointed to the chair of pathology in Oxford and continuing, among other objectives, to study Fleming's lysozyme.

Breaking through

Florey had wide interests but was determined to achieve results and not to dissipate his energies. His resources were grievously limited.[51] His laboratories were housed in an excellent building, provided by the generosity of Sir William Dunn's Trustees on the advice of Sir Walter Fletcher (see Chapter 4), but the morale of the staff was low. Florey's predecessor, Georges Dreyer (1873–1934) developed a vaccine against tuberculosis. The press published premature and optimistic reports, while trials of the vaccine showed that it was ineffective. Dreyer was unfairly discredited, and appears to have lost heart for further work. The story was no doubt well known to Florey and may have contributed to his own dealings with the press. There were no endowments to finance the upkeep of the building and the work inside it. A university steeped in classical, historical, and literary traditions was not unfriendly to science but had little awareness of the growing cost of high level scientific research, and even, perhaps, in some quarters reservations about its importance. Staff appropriate to Florey's aims had to be recruited and grants to pay them obtained from outside bodies. The Medical Research Council were helpful up to a point, but even Mellanby rapidly found Florey's demands tiresome and support from that quarter dwindled. When it failed, the Rockefeller Foundation contributed generously.[52] With the funds available, Florey collaborated with Chain, whose work on lysozyme, already mentioned, led naturally to a study of a wider range of antibacterial agents.

The mould *Penicillium* looked interesting, and was available: research workers in the department had been using it, just as Fleming was doing, to prevent the growth of unwanted organisms in bacterial cultures. Other antibiotics described in published research papers were considered. Eventually three were chosen for the first investigation—the products of *Bacillus subtilis*, *Pseudomonas pyocyanea*, and *Penicillium notatum*. Both Florey and Chain later emphasized that their study was directed to a better understanding of antibiotic action and not specifically to finding a therapeutic substance.[53,54]

Penicillin was an unpromising candidate for therapeutic use because of its instability and reputed inactivation by blood, but this did not reduce its scientific interest. Chain began working in the summer of 1938 on penicillin and pyocyanase, though by January 1939, when an application for funds was made to the Medical Research Council, penicillin and 'actinomycetin', an antibiotic which Waksman was studying, were mentioned. Evidently some of the progress made by Waksman attracted Florey's and Chain's attention. Sharpening the focus on penicillin appears to have been a gradual process. Various stories lend drama to the 'decision' to concentrate on penicillin.[54] A paper from the Dunn School in 1941 described three antibacterial substances from *Ps. pyocyanea*.[55]

Moulds do not usually grow fast, and conditions had to be found in which large quantities of *Penicillium notatum* could be produced as quickly as they were wanted. The solution to this problem was helped by N. G. Heatley, a young biochemist also from Hopkins's laboratory in Cambridge, who had been prevented by the outbreak of war from going to work in the Carlsberg laboratories in Copenhagen. His engineering skills and technical ingenuity were invaluable. Methods of extracting the active substance from the *Penicillium* cultures had to be found. Raistrick's work had been published[42] and provided a valuable starting point. A rapid method of assaying samples was essential in order to discover which extracts were giving the best yields, and the sooner a stable crude preparation was obtained which could act as a standard, the better.[56] Extractions were achieved without appreciable loss of activity (but not without discomfort) by working at temperatures below 5 °C. Freeze drying was valuable. This was a technique which had evolved during the 1930s and would hardly have been practical either for Craddock at St Mary's in 1929 or, perhaps, even for Raistrick in 1931. Florey's team produced material which stopped bacterial growth at the remarkable dilution of two parts per million. Moreover, it was not harmful when it was injected into mice. Later, it became clear that this material, despite its potency, contained only about 1 per cent of pure penicillin and that it was almost entirely composed of material which was *not* penicillin. Florey and Abraham wrote[57]

It must be considered extremely fortunate that the first preparations of the sodium salt of penicillin, which although of great antibacterial activity contained over 99 per cent of impurities, showed so little toxicity. If even one of the many impurities had been highly toxic, as it might well have been, the non-toxicity of penicillin might have been completely masked, with unpredictable effects on subsequent work.

By the time these results had been achieved, World War II had passed from the uneasy calm of its first winter into the violence of the European devastation of 1940. On 25th May, while the German army raced ruthlessly across Belgium and the inhabitants of England watched their skies for the first signs of airborne invasion, the scientists in the Oxford Pathology Department performed a crucial experiment. Chain and Heatley injected streptococci into mice and then treated some of the mice with penicillin. Later Chain went home. Heatley stayed until 3.30 a.m., when all of the untreated mice had died. The treated mice were alive.

When Chain came in on Sunday morning and saw the result, he is said to have danced. His colleagues 'all recognized that this was a momentous occasion. What they said is not recorded, but memory has supplied subsequent writers with various versions. One might suppose that Heatley said very little, that Chain was excited, and that Florey's reported comment "it looks quite promising" would be entirely in character'.[58] One wonders how the scene compared with that in Domagk's laboratory several years earlier, when the mice treated with Prontosil were alive and the controls had all died (see Chapter 8).

Penicillin arrives

In the experiment of 25th May 1940,[59] the treatment of four mice with penicillin required a substantial fraction of all the material so far extracted. A man weighs about as much as 3000 mice, and might need a bigger dose on the same scale. If penicillin was going to be tried in man, how on earth was enough going to be made? Chain had overcome the problems which had defeated every previous experimenter with penicillin, but only on a small scale. One cannot suddenly scale up a laboratory procedure by a thousandfold. Outside help was essential, but war-torn British industry was stretched to breaking capacity and the survival of a handful of mice was too thin a basis for disrupting existing essential activities to pursue what may well have appeared to be an academic will-o'-the-wisp. Somehow enough material had to be made to establish the worth of penicillin in man. Somehow, it was made. By making do and searching for essential materials—in a war-torn country—a production unit was created in the University Pathology laboratory and enough crude penicillin was extracted for further essential laboratory experiments and to treat several patients in the Radcliffe Infirmary half a mile away. The results were impressive, but marred by the death of at least one patient because there was insufficient material to continue treatment. In August 1941 a second paper appeared from Florey's group, now with clinical collaborators.[60] By then Florey and Heatley had gone to America to seek aid from a richer country not yet involved in war.

The difficulties which they met were formidable, and as much political as scientific. Florey had been supported by the Rockefeller Foundation and received guidance where to look. A. N. Richards (1876–1966) had worked in England with Florey and was now a scientific power in America and a source of valuable information. Of the many contacts explored, the most fruitful early collaboration came from the United States Department of Agriculture's North Regional Research Laboratory at Peoria, Illinois, where there was much experience of culturing micro-organisms. Here

Heatley stayed for some months, working with A. J. Moyer, a first-rate investigator of fermentation processes but also an anti-British isolationist,[61,62] in the development of deep fermentation (instead of surface culture). This process had already been invaluable to industry and proved to be so for penicillin. It was here that it was discovered that 'corn steep liquor', the material left over after extracting starch from maize, was a valuable nutrient for *Penicillium notatum*. When Heatley went to Peoria, culture methods were yielding $2 \, U \, ml^{-1}$; with these changes and new mutants of the mould, yields up to $900 \, U \, ml^{-1}$ were obtained.[63,64] Meanwhile Florey was, as he called it, 'carpetbagging' around American pharmaceutical businesses with varying success. In the end, several firms undertook penicillin production on a massive scale, but hardly any ever came to Florey himself for the clinical trials which he was desperate to extend.[65] The first patient to receive commercially produced penicillin in America was treated on the 14th March 1942. By August 1943 material was available at \$200 per million units, and a year later, just before the invasion of Europe on 6th June, the price was \$35. In 1950 it had dropped to 50 cents per million units.[66]

Although the American achievement in obtaining high yields of penicillin from mould cultures was crucial to the practical introduction of penicillin on a worldwide scale, manufacture by chemical synthesis was naturally an attractive alternative. First the chemical identity of penicillin had to be established: second, a method of making it had to be found. It was soon apparent that more than one penicillin existed. There was Florey's original material, which became known as Penicillin-I or Penicillin-F (for Florey); Penicillin-II or Penicillin-G (because G came after F) from the Squibb Laboratories; Penicillin-III or Penicillin-X from the laboratories at Peoria; and Penicillin-IV or Penicillin-K from the Abbot Laboratories. The central nucleus of all these penicillins was the same, but the exact structure was established only after much controversy.[67] Its identification was an early success of X-ray crystallography. Synthesis was achieved only after some years, and only by methods which did not justify expansion to an industrial scale. It is interesting to compare this fact with the ridicule to which some experts were exposed for developing fermentation methods of producing penicillin—'obviously to be a flop'.[68]

Hereafter the penicillins were studied in countless laboratories, academic and commercial, and discoveries were made which enabled them to be better understood[67] and used as the foundation for many new drugs. Inevitably, when penicillin was widely used, the least sensitive microbes proliferated and others were able to adapt by developing penicillin-splitting enzymes. Organisms resistant to penicillin, particularly staphylococci, became a major problem by about 1960, especially in hospitals, and accelerated the search for alternative antibiotics and for modifications of penicillin which were resistant to the bacterial penicil-linases. The discovery, in 1945 by Brotzu in Sardinia, of antibiotic activity in a species of *Cephalosporium* recovered from the sea near a sewage outfall was brought to Florey's attention by a former British public health officer and led to work at Oxford and elsewhere on this and similar organisms. Of the several active substances which were isolated, the cephalosporins have become particularly valuable.[69] The synthesis of penicillin on a com-mercial scale proved intractible, but enzymic and chemical modification of penicillins led to semi-synthetic substances, based particularly on economic ways of generating the nucleus, 6-aminopenicillanic acid, by fermentation. Chain, exasperated by the British failure to provide him with adequate research facilities, migrated to Rome, but was retained as a consultant by Beecham's at a time when the firm was rapidly developing its research facilities. He had much influence on the Beecham team who isolated 6-aminopenicillanic acid.[70-72]

From a therapeutic aspect, penicillin was marvellous because it cured many dangerous bacterial infections, but it was tiresome because it was inactive by mouth (it was destroyed by the acidity of the stomach contents), it acted for a very short time (it was rapidly excreted by the kidneys) and so was of little value unless given by injection at intervals of not longer than 3 hours, and because, after a time, the normal processes of evolution led to the appearance of resistant strains of the microbes which had previously been sensitive to penicillin. All of these problems were more or less overcome by ingenious chemical manipulation, so that a very wide range of semi-synthetic penicillins have become available. Resistance may be expected to develop in turn to these variants.

Publicity

As Fleming shared a Nobel Prize for the discovery of penicillin with Chain and Florey in 1945, one may reasonably ask what he was doing during the critical years 1938 to 1942. The matter has been handled in meticulous detail both by Hare[15-17] and Macfarlane,[22] and only the most salient points can be noted here. Fleming continued to work with penicillin as an aid to the isolation of bacteria in cultures all through the 1930s. When the sulphonamides came on the clinical scene, he gave much attention to them. In a letter[73] to the *British Medical Journal* in 1938 on the treatment of pneumonia he stated that sulphapyridine was active against pneumococci only when helped by leucocytes, and (true to the tradition of Almroth Wright) its action was improved if anti-pneumococcal serum was administered at the same time. He also recommended the use of vaccines to stimulate immunity to the pneumococci, but made no reference to the penicillin, a little of which, presumably, was in his reach while he wrote the letter. The first (1940) paper[59] from the Oxford workers attracted his attention. He arranged by telephone to visit Florey at the Oxford laboratory and did so on 2nd September.[74] He appears to have been pleased that someone was fostering the substance which so interested him, and went away with a sample of Oxford material, which was more potent than any he had himself prepared.[74] After this he began mentioning penicillin at the end of lectures on sulphonamides.[75]

The 1941 paper from Oxford led to some comments in the press, which included reference to Fleming and to Oxford, but Fleming's first substantial clinical observation with penicillin appears to have occurred in August 1942, when a patient, an employee of Fleming's brother's optical firm, was failing to respond to treatment for meningitis in St Mary's Hospital. Fleming succeeded in isolating a streptococcus from the cerebrospinal fluid of the patient, showed that it was sensitive to penicillin, and obtained a supply from Florey, because he himself had none of adequate potency. When intramuscular injections of penicillin failed to cure the patient, and after consulting Florey, Fleming took the heroic measure of injecting penicillin directly into the cerebrospinal fluid, where the streptococci still lurked. The treatment succeeded and the patient recovered rapidly. By arrangement with Fleming, the case was included in results of clinical trials published by the Oxford workers[76] in March 1943, and Fleming published a more detailed account[77] in October 1943. Each thanked the other in the two publications.

Before the patient left St Mary's Hospital, a leading article headed 'Penicillium' appeared in *The Times* and a few days later a letter was published from Sir Almroth Wright claiming Fleming as the discoverer of penicillin. This, of course, caught the attention of many journalists, and St Mary's did not fail to provide them with information. Another letter in *The Times*, from Sir Robert Robinson, drew attention to Florey's work and set enquirers off on a new tack to Oxford, but their journeys turned out to be unrewarding. Florey, for reasons already mentioned and for others,[78] had no wish to deal with the press. The reporters were sent away empty handed, while Florey told his staff to give no information whatsoever to journalists. For St Mary's his actions were a godsend. This was before the days of the National Health Service. The voluntary hospitals, of which St Mary's was one, depended on charity for their income, and were often desperately impoverished.[79] Any success that could be claimed to the credit of St Mary's Hospital improved its chances of acquiring much needed donations. The Dean, Lord Moran (1882–1977), had close connections with Lord Beaverbrook (1879–1964), already a benefactor and, more important, proprietor of the *Daily Express* newspaper. The story of Fleming's discovery was a superb one and gained delightful embellishments as it dwindled in truthfulness.[80] Fleming himself, to his credit, was at first distressed by the publicity and wrote accordingly to Florey.[80] But the publicity machine had come into operation and nothing would stop it. Later, Fleming evidently came thoroughly to enjoy the fame which was thrust upon him, and to travel across the world to receive prizes and honours of many kinds. Most people believe what they are told most often, and the myth has become unshakeable.

A few people, very few, had a healthy scepticism. J. Brunel, professor of mycology in Montreal in 1944, alerted by a popular article on the wonderful new drug, looked briefly at Fleming's original publication on *Penicillium*, and at other publications, and wrote an article very properly drawing attention to some of the literature which Fleming had omitted to quote.[81] Florey himself, in a lecture[82] given in 1945, made a similar review, which was amplified in the great book on antibiotics written by the Oxford group in 1949.[39] But the myth was more entertaining when dull but inescapable facts were omitted. Some of Fleming's biographers have been writers without medical or scientific training. They have tried to make a personal and human drama out of what was, in fact, a sequence of astonishing scientific events. Some of these events are still not well understood; although Hare and Macfarlane have done much to bring the most important medical discovery of this century into proper perspective. Fleming discovered the important strain of the mould later identified as *Penicillium notatum*. He coined the word 'penicillin'. But he did not isolate or identify the substance, and for 12 years or so confined his work to the use of crude extracts in selective culture media.

It may seem sad, or even tragic, that so much credit has gone to Fleming and so little to Florey and his team at Oxford. All the early supplies of penicillin which came to the public were manufactured in America, and, in America at least, it was natural to assume that penicillin was an American discovery. If the trumpets of Britain had not been blown loud and long from St Mary's, a different myth would have grown giving all the credit to America. There was plenty of substance for such an idea. Without the work at Peoria and in many American industrial laboratories, many more months or years would have passed before penicillin became readily available. From another angle, one may note that the receipt of publicity, as Florey knew well, is an astonishing waste of time which can be spent on more worthy scientific activities. While Fleming travelled the world receiving plaudits, Florey was continuing to direct 'the best school of experimental pathology in the world'[83] and pursuing the discovery of more antibiotics. He also became President of the Royal Society, so that for 5 years he led its activities in nurturing British science and ripening its fruits. He received a life peerage and the Order of Merit; headed an Oxford College with distinction, and promoted the birth of the Australian National

University at Canberra. It was not a bad set of compensations for unawarded praise.

Streptomycin and onwards

The next antibiotic to be discovered, streptomycin, was of great value because it attacked microbes which were insensitive to penicillin, especially those which caused tuberculosis. Its discovery was the result of prolonged studies of the microbes which exist in soil and survive by breaking down the remains of plants and animals, and by competition with each other. This was the real world of antibiosis, which occupied much of the life of S. A. Waksman. The scavenging activities of soil microbes are of great importance to human welfare, but Waksman's studies were not orientated to medical applications.

Waksman was born in the Ukraine, the son of a small trader in a small town, and suffered the difficulties common to the Jewish community in that environment. He has written a memorable account[48] of his spiritually rich but socially oppressed childhood, his struggles for education, and his emigration to the United States in 1913. In America he made good. He obtained a Ph.D. in biochemistry at the University of California and spent his working life at the Agriculture Experimental Station of the State University of New Jersey, where he gave strength to the science of soil microbiology. His studies inevitably included competition between microbes living free in the soil. The organisms which were his special interest, the actinomycetes or ray fungi, were particularly vigorous in producing substances which inhibit the growth of other microbes.

Waksman is said to have been shown a culture of tubercle bacilli which had been killed by a fungus in 1932, but he was not tempted to pursue the implications of this observation.[84] The general lack of interest in antibiosis at this time was no doubt discouraging, but it did not prevent his former pupil Dubos from bringing tyrothricin to clinical trial. Waksman records that the great American pharmaceutical firm of Merck established a fellowship in his department in 1938 for the study of industrial fermentation processes, and made grants to him from 1939 onwards for research into antibiotics.[85] In 1942 his son, then a medical student, urged Waksman senior to isolate strains of *Actinomyces* active against human tubercle bacilli, but he replied that the time had not come yet.[86] However, in 1944 a paper appeared describing the properties of streptomycin,[87] a substance obtained from a kind of *Actinomyces* recently reclassified and named *Streptomyces*. The short report included a table showing the concentration of streptomycin which stopped the growth of various organisms. Tubercle bacilli are included in this table, in a single line, without information about the strain used and whether it was virulent or not. The paper is often quoted as the original report of a major discovery, but it does not look like it. It is difficult to believe that the authors saw the implication of their findings for human medicine.[84]

The report, or news of it,[88] attracted the attention of W. H. Feldman and H. C. Hinshaw, who were studying drugs for the treatment of tuberculosis at the Mayo Clinic in Rochester, Minnesota. They made contact with Waksman and with Merck, and obtained supplies of the drug for trials in infected guinea pigs, and, when these were successful, in man.[89] Once clinical investigations began, events moved fast. Economical methods of manufacture were devised at Merck, while clinical trials were organized by the Committee on Chemotherapeutics of the USA National Research Council, and the drug was distributed without charge to selected institutions. By the beginning of 1947 the drug began to be available commercially. The cost, like that of penicillin, was reduced about seventyfold between 1946 and 1950 by improvements in large scale production.[48,89]

Streptomycin is more stable than penicillin and so it was easier to isolate and manufacture. It is also more toxic, and from an early stage a further disadvantage gradually became apparent: tubercle bacilli become resistant to it remarkably quickly. Later other drugs were discovered which mitigated this handicap. Without them the useful life of streptomycin might have been confined to a few years. But it was the first drug generally recognized as life saving in tuberculosis. Miliary tuberculosis (tuberculous septicaemia) and tuberculous meningitis, hitherto consistently fatal conditions, yielded to it. It is not surprising that a Nobel prize was awarded in 1952 to Waksman, though there are many reasons why other contributors were also worthy of honour.[88]

The successes of penicillin and streptomycin naturally led to a search for further antibiotics. The resources of many pharmaceutical firms were applied to the search for new moulds and other micro-organisms. Innumerable materials were screened for antibacterial activity, and, incidentally, for any other kind of activity which screens (see Chapter 6) might detect. Active principles were purified and identified chemically, and, when good fortune attended the search, developed into therapeutically useful agents. Duplication of discovery was not surprising. One agent, polymyxin, (later found to comprise a family of closely related substances) was reported[90–93] independently from three separate laboratories in Britain and America in 1947. The material originated from a widely distributed soil organism and was active against various bacteria, some of which at that time still awaited effective therapy. The polymyxins had various limitations, but they were followed by the tetracyclines, chloramphenicol, and many other antibiotics. Some were more active than any predecessor against some particular organism, and some had special merits in use, such as working when given by mouth, or acting for a longer period than usual. Some were active against strains of organisms which had become resistant to older antibiotics. But the pathway to discovery was established and the scale on which resources were expended to make these discoveries represents a fantastic extravagance when it is compared with the few thousands of pounds available to Florey and his colleagues, and even with the subsequent development of the original penicillins in the United States.

Although it was more economical to use living organisms to produce penicillin, and also streptomycin, there was no fundamental reason why antibiotics should be regarded differently from drugs made in a laboratory or industrial plant by chemical processes. Chloramphenicol was the first antibiotic to be prepared synthetically on a large scale,[94] and no difference has been recognized between the therapeutic properties of material made by microbes and by man. No mystique need surround antibiotics because of their biological origin.

Now there are remedies for most bacterial infections. Viruses are more difficult, and are discussed in the next chapter. However, it remains to be seen how completely the principle of antibiosis, or chemical warfare between species, has been exploited, and whether the capacity of microbes to adapt to their environment will outstrip the capacity of man to find means of destroying not only those which are his enemies but also those which are his friends and which he unwittingly eliminates by the over-zealous distribution of his new agents.

Notes

1. Selwyn, S. (1979). Pioneer work on the 'penicillin phenomenon' 1870–1876. *Journal of Antimicrobial Chemotherapy* 5, 249–55.
2. Fraser-Moodie, W. (1971). Struggle against infection. *Proceedings of the Royal Society of Medicine* 64, 87–94.
3. Florey, H. W. (1945). The use of micro-organisms for therapeutic purposes. *British Medical Journal* ii, 635–42.
4. Brunel, J. (1951). Antibiosis from Pasteur to Fleming. *Journal of the History of Medicine* 6, 287–301.
5. Emmerich, R. and Löew, O. (1899). Bakteriologische Enzyme als Ursache der erworbenen Immunität und die Heilung von Infectionskrankheiten. *Zeitschrift für Hygiene und Infectionskrankheiten* 31, 1–65.
6. Wagner, W. (1929). Untersuchungen der bakteriziden Bestandteile des *Bac. pyocyaneus*. *Zeitschrift fur Immunitätsforschung* 63, 483–91.
7. d'Herelle, M. F. (1917). Sur un microbe invisible antagoniste des bacilles dysentériques. *Comptes rendus des séances de l'Académie des sciences* 165, 373–5.

8. d'Herelle, M. F. (1918). Sur le rôle du microbe filtrant bactériophage dans la dysenterie bacillaire. *Comptes rendus des séances de l'Académie des sciences* 167, 970–2.

9. Colebrook, L. (1947). Edward Almroth Wright 1861–1947. *Obituary Notices of Fellows of the Royal Society* 6, 297–314.

10. Macfarlane, G. (1984). *Alexander Fleming. The man and the myth*, chapters 6–9. Chatto and Windus, The Hogarth Press, London.

11. Oakley, C. L. (1971). Leonard Colebrook 1883–1967. *Biographical Memoirs of Fellows of the Royal Society* 17, 91–138.

12. Noble, W. C. (1973) *Coli: Great healer of men. The biography of Dr. Leonard Colebrook F.R.S.* Heinemann, London.

13. Colebrook, L. (1956). Alexander Fleming 1881–1955. *Biographical Memoirs of Fellows of the Royal Society* 2, 117–27. Numerous other biographies of Fleming have been written, some by non-scientists. The biography which appears to have carried most influence was written by André Maurois (note 14) at the request of Fleming's second wife. Hare (1970, note 15), who worked in Wright's laboratories until he went with Colebrook to Queen Charlotte's Hospital, has reviewed the whole discovery of penicillin and Fleming's contribution to it: see also Hare (1982, note 16), Hare (1983, note 17) and Macfarlane (1984, note 10 above).

14. Maurois, A. (1959). *The Life of Sir Alexander Fleming, discoverer of penicillin*. Transl. from the French by Gerard Hopkins, Jonathan Cape, London.

15. Hare, R. (1970). *The birth of penicillin*. Allen & Unwin, London.

16. Hare, R. (1982). New light on the history of penicillin. *Medical History* 26, 1–24.

17. Hare, R. (1983). The scientific activities of Alexander Fleming, other than the discovery of penicillin. *Medical History* 27, 347–72.

18. Exactly how the discovery was made is not as simple as appears from Fleming's paper. See Macfarlane, note 10, pp. 98–103.

19. Fleming, A. (1922). On a remarkable bacteriolytic element found in tissues and secretions. *Proceedings of the Royal Society, ser. B* 93, 306–317.

20. Macfarlane, G. (1984). *Alexander Fleming. The man and the myth* p. 98. Chatto and Windus, The Hogarth Press, London.

21. Fleming, A. (1929). On the antibacterial action of cultures of a *Penicillium* with special reference to their use in the isolation of *B. influenzae*. *British Journal of Experimental Pathology* 10, 226–36.

22. Abraham, E. P. (1971). Howard Walter Florey—Baron Florey of Marston and Adelaide. *Biographical Memoirs of Fellows of the Royal Society* 17, 255–302.

23. Macfarlane, G. (1979). *Howard Florey. The making of a great scientist*. Oxford University Press, Oxford.

24. Williams, T. I. (1984). *Howard Florey. Penicillin and after*. Oxford University Press, Oxford.

25. Ibid., p. 29.

26. Goldsworthy, N. E. and Florey, H. (1930). Some properties of mucus, with special reference to its antibacterial properties. *British Journal of Experimental Pathology* 11, 192–208.

27. Florey, H. W. (1930). The relative amounts of lysozyme present in the tissues of some mammals. *British Journal of Experimental Pathology* 11, 251–61.

28. Carleton, H. M. and Florey, H. (1931). Birth control studies. 2. Observations

on the effects of common contraceptives on the vaginal and uterine mucosae. *Journal of Obstetrics and Gynaecology of the British Empire* 38, 555–64.

29. Florey, H. W., Harding, H. E., and Fildes, P. (1934). The treatment of tetanus. *Lancet* ii, 1036–41.
30. Williams, T. I. (1984). *Howard Florey, penicillin and after*, p. 43. Oxford University Press, Oxford.
31. Abraham, E. P. (1983). Ernst Boris Chain (1906–1979). *Biographical Memoirs of Fellows of the Royal Society* 29, 43–91.
32. Clark, R. W. (1985). *The life of Ernst Chain. Penicillin and beyond*. Weidenfeld and Nicolson.
33. Abraham, E. P. and Robinson, R. (1937). Crystallization of lysozyme. *Nature* 140, 24.
34. Epstein, L. A. and Chain, E. (1940). Some observations on the preparation and properties of the substrate of lysozyme. *British Journal of Experimental Pathology* 21, 339–55.
35. Hare, R. (1970). *The Birth of Penicillin*, pp. 54–80. Allen and Unwin, London.
36. *Ibid.*, pp. 81–7; Hare, R. (1982). New light on the history of penicillin. *Medical History* 26, 1–24, at p. 5.
37. Wainwright, M. and Swan, T. (1986). C. G. Paine and the earliest surviving clinical records of penicillin therapy. *Medical History* 30, 42–56.
38. Wainwright, M. (1987). The history of the therapeutic use of crude penicillin. *Medical History* 31, 41–50.
39. Florey, H. W. *et al.* (1949). *Antibiotics. A survey of penicillin, streptomycin, and other antimicrobial substances from fungi, actinomycetes, bacteria, and plants*, p. 634. Oxford University Press, Oxford.
40. Birkinshaw, J. H. (1972). Harold Raistrick (1890–1971). *Biographical Memoirs of Fellows of the Royal Society* 18, 489–509.
41. *Ibid.*, p. 497–8.
42. Clutterbuck, P. W., Lovell, R. and Raistrick, H. (1932). Studies in the biochemistry of micro-organisms. XXVI. The formation from glucose by members of the *Penicillium chrysogenum* series of a pigment, an alkali-soluble protein and penicillin—the antibacterial substance of Fleming. *Biochemical Journal* 26, 1907–18.
43. Raistrick, H. (1943). Patulin in the common cold. Collaborative research on a derivative of *Penicillium patulum* Bainier. *Lancet* ii, 625.
44. Oxford, A. E., Raistrick, H. and Simonart, P. (1939). Studies in the biochemistry of micro-organisms. LX. Griseofulvin, $C_{17}H_{17}O_6Cl$, a metabolic product of *penicillium griseofulvum* dierckx. *Biochemical Journal* 33, 240–8.
45. Macfarlane, G. (1984). *Alexander Fleming. The man and the myth*, pp. 146–7. Chatto and Windus, The Hogarth Press, London.
46. Dubos, R. J. (1939). Studies on a bactericidal agent extracted from a soil bacillus. I. Preparation of the agent. Its activity in vitro. *Journal of Experimental Medicine* 70, 1–10.
47. Dubos, R. J. (1939). Studies on a bactericidal agent extracted from a soil bacillus. II. Protective effect of the bactericidal agent against experimental pneumococcus infection in mice. *Journal of Experimental Medicine* 70, 11–17.
48. Waksman, S. A. (1954). *My life with the microbes*. Simon and Schuster, New York. (British edition, Hale, London, 1958).

49. Waksman, S. A. and Woodruff, H. B. (1940). The soil as a source of microorganisms antagonistic to disease-producing bacteria. *Journal of Bacteriology* 40, 581–600.
50. Waksman, S. A. (1946). Antibiotic substances—contribution of the microbiologist. *Annals of the New York Academy of Sciences* 48, 35–9.
51. Macfarlane, G. (1979). *Howard Florey. The making of a great scientist*, p. 239 f. Oxford University Press, Oxford.
52. *Ibid.*, pp. 299–303.
53. Florey, H. W. *et al.* (1949). *Antibiotics. A survey of penicillin, streptomycin, and other antimicrobial substances from fungi, actinomycetes, bacteria, and plants*, p. 636 f. Oxford University Press, Oxford.
54. Macfarlane, G. (1979). *Howard Florey. The making of a great scientist*, p. 284–9. Oxford University Press, Oxford.
55. Schoental, R. (1941). The nature of the antibacterial agents present in *Pseudomonas pyocyanea* cultures. *British Journal of Experimental Pathology* 22, 137–47.
56. Cf. chapters 3, 5 and 6 (Biological standardization).
57. Florey, H. W. and Abraham, E. P. (1951). The work on penicillin at Oxford. *Journal of the History of Medicine* 6, 302–17.
58. Macfarlane, G. (1979). *Howard Florey. The making of a great scientist*, p. 315. Oxford University Press, Oxford.
59. Chain, E. *et al.* (1940). Penicillin as a chemotherapeutic agent. *Lancet* ii, 226–8. The initial results were published within three months of the first dramatically successful experiment in mice.
60. Abraham, E. P. *et al.* (1941). Further observations on penicillin. *Lancet* ii, 177–89.
61. Macfarlane, G. (1979). *Howard Florey. The making of a great scientist*, p. 340. Oxford University Press, Oxford.
62. Williams, T. I. (1984). *Howard Florey. Penicillin and after*, p. 133. Oxford University Press, Oxford.
63. Richards, A. N. (1964). Production of penicillin in the United States (1941–1946). *Nature* 201, 441–5.
64. Hobby, G. (1985). *Penicillin. Meeting the challenge*, p. 94 f. Yale University Press, New Haven.
65. Macfarlane, G. (1979). *Howard Florey. The making of a great scientist*, p. 342. Oxford University Press, Oxford.
66. Williams, T. I. (1984). *Howard Florey. Penicillin and after*, pp. 148–9. Oxford University Press, Oxford.
67. Sheehan, J. C. (1982). *The enchanted ring. The untold story of penicillin*. MTP Press, Cambridge, Mass.
68. Williams, T. I. (1984). *Howard Florey. Penicillin and after*, p. 135 f. Oxford University Press, Oxford.
69. Abraham, E. P. (1962). The cephalosporins. *Pharmacological Reviews* 14, 473–500.
70. Batchelor, F. R., Doyle, F. P., Nayler, J. H. C. and Rolinson, G. N. (1959). Synthesis of penicillin: 6-aminopenicillanic acid in penicillin fermentations. *Nature* 183, 257–8.
71. Chain, E. (1971). Thirty years of penicillin therapy. *Proceedings of the Royal Society, ser. B* 179, 293–319.

72. Lazell, H. G. (1975). *From pills to penicillin. The Beecham Story*. Heinemann, London.
73. Fleming, A. (1938). Letter to the editor. *British Medical Journal* ii, 37–8.
74. Macfarlane, G. (1984). *Alexander Fleming. The man and the myth*, pp. 178–9. Chatto and Windus, The Hogarth Press, London.
75. Fleming, A. (1940). Letter to the editor. *British Medical Journal* ii, 715.
76. Florey, M. E. and Florey, H. W. (1943). General and local administration of penicillin. *Lancet* i, 387–97.
77. Fleming, A. (1943). Streptococcal meningitis treated with penicillin. *Lancet* ii, 434–8.
78. Macfarlane, G. (1979). *Howard Florey. The making of a great scientist*, pp. 349–50. Oxford University Press, Oxford.
79. Rivett, G. (1986). *The Development of the London Hospital System 1823–1982*, pp. 186–91. King Edward's Hospital Fund for London.
80. Macfarlane, G. (1984). *Alexander Fleming. The man and the myth*, p. 200. Chatto and Windus, The Hogarth Press, London.
81. Brunel, J. (1944). Qui a dècouvert la pènicilline? *Revue Canadienne de Biologie* 3, 333–43.
82. Florey, H. W. (1945). The use of micro-organisms for therapeutic purposes. *British Medical Journal* ii, 635–42.
83. Macfarlane, G. (1979). *Howard Florey. The making of a great scientist*, p. 273. Oxford University Press, Oxford.
84. Comroe, J. H. (1978). Retrospectroscope. Pay dirt: the story of streptomycin. Part I. From Waksman to Waksman. *American Review of Respiratory Diseases* 117, 773–81.
85. Waksman, note 48, p. 184f. of British (1958) edition.
86. Waksman, note 48, pp. 211–12 of British (1958) edition.
87. Schatz, A., Bugie, E., and Waksman, S. A. (1944). Streptomycin, a substance exhibiting antibiotic activity against Gram-Positive and Gram-Negative bacteria. *Proceedings of the Society for Experimental Biology and Medicine* 55, 66–9.
88. Comroe, J. H. (1978). Retrospectroscope. Pay dirt: the story of streptomycin. Part II. Feldman and Hinshaw: Lehmann. *American Review of Respiratory Diseases* 117, 957–68.
89. Feldman, W. H. (1954). Streptomycin: some historical aspects of its development as a chemotherapeutic agent in tuberculosis. *American Review of Tuberculosis* 69, 859–68.
90. Ainsworth, G. C., Brown, A. M., and Brownlee, G. (1947). 'Aerosporin', an antibiotic produced by *Bacillus aerosporus* Greer. *Nature* 160, 263.
91. Benedict, R. G. and Langlykke, A. F. (1947). Antibiotic activity of *Bacillus polymyxa*. *Journal of Bacteriology* 54, 24–5.
92. Brownlee, G., Bushby, S. R. M., and Short, E. I. (1952). The chemotherapy and pharmacology of the polymyxins. *British Journal of Pharmacology* 7, 170–88.
93. Stansly, P. G., Shepherd, R. G., and White, H. J. (1947). Polymyxin: a new chemotherapeutic agent. *Johns Hopkins Hospital Bulletin* 81, 43–54.
94. Controulis, J., Rebstock, M. C., and Crooks, H. M. Jr. (1949). Chloramphenicol (Chloromycetin). V. Synthesis. *Journal of the American Chemical Society* 71, 2463–8.

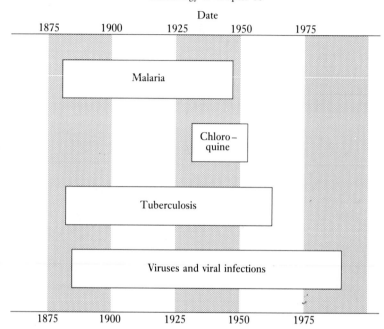

Chronology of Chapter 10

NB The blocks show periods discussed in the text, and do not mean that there was no activity about the subject at other times.

10

Towards the conquest of infectious diseases

Malaria: discovery in the laboratory

Long before the sulphonamides were used as medicines, quinine was known to be a specific drug against malaria. It did not prevent infection; it only checked the later multiplication of the parasite after infection had taken place. Nearly all the world's quinine came from Java, and, if Java fell into enemy hands (as it did in 1942), supplies would virtually cease. Chemists had attempted to synthesize quinine for the previous hundred years (see Chapter 2) but all they had achieved was to discover the extreme complexity of the problem. In the 1930s, with the threat of war and the knowledge that it might be fought in territories infested with malaria, the need for additional remedies was evident. Little was known in English-speaking countries about the new synthetic drugs (see Chapter 8) discovered in Germany in the 1920s and 1930s. In these circumstances military disaster from the commonest of all causes, sick troops, was a serious possibility, and a number of scientists gave urgent thought to finding better drugs to control, and preferably prevent the disease.

The causal parasite, a one-celled organism which thrived in red blood cells and was named *Plasmodium malariae*, was first seen under a microscope in 1880 by the French physician Alphonse Laveran (1845–1922) while he was serving in Algeria.[1] Improved microscopic methods and new synthetic staining materials from the German chemical industry made it possible to see the parasite in films of blood from malarial patients. Within a few years four species of *Plasmodium* had been recognized, responsible for clinically distinct variants of malaria. Now over a hundred are known, mostly infecting species other than man. Malaria was known to be associated with marshes: it was often called marsh fever, but the connection was not understood until certain mosquitoes were recognized as carriers of the disease and as alternative hosts of plasmodia. Stage by stage, the whole complex life cycle of the parasite was worked out.

Many points were still obscure in the 1930s, notably how relapses could occur in subjects long after they had left malarial regions and remained healthy for months with no sign of parasites in their blood. The period between the bite of an infected mosquito and the first attack of fever was also puzzling. Again no parasites could be found in the blood, but they appeared later, even if a course of quinine had been given in the meantime. So a drug which acted during this phase, a 'true causal prophylactic', was exactly what was wanted, but how was it to be found? Several workers suggested that there must be a tissue phase, in which the *Plasmodium* went to ground, but it was a long time before objects, evidently forms of the parasite, were recognized in microscopic sections of the spleen and liver of infected domestic fowl, monkeys and finally man.[2,3]

Very soon after Prontosil and sulphanilamide became available they were shown to be active against malaria. Their activity created some initial enthusiasm, but was found to be poor compared with quinine. Unexpectedly, when sulphanilamide was tried in bird malaria (see Chapter 8), at that time the standard procedure for discovering new antimalarial agents,[4] it did not work. This raised questions about the methods of testing. The support of a British parasitologist Warrington Yorke (1883–1943), at the Liverpool Institute of Tropical Medicine, who foresaw the wartime problems well in advance, was most important to a programme of research into antimalarial drugs developed at the Imperial Chemical Industries Research Laboratories at Manchester.

Much of the work centred on the complex biology of the malaria parasite[5] and on improvements in the methods of testing potential antimalarial drugs. Up to five stages in the life cycle could be identified, each of which might respond to a different agent. As a result of this work it could be declared that 'All the drugs now used against malaria were found by using experimental laboratory infections: the design of the tests employed to find them, or to uncover their particular attributes and imperfections, was based on knowledge of the life cycle'.[6]

Application of the principle of competitive antagonism (see Chapter 8) to the discovery of new drugs required some knowledge of the essential metabolic pathways in the infecting organism. This was particularly difficult for malaria, because the parasites could not be grown *in vitro*, and so were not easily available. There was no obvious starting point for experiments designed to discover the details of its biochemistry.[7] All that could be done was to work by analogy. Knowledge that plasmodia were vulnerable to sulphonamides was useful, particularly when they were shown to be protected, like bacteria, by p-aminobenzoic acid.[8] Chemically, the antimalarial drug Atebrin or mepacrine had some resemblance to riboflavin. Experiments with bacteria suggested that mepacrine competed with riboflavin, i.e. that mepacrine was a false building block which blocked the place belonging to riboflavin. It was an attractive idea, but it led nowhere. Progress came instead from testing a long sequence of compounds leading from the most potent sulphonamides to pyrimidines, a family of moderately complex organic compounds some of which are constituents of nucleic acids (Chapter 11), and later to guanidines. A sequence of antimalarial drugs of increasing power were produced, culminating in the drug 'Paludrine', alias proguanil or chloroguanide, which became publicly known in 1945.[9,10] It was rated very highly as a causal prophylactic. For suppressing attacks, especially those due to *Plasmodium falciparum*, it was rather slow in action and inferior to mepacrine. Furthermore, after many years of use, strains of resistant *Plasmodium* emerged, and, like nearly all antimicrobial agents, its power began to wane. But the main objective at the time, a true causal prophylactic, had been achieved.

This work depended on intimate collaboration between laboratory chemists, laboratory parasitologists, and clinicians expert in the disease in question. It was greatly fortified by an army research unit based in the United Kingdom, and by a unit established in Australia where antimalarials were systematically evaluated in controlled clinical trials in army volunteers. This unit provided the necessary information which allowed the most effective system of dosage with mepacrine and Paludrine to be established.[11] No doubt the availability of a population of soldiers willing to volunteer for such trials was an incidental benefit of the conditions of war: no such opportunities are normally available for evaluating and learning quickly and accurately how to use new drugs.

Malaria: discovery by committee

Perhaps the best known of all recent antimalarial drugs is chloroquine. As antimalarial drugs were desperately needed during the war, it is startling to realize that chloroquine was first synthesized several years before the war and recognized *at that time* as having antimalarial activity. A delightfully frank account of its development[12] shows some of the less scientific problems of discovery.

In 1934, H. Andersag, a chemist at the Bayer Laboratories in Elberfeld, prepared a compound which was first known as Resochin. It might be described as a simplified mepacrine, although it belonged to a substantially different chemical class. For 10 years few people knew anything about it. Some limited laboratory and clinical tests undertaken at Bayer led to the belief that it was slightly too toxic to be acceptable. A closely related substance, which was given the name Sontochin, was also laid aside. Under cartel agreements, both the Winthrop Chemical Company in the United

States and, later, the French firm of Specia (Société Parisienne d'Expansion Chimique, the pharmaceutical branch of the great chemical company, Rhone–Poulenc) were informed about Resochin and Sontochin. By this time, 1940, the German army had overrun France, and in Europe the normal assessment of new drugs was made subordinate to military needs.

In the USA, the need for research into antimalarial drugs had naturally been foreseen. A program on the synthesis of new antimalarial drugs was initiated at the National Institutes of Health in 1939 and formed the basis of a war program organized by the Committee on Medical Research of the Office of Scientific Research and Development, National Research Council. The program involved scientists from the universities and industry, private individuals, the US Army, the Navy and the Public Health Service, and included liaison with Great Britain and Australia. It was co-ordinated by a group of conferences, subcommittees, and, from November 1943, by the Board for the Co-ordination of Malarial Studies. The overall search for new antimalarial agents involved the screening of some 16,000 compounds, most of them for both suppressive and prophylactic activity against several avian malarias, plus a thorough study of the toxicology and pharmacology of many of the preparations in lower animals. Finally the appraisal was undertaken of some 80 compounds against the malarias of man.[12]

A great deal of useful information was assembled.[13,14] Some people will regard such a massive exercise in organization with awe. To others it may seem like a recipe for disaster rather than for successful research. It seems that the machinery did not work very well in discovering chloroquine. Under cartel agreements, the identity and properties both of Sontochin and of Resochin had been disclosed to the Winthrop Chemical Company. The reports created sufficient interest for Sontochin to be synthesized and tested by the American company, and it was found to be active against malaria in canaries. Resochin was not pursued at that time, but both compounds were duly registered for purposes of patenting. The information very properly reached the files of the survey for antimalarial compounds under the Survey Number SN-183 and was available to various conferences, committees, and subcommittees on the Co-ordination of Malarial Studies. History does not record what the members of these committees thought about it, or whether they had time to read their papers.

Meanwhile, P. Decourt, a French clinical consultant to Specia, took samples of the drug to Tunisia for human trials. There he was helped by Jean Schneider, later a professor in the medical faculty at Paris. Schneider's trials were interrupted by the Allied invasion of North Africa in November 1942, and after the capture of Tunis, Schneider turned over his promising results and his remaining samples of Sontochin to the US Army.

In due course some of the drug reached the United States. Its chemical composition was determined at the Rockefeller Institute in New York and found to be identical with the material synthesized at the Winthrop laboratories 3 years earlier. According to Coatney,[15] this discovery 'created havoc bordering on hysteria. We had "dropped the ball" and in doing so had lost valuable time in the search for a reliable synthetic antimalarial'. Naturally the lapse had to be covered up. The compound was given a new number and the biological data declared secret.

Further trials on Sontochin, now SN-6911, confirmed its considerable effectiveness, and also generated several ideas which resulted in the synthesis of a compound which was named SN-7618. It was made, but it was also found to be known already and to have been patented in the USA, along with Sontochin, by Winthrop. It was, in fact, a different salt of the same base as the original Bayer Resochin, and after much comparison with other related compounds it became established and was recognized formally in February 1946 under the name chloroquine. It had been synthesized and first tested in animals about 12 years earlier, and rejected, or ignored, twice. It has become the drug of choice for the American armed forces and for the World Health Organization, and has survived against the threat of resistant strains for a surprisingly long time.[16]

Since this time much progress has been made in understanding the biology of plasmodia and new antimalarial drugs have been discovered. Primaquine, chemically in the same family as the pre-war German drug pamaquin, emerged as the most notable success of the great US wartime project. Daraprim (pyrimethamine), a very different substance, evolved some years later from research of more general significance, to which we shall return in the next chapter. There are also more recent drugs and a radically different approach, based on advances in immunology, is leading towards the development of vaccines.[17] Much has been done to control mosquitoes too, but they, like the malaria parasites, have adapted to survive in the face of new enemies. The great hope of eradicating malaria has faded: new battles must be fought and new weapons devised if even the present level of control is to be maintained.[18]

Tuberculosis

For another widespread disease the biochemical approach, and particularly the idea of competitive antagonism, proved to be more rewarding. Every disease presents its own peculiar problems which fascinate and challenge investigators, but tuberculosis has more than the usual range of difficulties.

Before Koch, the cause of tuberculosis was unknown, but was suspected to be hereditary. After the tubercle bacillus was identified, accurate diagnosis of tuberculosis, of the lungs and of other organs, became possible. Controversy continued for some time about the identity or otherwise of human and bovine tuberculosis, until the infecting organisms were shown clearly to belong to different species of the large genus Mycobacteria, most of which are harmless to man. *Mycobacterium tuberculosis* and *Mycobacterium bovis* both grow slowly and require complicated rich media, including substances of unknown or uncertain chemical composition. They have waxy cell walls which perhaps protect them against many antibacterial substances. Furthermore, immune responses to tubercle bacilli are extraordinarily complicated. Koch's attempt to promote immunity with extracts of the organism (tuberculins) were not crowned with therapeutic success. A tuberculin remains as a useful diagnostic agent for detecting the state of immunity to tuberculosis, and so giving evidence of the presence, or past presence, of the disease.

Tissues respond to the presence of tubercle bacilli by forming a fibrous wall round the organisms and encasing them in the small nodules called tubercles. Sometimes the control is effective; sometimes damage continues within the enclosed areas and great destruction of tissues occurs, with cavities and haemorrhages. The fibrous tissues round the infected area impedes access of antituberculous substances. So the tubercle bacillus is a particularly difficult target for chemical attack.

In the medical climate of the 1920s, when the possibility of chemotherapy was looked on so unfavourably, mycobacteria were among the least promising organisms to choose for such treatment. A comprehensive review[19] written in America in 1932 concluded that, of the many substances which had been tried, more or less empirically, against tuberculosis, none was effective. One of the authors summarized the position in the words: 'Since 1922 we have come to recognise more and more that chemotherapy in the sense in which Ehrlich introduced the term, is more of a dream than a reality'.[20] In England Almroth Wright was expressing similar views. Happily, such pessimism was dispelled a decade later, when the study of bacterial chemistry began to bear fruit.

For anyone studying the biochemistry of microbes, mycobacteria must have offered a mixture of attraction and discouragement. The genus was sufficiently unlike other bacteria to arouse curiosity. The disease both of man and cattle was enormously important, but working with the human tuberculosis organism carried considerable risk of contracting an often incurable and sometimes fatal infection. There was therefore good reason for using related but innocuous species and accepting the chance of missing key properties which made the virulent organisms dangerous. The slow growth of the bacteria did not help. Experiments which take weeks are tiresome to arrange, more likely to go wrong, and less fun to do than those which given an answer at once, or within a day or so. Also, the slow course of the disease implied that any drug would have to be given for a long time in order to be effective and so would have to be particularly harmless to patients.[21,22]

After Domagk's discovery of Prontosil (see Chapter 8), its active principle, sulphanilamide, was tried against tuberculosis. In large doses it was effective against experimental infections in animals, and some related compounds called sulphones were better.[23,24] But their value in human tuberculosis was slight. They were more beneficial in a disease caused by another *Mycobacterium*, leprosy.[25] Useful results were achieved by studying the chemical needs of Mycobacteria. In Duke University, in North Carolina, F. Bernheim studied the effect of various substances on tubercle bacilli.[26] He worked with the bovine strain, but confirmed his positive findings with virulent human organisms. He was not primarily a bacteriologist, and instead of using the standard criterion of survival of the organism, growth in culture, he measured the uptake of oxygen by the cultivated organisms over short periods and so obtained results more quickly. Among numerous negative results he found some quite simple compounds, aldehydes and fatty acids, which stimulated oxygen uptake. Benzoic acid doubled it and salicylic acid increased it five-fold. The closely related *p*-aminobenzoic acid (PABA) had just become famous because of the work of Fildes and Woods. Bernheim tried it, but it had no effect. Another close relative, acetylsalicylic acid, i.e. aspirin, was active only if it was broken down to release salicylic acid.

Bernheim's results suggested that he had found some substances which were good for tubercle bacilli, and, by applying the principle of competitive antagonism, he sought closely related compounds which would also be taken up by them but would block the pathway of the stimulants. He had slight success with tri-iodobenzoic acid,[27,28] but it was J. Lehmann, in Sweden, who found that, among some fifty or so compounds tested, *p*-aminosalicylic acid, commonly known as PAS, prevented the growth of tubercle bacilli.[29,30] The compound was tried clinically, and proved to be of considerable benefit to a number of patients, but Swedish physicians were somewhat sceptical.[31] The reduction of fever is an early sign of improvement in tuberculosis. The aminosalicylate reduced the fever, but as salicylates are known to do this it was thought that the effect might be non-specific. Indeed, as gradually became clear, PAS alone is not a powerful drug against tuberculosis. But it was found to be valuable in conjunction with the newly-discovered antibiotic streptomycin (see Chapter 9).

The appearance of PAS on the clinical scene coincided with the introduction of streptomycin in the United Kingdom. Very small amounts were available, quite insufficient to treat all the patients who might benefit. The opportunity was taken to devise comparative trials, in which some patients were treated with streptomycin and others served as controls. In the earliest of these trials, as also in the USA, the rapid development of resistance was observed.[32,33] A further trial was promptly conducted in which streptomycin was compared with PAS and with the combination of both drugs. The results were unequivocal, that streptomycin was better than PAS, but that the addition of PAS to streptomycin virtually prevented the appearance of resistant strains.[34] So the value of PAS was established, and streptomycin was saved from the loss of efficacy which would otherwise soon have rendered it useless.

With the idea of competitive antagonism as a guide and the success of PAS as an encouragement, many other compounds were investigated in many laboratories. A French study showed that nicotinamide, a member of the B-complex of vitamins that prevents the disease pellagra in man (see Chapter 13), had some activity against the development of leprosy in rodents. As leprosy and tuberculosis are caused by related organisms, nicotinamide was tested against tubercle bacilli. It was found to be active but only in concentrations too high to be clinically acceptable. This discovery was apparently overlooked in the United States, and was made again 3 years later during the screening of large series of compounds. In Germany, Domagk and his colleagues investigated the potential of sulphonamides, gradually moving towards compounds which had no connection with his original antistreptococcal drug Prontosil. From such work and from studies on the biochemistry of tubercle bacilli, on substances which prevented their growth, and on potential therapeutic agents, extensive research, especially in the USA, led to the substance named isoniazid[35] which was discovered almost simultaneously in the laboratories of E. R. Squibb and of Hoffmann-La Roche and publicized in 1952. It has remained the most effective and least toxic drug against tuberculosis.

Like malaria, tuberculosis has been contained but not eliminated. Several more drugs have been discovered, though their activity is limited by their toxicity.[36] Resistance is a major problem, particularly in countries where the supply of drugs is limited and their use is ill-controlled and haphazard. In such conditions, where drug combinations are not used for an adequate length of time, resistant strains have every chance of profiting from the destruction of their more sensitive relatives, and establishing a reservoir of intractible organisms. The patient, rendered incurable, is then particularly liable to travel in search of treatment to better equipped countries, and there to disseminate the resistant organisms which he carries. Investigating and controlling the spread of tuberculosis socially[37] is as difficult as understanding the body's internal defences against the disease. So the need for new antituberculous drugs grows, and the older remedies become less valuable. New synthetic chemicals and new antibiotics have been discovered and brought into use, but at present

(1990) it does not seem likely that tuberculosis will be eradicated simply by the discovery of effective drugs. The behaviour of mycobacteria and of humans is too complex. However the pessimism of the 1920s about chemotherapy[20] is an awful warning about the hazards of prediction!

At the frontiers of life: viruses

After the great innovative period of bacteriology had achieved its ends, there remained a number of diseases, evidently infectious, for which no causative organisms could be found. They included familiar childhood fevers—measles, chicken pox, and mumps—and some much more dangerous conditions, including smallpox and rabies, as well as various diseases of animals, including the distemper of dogs and the foot-and-mouth disease of cattle. Similar infections were also observed in plants. A paper by the Russian scientist D. Ivanovsky,[38] published in 1892 and often regarded as the beginning of the science of virology, describes how a disease of tobacco plants can be transmitted by the sap after it has passed through a filter capable of retaining bacteria and other particles. Whatever made the sap infectious was not visible with the conventional microscope. Loeffler showed another filter-passing material to be the cause of foot-and-mouth disease in cattle.[39] These agents were given the old familiar name for obscure virulent materials, virus, and presented considerable difficulties because they could be detected only by infecting susceptible species and observing whether disease resulted. As many of the viruses, including those of man, attacked only a single species of host, investigation was a problem.

Animal viruses became easier to study when methods of growing them were devised. For a decade, fertile hen's eggs provided the most suitable medium. Later cultures of, usually embryonic, animal cells were used. The presence of virus could be recognized by microscopic changes in the infected cells, and means of assay were devised to detect the presence, and preferably the amount, of virus present. Advances followed the development of the electron microscope. Virus particles placed in the beam of electrons cast shadows which could then be photographed enabling them to be seen. Until then it was difficult to decide the status of these invisible materials. I remember, about 1946 or 1947, a somewhat heated debate among microbiologists about the nature of life and whether viruses were alive or not. As usual, ideas associated with 'life' generated much emotion, especially in those who were not anxious to define their words accurately. The debate had no practical importance, nor did it influence the direction and methods of further research. Organic chemists had been troubled and had had to overcome the same semantic problems a century earlier (see Chapter 2).

As common virus diseases produced good immunity the prospects for inventing vaccines seemed encouraging; indeed the whole process of vaccination began with smallpox. But there were no obvious ways of producing attenuated strains of virus until a means of cultivation was established. The greatest progress during the 1930s was made in investigations, sponsored by the Rockefeller Foundation in West Africa and in South America, of the tropical disease called yellow fever. By selective cultivation of strains of the virus causing yellow fever a vaccine was produced shortly before World War II. It made an immense difference to military operations in Africa, stimulated research for other virus vaccines, and led to the award, in 1951, of a Nobel prize[40] to the medical scientist Max Theiler (1899–1972).

The investigations in the 1930s also gave one of the earliest hints of a new kind of cellular defence mechanism against viruses. Quite simply, when monkeys had been infected with one strain of yellow fever virus, they became transiently less easy to infect with another strain.[41] Their resistance developed too rapidly and disappeared long before ordinary immune reactions could appear. Some other cellular event was involved, but its nature remained obscure and unexplained for another 30 years.

After World War II, progress was made towards understanding the nature of viruses, the mechanisms by which viral infections were rendered harmless, and discovering suitable virus strains from which vaccines could be made. Early in the 1950s both killed and attenuated living vaccines became available against poliomyelitis[43] and the disease was almost eliminated from those parts of the world where mass immunization campaigns were undertaken. However deaths are still recorded from this disease and children are crippled by it. But the problem of control of poliomyelitis is now one of social organization and, until new strains and new viruses appear, not principally one of medical discovery. Similarly the control of rubella is complex and important, because rubella in pregnancy is a cause of congenital defects.[44] Effective vaccines prevent such hazards, but only if a social organization ensures that all potential mothers are vaccinated in good time.[45]

The interference between one kind of virus and another was investigated more fully late in the 1950s by an outstanding research worker, Alick Isaacs[46] (1921–1967), at the National Institute for Medical Research. He exposed cells in culture to influenza virus, showed that they were resistant to infection with certain other viruses, and isolated from his cultures a protein which conferred the same resistance on fresh cells.[47] Once again progress had been made by identifying a kind of messenger or transmitter substance. Isaacs named the substance interferon, and for some years vigorous efforts were made to develop it into a practical therapeutic agent.[48] But it soon became evident that there were many kinds of interferon, all produced when cells were exposed to particular viruses but varying in their exact nature according to the cells which produced them and the species of animal from which the cells originated. As the interferons were specific for their species, it was essential to prepare interferon from *human* cells for the treatment of humans. To multiply human cells in tissue culture on the scale necessary presented considerable difficulties, but enough material has been made for a variety of clinical trials against many different viral diseases and certain cancers. Here too, the difficulties are formidable, with much disappointment and rare success.[49] Both research and the public have suffered greatly from premature publicity which has provoked excited demands for the material. In the absence of supplies of the curative substance and of any prospect of making it within months or years, such publicity is a tragic source of false hope and disillusion, and would be better avoided. However, advances in recombinant DNA technology have helped and interferon is now available for certain licensed uses.

The alternative approach to treating an infectious disease, by means of chemotherapy, made little progress until viruses could be grown and studied in cultures outside living organisms.[50] In some early post-war work on tuberculosis, compounds under investigation were screened for antiviral activity, and some active compounds were found[51] The observation was pursued and many compounds were examined for activity against viruses in culture, and in infected animals. One derivative, later named methisazone, attracted interest because it was shown in clinical trials in Madras, in 1963, to prevent people exposed to smallpox during an epidemic from developing the disease, even though it did not hasten recovery once the disease had developed.[52] Many other substances were screened for antiviral activity and a few drugs with limited clinical application were found.[53]

Progress, as always, in the practical discovery of new drugs depended on basic advances in understanding the chemical identity of the living system. Gradually it was established that each of the many kinds of virus consisted of a core of a nucleic acid, either ribonucleic acid (RNA) or deoxyribonucleic acid (DNA) (see Chapter 11), with a protein coat. At the same time, during the 1950s and 1960s, great advances were made in determining the molecular structures of RNA and DNA, and in recognizing their functions in controlling the activities of living cells, including the power to reproduce themselves and to become immune to infecting organisms. Advances in virology became inseparably linked with the new science of molecular biology, and the discovery of antiviral drugs began to have a deeper and stronger foundation.

A drug named idoxuridine, synthesized as a possible agent against cancer, proved to be effective in treating eye infections with a herpes virus.[54,55] It was closely related to a naturally occurring substance. Its activity depended on its being sufficiently similar to be taken up by the chemical processes of the virus but sufficiently different to be useless to the virus and to jam its works. Once again advantage was being taken of the false building block principle which had been discovered for sulphanil-amide (see Chapter 8).

The differences between virus nucleic acids and those of the hosts which they infect are very subtle. Thus it is critically important that an antiviral agent does not become incorporated into the host's nucleic acids,[56,57] or those works too could be jammed, with devastating consequences. The newer antiherpes agent, acyclovir,[58] is of great interest, because it acts in virally-infected cells, but only to a negligible extent in healthy human cells. Thus it has an unusual and invaluable safety factor among its properties.

It remains to be seen whether vaccines or chemotherapy will play a greater part in controlling the human immunodeficiency virus.[59] What is quite clear is that any success will rest on the basic sciences of virology and molecular biology, and that, without research into obscure aspects of these subjects the direct hunt for new cures would be no better than shooting in the dark with unfamiliar weapons at an undefined target.

The new look in science

In surveying the discovery of penicillin, we moved from one domain of science to another. Fleming, a classical, conventional scientist, saw comprehensible objects and made obvious inferences about the things he saw. You and I might have done the same if we had been there, though we would have had none of the technical skill and detailed knowledge of Fleming. On the other hand, Chain and Florey were armed with apparatus which means little or nothing to the layman. When he arrived at Florey's laboratory, Chain was delighted to see a Sohxlet apparatus in a cupboard, but caused consternation by saying 'I shall want six of those'. Yes, but what *is* a Sohxlet apparatus, and why isn't one enough? Would we be wiser if we were told? And Chain isolated substances which could be recognized as novel and peculiar only with a considerable knowledge of chemistry. One feels a certain loss of romance. The antibiotic mould supposedly drifting in on the wind becomes prosaic when its magic is revealed only by tedious operations in a laboratory or factory.

What would have happened if penicillin had not been discovered, or if years of patient research had failed to explain why sulphonamides acted only on a limited selection of microbes? Would the idea of drugs useful against germs once again have faded into obscurity? Would many firms in the pharmaceutical industry have expended large resources in the pursuit of moulds which might produce useful drugs? Would streptomycin have become available if the example of penicillin had not given impetus to its investigation? And if there had been no synthetic drugs to control the resistant strains, would the decline of streptomycin have been taken to prove that antibiotics were a waste of time because organisms adapted to them too easily?

However, the medical advances *were* made, and so were many other scientific discoveries, impelled by the necessities of war. The obvious achievements—radar, jet propulsion, guided missiles, and the atomic bomb which brought the war to a sudden and unexpectedly early end at a price only less horrible than the continuation of war—depended on much basic research in physics and chemistry, particularly in the fields of radio-activity and nuclear energy, and of electronics. The same kind of basic research continued after the war and provided the groundwork for endless discoveries which could be applied to peaceful ends. The material amenities of life increased, at least in Western society, and prosperity came to many who had never known it before. Public appreciation of science was high, and funds for research became available on a previously unheard-of scale.

So there was plenty of enthusiasm, and plenty of opportunity. Leader-ship in science, which had moved from France to Germany in the late nineteenth century and from Germany to Great Britain early in the twentieth, now went to the United States, where immense resources had been developed during the war and were ready to be applied to more beneficial ends. The expansion of American medical research, much of the finest quality, attracted European doctors who aspired to a scientific career, and they did well to spend a year or two in an American laboratory, or even to emigrate there altogether. The National Institutes of Health of the United States Public Health Service were extended[60] far beyond the size of any comparable European centre. Laboratories and institutes established by charitable foundations became bigger: the Rockefeller Institute in New York was renamed the Rockefeller University in 1977 to reflect the scale of its research and education.[61] Industry, especially chemical and pharmaceutical, was no less involved in the great expansion, and the opportunities for research were on an unprecedented scale.

Expansion was not confined to the USA. In Europe the ravages of war had to be made good and then came the enlargement of universities, government institutes, and industrial establishments. The British Medical Research Council expenditure increased twenty-fold between 1948 and 1971.[62] The Wellcome Trust, set up by the will of Sir Henry Wellcome to devote the profits of the Burroughs Wellcome businesses to medical and veterinary research, became a major sponsor, second in Britain only to the Medical Research Council itself, and with notable influence elsewhere in the world.[63] The growth of the trust owed much to its first chairman, Sir Henry Dale (see Chapter 4). By the time of Wellcome's death in 1936,

Dale was long established as Director of the National Institute of Medical Research, and soon afterwards (1940–1945) became President of the Royal Society and a principal scientific advisor of government during the war. Both in these positions and as Chairman of the Wellcome Trust for 22 years, he had great influence on the development of medical research during and after the war. Other trusts based on the profits of industry gave increasing support to medical research, and the pharmaceutical industry, rapidly becoming international in its outlook and activities, promoted its own laboratories directed towards the discovery of new drugs. Public hopes for medical science, coupled with increasing prosperity, encouraged the formation of charities which collected funds for research, largely from individual donations and legacies. Cancer research had a long tradition of support from private contributions, but new charities devoted to particular diseases, such as arthritis and rheumatism, leukaemia, and muscular dystrophy, were founded and became a great source of strength to workers whose interests had or might have application to the desired objectives.[64]

The style of discovery changed after World War II. The lone worker and thinker became a rarity. There was more team work, and the teams were larger. Their members were more specialized: a growing proportion of them had no medical background. The range of sciences on which they drew was greater. The significance of many particular facts or operations could be perceived only with some background knowledge, which the uninitiated would have little wish or reason to acquire. Reports of original work, headed often by the names of many joint authors, became too full of jargon to be understood even by trained scientists who were not working in the particular field. This situation persists today, though strong movements towards interdisciplinary research help to avoid total fragmentation of scientific understanding.

In such a climate, it is ambitious to attempt to explain scientific problems in simple terms. But the gap between amateur and expert must be bridged if society is not to be split between two cultures. It can be done, though it involves going into a little detail from time to time about points of central importance, and it needs simplifications which trouble an author's respect for exact truth and may appal an unsympathetic expert. In spite of specialization, the *general* outline of the discovery of many recent drugs is remarkably similar[65,66] and indeed has been said to follow a text-book pattern.[67] To avoid too many technicalities, we will confine the story to a small number of topics: cancer, high blood pressure, and mental disorders will serve to illustrate the ways in which problems have been solved.

Notes

1. Laveran, A. (1907). Protozoa as causes of disease. *Nobel lectures physiology or medicine, 1901—1921* 264–71. Published for the Nobel Foundation, 1967, Elsevier, Amsterdam.
2. James, S. P. and Tait, P. (1937). New knowledge of the life-cycle of malaria parasites. *Nature* 139, 545.
3. Shortt, H. E. and Garnham, P. C. C. (1948). Demonstration of a persisting exoerythrocytic cycle in *Plasmodium cynomolgi* and its bearing on the production of relapses. *British Medical Journal* i, 1225–8.
4. Davey, D. G. (1963). Chemotherapy of malaria. Part I. Biological basis of testing methods. In *Experimental Chemotherapy*, eds R. J. Schnitzer and F. Hawking. Academic Press, New York.
5. Davey, D. G. (1944). Biology of the malaria parasite in the vertebrate host. *Nature* 153, 110–11.
6. Davey, note 4, p. 488.
7. Curd, F. H. S., Davey, D. G., and Rose, F. L. (1945). Studies on synthetic antimalarial drugs. II—General chemical considerations. *Annals of Tropical Medicine and Parasitology* 39, 157–64.
8. Maier, J. and Riley, E. (1942). Inhibition of antimalarial action of sulfonamides by *p*-aminobenzoic acid. *Proceedings of the Society for Experimental Biology and Medicine* 50, 152–4.
9. Editorial notice (1945). Malaria research leading to paludrine. *Annals of Tropical Medicine and Parasitology* 39, 137–8.
10. Fairley, N. H. (1945). Chemotherapeutic suppression and prophylaxis in malaria. *Transactions of the Royal Society of Tropical Medicine and Hygiene* 38, 311–55.
11. Fairley, N. H. (1946). Researches on paludrine in malaria. *Transactions of the Royal Society of Tropical Medicine and Hygiene* 40, 105–51.
12. Coatney, G. R. (1963). Pitfalls in a discovery: the chronicle of chloroquine. *American Journal of Tropical Medicine* 12, 121–8. Coatney's account includes reference to numerous wartime reports, not readily accessible.
13. Wiselogle, F. Y. (ed.) (1946). *A survey of antimalarial drugs, 1941—1945*. J. W. Edwards, Ann Arbor, Michigan.
14. Peters, W. and Richards, W. H. G. (eds) (1984). Antimalarial drugs. I. Biological background, experimental methods, and drug resistance. *Handbook of Experimental Pharmacology*, vol. 68. Springer-Verlag, Berlin.
15. Coatney, G. R. (1963). Pitfalls in a discovery: the chronicle of chloroquine. *American Journal of Tropical Medicine* 12, 121–8 at p. 125.
16. Bruce-Chwatt, L. J. (1982). Chemoprophylaxis of malaria in Africa: the spent 'magic bullet'. *British Medical Journal* 285, 674–6.
17. Bruce-Chwatt, L. J. (1987). The challenge of malaria vaccine: trials and tribulations. *Lancet* i, 371–3.
18. W.H.O. (1984). Advances in malaria chemotherapy. *W.H.O. Technical Report Series no. 711*.
19. Wells, H. H. and Long, E. R. (1932). The chemistry of tuberculosis. Ballière, Tindall and Cox, London.
20. Wells. H. H. (1931). The chemotherapy of tuberculosis. *Yale Journal of Biology and Medicine* 4, 611–26; 625.

21. Brownlee, G. (1953). The wider aspects of the chemotherapy of tuberculosis. *Pharmacological Reviews* 5, 421–50.
22. Robson, J. M. and Sullivan, F. M. (1963). Antituberculosis drugs. *Pharmacological Reviews* 15, 169–223.
23. Rich, A. R. and Follis, R. H. (1938). The inhibitory effect of sulfanilamide on the development of experimental tuberculosis in the guinea-pig. *Bulletin of the Johns Hopkins Hospital* 62, 77–84.
24. Feldman, W. H., Hinshaw, H. C. and Moses, H. E. (1941). Treatment of experimental tuberculosis with promin (sodium salt of *p,p'*diamino-diphenyl sulphone-N,N'-dextrose sulfonate): preliminary report. *Proceedings of the Staff Meetings of the Mayo Clinic* 18, 118–25.
25. Bushby, S. R. M. (1958). The chemotherapy of leprosy. *Pharmacological Reviews* 10, 1–42.
26. Bernheim, F. (1941). The effect of various substances on the oxygen uptake of the tubercle bacillus. *Journal of Bacteriology* 41, 387–95.
27. Saz, A. K. and Bernheim, F. (1941). The effect of 2,3,5,triiodobenzoate and various other compounds on the growth of the tubercle bacillus. *Journal of Pharmacology and Experimental Therapeutics* 73, 78–84.
28. Saz, A. K., Johnston, F. R., Burger, A. and Bernheim, F. (1943). Effect of aromatic iodine compounds on the tubercle bacillus. *American Review of Tuberculosis* 48, 40–50.
29. Lehmann, J. (1946). *Para*-aminosalicylic acid in the treatment of tuberculosis. *Lancet* i, 15–16.
30. Lehmann, J. (1964). Twenty years afterward. Historical notes on the discovery of the antituberculous effect of para-aminosalicylic acid (PAS) and the first clinical trials. *American Review of Respiratory Diseases* 90, 953–6.
31. Birath, G. (1969). Introduction of para-amino-salicylic acid and streptomycin in the treatment of tuberculosis. *Scandinavian Journal of Respiratory Diseases* 50, 204–9.
32. Medical Research Council (1948). Streptomycin treatment of pulmonary tuberculosis. *British Medical Journal* ii, 769–82.
33. Crofton, J. and Mitchison, D. A. (1948). Streptomycin resistance in pulmonary tuberculosis. *British Medical Journal* ii, 1009–15.
34. Medical Research Council (1950). Treatment of pulmonary tuberculosis with streptomycin and para-amino-salicylic acid. *British Medical Journal* ii, 1073–85.
35. Fox, H. H. (1953). The chemical attack on tuberculosis. *Transactions of the New York Academy of Science* 15, 234–42.
36. Mitchison, D. A., Ellard, G. A., and Grosset, J. (1988). New antibacterial drugs for the treatment of mycobacterial disease in man. *British Medical Bulletin* 44, 757–74.
37. Sutherland, I. (1988). Research into the control of tuberculosis and leprosy in the community. *British Medical Bulletin* 44, 665–78.
38. Ivanovsky, D. (1892). Ueber die Mosaikkrankheit der Tabakspflanze. *Bulletin of the Academy of Imperial Science, St. Petersburg* 3, 67. Quoted by F. M. Burnet (1960). *Principles of animal virology*, 2nd edn. Academic Press, New York.
39. Loeffler, F. and Frosch, P. (1898). Berichte der Kommission zur Erforschung der Maul- und Klauenseuche bei dem Institut für Infektionskrankheiten in Berlin. *Zentralblatt für Bakteriologie*, Abt. I Orig., 23, 371. Quoted by F. M.

Burnet (1960). *Principles of animal virology*, 2nd edn. Academic Press, New York.

40. Theiler, M. (1951). The development of vaccines against yellow fever. *Nobel lectures physiology or medicine, 1942–1962*, 351–9. Published in 1964 for the Nobel Foundation, Elsevier, Amsterdam.

41. Hoskins, M. (1935). A protective action of neurotropic against viscerotropic yellow fever virus in Macacus rhesus. *American Journal of Tropical Medicine and Hygiene* 15, 675–80.

42. Findlay, G. M. and MacCallum, F. O. (1937). An interference phenomenon in relation to yellow fever and other viruses. *Journal of Pathology and Bacteriology* 44, 405–24.

43. Enders, J. F. (1954). The cultivation of the poliomyelitis viruses in tissue culture. *Nobel Lectures Physiology or Medicine, 1942–62*, 448–67. Published in 1964 for the Nobel Foundation, Elsevier, Amsterdam.

44. Gregg, N. McA. (1941). Congenital cataract following German measles in the mother. *Transactions of the Ophthalmological Society of Australia* 3, 35–45.

45. Badenoch, J. (1988). Big bang for vaccination. *British Medical Journal* 297, 750–1.

46. Andrewes, C. (1967). Alick Isaacs 1921–1967. *Biographical Memoirs of Fellows of the Royal Society* 13, 205–21.

47. Isaacs, A. and Lindemann, J. (1957). Virus interference. I. The interferon. *Proceedings of the Royal Society, ser. B* 147, 258–67.

48. Ho, M. and Enders, J. F. (1959). An inhibitor of viral activity appearing in infected cell cultures. *Proceedings of the National Academy of Sciences, USA* 45, 385–9.

49. Sehgal, P. B., Pfeiffer, L. M. and Tamm, I. (1982). Interferon and its inducers. In Chemotherapy of viral infections. *Handbook of Experimental Pharmacology*, vol. 61, eds. P. E. Came and L. A. Caliguiri, pp. 205–311. Springer-Verlag, Berlin.

50. Came, P. E. and Caliguiri, L. A. (eds) (1982). Chemotherapy of viral infections. *Handbook of Experimental Pharmacology*, vol. 61. Springer-Verlag, Berlin.

51. Hamre, D., Bernstein, J. and Donovick, R. (1950). Activity of *p*-aminobenzaldehyde, 3-thiosemicarbazone on vaccinia virus in the chick embryo and in the mouse. *Proceedings of the Society for Experimental Biology and Medicine* 73, 275–8.

52. Bauer, D. J., St. Vincent, L., Kempe, C. H. and Downie, A. W. (1963). Prophylactic treatment of smallpox contacts with N-methylisatin β-thiosemicarbazone. *Lancet* ii, 494–5.

53. Hay, J. and Bartkoski, M. J. Jr. (1982). Pathogenesis of viral infections. In *Chemotherapy of viral infections. Handbook of Experimental Pharmacology*, vol. 61, eds. P. E. Came and L. A. Caliguiri, pp. 1–91. Springer-Verlag, Berlin.

54. Prusoff, W. H. (1959). Synthesis and biological activities of iododeoxyuridine an analog of thymidine. *Biochimica Biophysica Acta* 32, 295–6.

55. Kaufmann, H. E. (1962). Clinical cure of herpes simplex keratitis by 5-iodo-2′deoxyuridine. *Proceedings of the Society for Experimental Biology and Medicine* 109, 251–2.

56. Albert, A. (1985). *Selective Toxicity* 7th edn., pp. 125–47. Chapman and Hall, London.

57. Elion, G. B. (1985). Selectivity—key to chemotherapy: presidential address. *Cancer Research* 45, 2943–50.
58. Elion, G. B. *et al.* (1977). Selectivity of action of an antiherpetic agent, 9-(2-hydroxyethoxymethyl)guanine. *Proceedings of the National Academy of Sciences, USA* 74, 5716–20.
59. Mitsuya, H. *et al.* (1986). 3'-Azido-3'-deoxythymidine (BW A509U): an antiviral agent that inhibits the infectivity and cytopathic effect of human T-lymphotropic virus type III/lymphadenopathy-associated virus *in vitro*. *Proceedings of the National Academy of Sciences, USA* 82, 7096–100.
60. Shannon, J. A. (1987). The national institutes of health: some critical years, 1955–1957. *Science* 237, 865–8.
61. Rockefeller Foundation (1977). *Institute to University. A Seventy-fifth Anniversary Colloquium*. The Rockefeller University, New York.
62. Thomson, A. L. (1973). *Half a century of medical research, Vol. 1: Origins and policy of the Medical Research Council (UK)*, Her Majesty's Stationery Office, London.
63. Hall, A. R. and Bembridge, B. A. (1986). *Physic and Philanthropy. A History of the Wellcome Trust 1936—1986*. Cambridge University Press, Cambridge.
64. Association of Medical Research Charities (1988). *Handbook of British Medical Research Charities*. AMRC, London.
65. Gross, F. (ed.) (1983). *Decision Making in Drug Research*. Raven Press, New York.
66. Burley, D. M. and Binns, T. B. (eds) (1985). *Pharmaceutical medicine*. Edward Arnold, London.
67. Williams, M. and Malick, J. B. (eds) (1987). *Drug discovery and development*. Humana Press, Clifton, New Jersey.

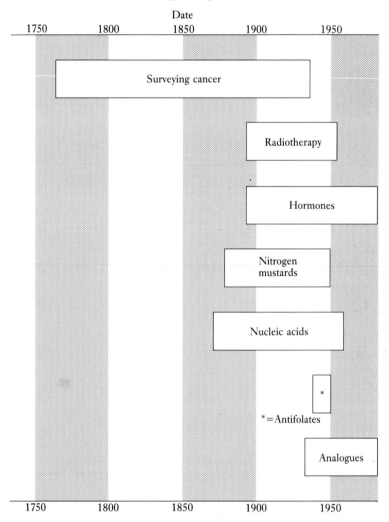

Chronology of Chapter 11

Date

| 1750 | 1800 | 1850 | 1900 | 1950 |

Surveying cancer

Radiotherapy

Hormones

Nitrogen mustards

Nucleic acids

*

* = Antifolates

Analogues

| 1750 | 1800 | 1850 | 1900 | 1950 |

NB The blocks show periods discussed in the text, and do not mean that there was no activity about the subject at other times.

11

Cancer

The background

Cancer probably evokes more fears and grievously false hopes than most diseases. The causes of cancer are still obscure, but it is not always necessary to know the cause of a disease in order to cure it or even to prevent it. Salvarsan and later the sulphonamides could hardly have been invented without the bacteriological advances made between 1880 and 1910, and bacterial test systems were essential for isolating pure penicillin from mould juice. On the other hand, Jenner introduced vaccination when he neither knew nor had any means of knowing anything about viruses. Snow associated the occurrence of cholera with the source of the victim's water supply 20 years or so before bacteria were identified as causing the disease. Lime juice was recognized (though not widely established) as preventive for scurvy, and cod liver oil for rickets long before specific nutritional deficiencies began to be understood or vitamins were discovered. So seeking 'the cause' of cancer is not the only route to finding drugs which will relieve or cure it.

Research directed towards preventing and curing cancer repeatedly illustrates the dilemma. Should one seek the causes, eliminate them and so prevent the disease? Should one seek drugs which will stop the disease, and, if so, what strategy should guide the search? Or should one take a middle road, trying to understand what goes wrong and then deciding whether it is easier to prevent or to cure?

As we shall see, the real advances have come not from research specifically directed towards cancer, but from discoveries made in quite different fields, as wide ranging as electrical discharges in gases, heredity in banana flies, analysis of the spermatozoa of salmon, and the development of instruments of chemical warfare. It is unlikely that anyone thought of any of these possibilities when, in 1801, a society was formed to investigate the nature and cure of cancer.[1] The society appointed a medical committee, which outlined the essential problems in a series of simple questions. They asked:

1. What are the diagnostic signs?
2. Are there any characteristic changes in a tissue which precede the growth of cancer?
3. Is a cancer a degenerative change which arises from some other disease?
4. Is cancer inherited?
5. Is cancer contagious?
6. Is cancer associated with other diseases?
7. Is it a localized condition or the sign of a generalized bodily disorder?
8. Is it produced by an unfavourable climate or environment?
9. Is cancer associated with a particular temperament?
10. Does it occur in 'brute creatures'?
11. Are there exempt periods of life?
12. Is it ever susceptible of natural cure?

Several books and treatises were written by members of the committee in their attempt to answer these questions. An enormous amount of information has been gathered since. It shows that no simple answer can be given to any of these questions, and that the questions, though reasonable, are themselves over-simplified. Cancer is not one disease but many, and the answers are unlikely to be the same for all of them.

Many people have also asked 'Can any substance be given to a patient with cancer to stop the growth and dissolve the cancer?' And many a quack has traded on the anguish of cancer sufferers and the credulity of their friends.[2-4] Laws have been made to check such deceits, but sometimes remedies have been honestly although misguidedly trusted until experience gradually showed their worthlessness.

It is difficult to test a potential cure when a disease is ill-defined. Cancers of all kinds are now recognized as the uncontrolled multiplication of a particular cell type, either localized in simple tumours or invading surrounding tissues and spreading via the blood stream to create secondary growths in distant organs. Recognizing the cellular origin of cancers was a natural development from the cell theory of bodily structure and was established by the great labours of the nineteenth-century descriptive pathologists. Virchow himself, the founding father of the subject, at first regarded cancers as arising from formless material which would later develop into cancer cells. This was the time when theories of the spontaneous generation of living cells were receiving hard treatment from Pasteur, and Virchow came to assent without doubts that cancer cells had their origin in other cells.[5]

There are many problems for seekers after cures for cancer. Advances in pathology brought accurate descriptions of the different kinds of new growth and allowed some sort of classification, but gave little help in suggesting remedies. The most obvious treatment was to cut out the growth. When anaesthesia and antisepsis were both well established, surgery became a bearable procedure. But even surgery was helpless when there were widespread and multiple secondary deposits.

Epidemiological studies sometimes threw light on preventable causes of cancer. In 1761 J. Hill associated cancer of the nasal passages with the immoderate use of snuff. In 1775 Percival Pott recognized that scrotal cancers rarely occurred except in chimney-sweeps, an observation which has become famous as one of the earliest records of an industrial cancer. But doctors who had a taste for epidemiology tended to study diseases which spread rapidly and killed more quickly, such as cholera and typhoid fever. New industrial processes occasionally led to the discovery of new causes of cancer, such as the 'paraffin cancer' of workers in the Scottish shalefield and the 'mulespinner's cancer' of Lancashire cotton operatives.

Once it was recognised that any kind of cancer had an identifiable cause, it was natural to try and determine the exact chemical identity of the substances responsible (the 'carcinogens'). Little progress could be made until these cancers could be reproduced by experiments in animals. How else, short of applying it to humans, could one test whether a particular extract contained the carcinogenic material? Success in this field was elusive until in 1915, Yamagiwa and Ichikawa produced tumours by painting coal tar on the ears of rabbits. Other means of detecting carcinogens followed, but the chemical problems of isolating minute amounts of active material from large quantities of tar were also formidable. Advances in physical and chemical methods enabled Kennaway and Hieger in 1930 at the Chester Beatty Institute in London[7] to isolate a substance with the chemical name 1:2:5:6-dibenzanthracene, 'the first known pure chemical compound manifesting pronounced carcinogenic properties.[6,8]

Many more carcinogens have been identified since then. Some are mainly of labortory interest but others are clearly related to substances in more or less widespread use. Epidemiologists have incriminated substances and circumstances, and identified potentially carcinogenic material, such as tobacco smoke and tobacco tar, showing how a substantial burden of cancer could be prevented by avoiding exposure to the sources of carcinogens. But it has taken many years for such discoveries to be accepted and acted upon, especially when they conflict with public convenience and related commercial interests, as the sad story of tobacco promotion and consumption shows.

A young American doctor named Peyton Rous turned to laboratory work because he was not fit enough for the stress of medical practice. In 1911, when he was working at the Rockefeller Institute in New York, he described the transmission of a malignant new growth by means of a cell-free filtrate. The tumour came from a hen, and the material transmitted caused tumours in other fowl, so it had no obvious importance to human medicine. The microbes known at the time did not pass through filters, so the nature of the material was obscure. The paper did not cause much excitement and Rous turned to other subjects. Forty years later, when viruses were better known and several people had attributed viral origins to some tumours and perhaps leukaemia, the paper was rediscovered. In 1966 Peyton Rous was duly honoured with a Nobel prize.[9] But the possible role of antiviral drugs in the treatment of such cancers remains unknown.

The beginning of radiotherapy

There is no reason to suppose that either Wilhelm Roentgen (1845–1923) or Pierre Curie (1859–1906) had any thought of discovering a treatment for cancer when they first encountered X-rays and radium. Roentgen was experimenting with electrical discharges in evacuated tubes when he observed the emission of rays which made nearby materials fluoresce, i.e. emit light when the rays fell upon them. His article in the *Transactions of the Würzburg Physical Medical Society* was promptly reported in England and America and, as many scientists had been working in the field, was widely taken up.[10] The X-rays were quickly (and accidentally) found to penetrate living tissue and leave shadows of the less penetrable denser parts, allowing X-ray pictures to be made of the skeleton. Later, and also accidentally , they were found to damage and destroy living tissues, so they were applied as a kind of cautery to ulcerating cancerous growths as well as to other conditions such as inflammations and tuberculosis.

Antoine Becquerel (1852–1908) observed that similar rays are given off by uranium. Soon afterwards radium, a considerably more powerful emitter of such rays, was discovered by Pierre and Marie Curie (1867–1934). Much work was needed to prepare enough radium to treat patients, but once it was made, it was a more convenient source of radiation than the apparatus, extremely primitive by modern standards, which produced X-rays from electric discharges.

X-rays were greeted with uncritical enthusiasm. It was soon shown that the waters of many spas emitted radiations. Their supposed merits in restoring health were promptly attributed, without evidence, to the new magic. Commercial interests promoted the dispensing of radioactive solutions for therapeutic purposes, and some members of the medical profession supported their use. According to an abstract of the time, 'Half a pint a day six days a week for six weeks for patients suffering from rheumatic gout and similar affections was advised by the late Sir F. Treves.

Two such courses generally effect a cure'.[11] Methods of clinical investigation were in their infancy, and it was open to any distinguished consultant to adopt a new treatment, profess himself satisfied by its results and enhance his reputation and his practice accordingly. Methods for measuring the amount or intensity of radiation had not been developed, and the doses to which different patients were exposed probably varied from the imperceptible to the lethal. Deaths from overexposure would not have occurred until years later and so caused no immediate alarm.

The early use of X-rays and radium for the treatment of cancer was very much a matter of trial and error. Experiments on the biological effects of radiation were similarly hampered by the problems of measuring the intensity of radiation and the proportion of it absorbed by living tissues. A unit of radiation dosage, the 'roentgen',[12] was not defined until 1928, and it took time for it to be universally accepted. Basic questions about how X-rays affect living cells could hardly be answered, or even investigated with any precision until the doses absorbed could be measured. None the less, various biological effects of X-rays were noted and explored. Just as drugs were often used, even when their properties were poorly understood, to investigate physiological processes which they evidently modified, so X-rays were found to be a valuable tool in certain fields of study.[13,14]

Cell division, and the processes by which the characteristics of parent cells are reproduced in their offspring, was one such field. At the beginning of the twentieth century, two approaches converged. One was the study of inheritance by direct observation of specific properties of parents and offspring. Mendel's work had been rediscovered and repeated. The other was the search in living tissues for the carriers or agents which determined these properties, i.e. the genes responsible for producing them in successive generations. The discovery that suitable doses of X-rays sometimes cause a mutation, that is, a change in the property produced by a specific gene, was a valuable tool in relating the observed features of heredity to the cellular changes which control them. In the hands of the American zoologists Jacques Loeb (1859–1924), T. H. Morgan (1866–1945), and H. J. Muller (1890–1967) from the early 1900s onwards, the study of mutations induced by the radiations emitted by radium led to major advances in the science of genetics.[15,16] Morgan received a Nobel prize in 1933, and Muller in 1946.

It was shown that the effects of irradiation are greatest on dividing cells. The ovaries and testes are particularly sensitive, and sterilization (sometimes therapeutically desirable) is easily produced by a beam of X-rays of sufficient intensity. As a cancer consists of cells dividing uncontrolledly, there were good prospects of finding a dose of radiation which would stop the cancer cells without damaging healthy cells. Some tumours are indeed radio-sensitive, and some progress was made in therapy with x-rays and also with radium, which could be inserted into a tumour so that its effect was concentrated. When cancers did respond, radiotherapy was particularly useful in dealing with secondary growths which were too widespread for surgical relief. But most cancers failed to respond at all to the largest doses which were tolerated by healthy tissues.[17]

At the same time, people began to discover that irradiation not only arrests the growth of some cancers, but can have the opposite effect and generate new cancers in tissues which received excessive exposure. Early workers with X-rays, who were exposed to what by modern standards are horrifying doses, developed cancers at an alarming rate. Some girls, who worked in a factory where luminous radioactive paint was applied to the dials of watches, developed necrosis of the jaw and fatal anaemia. Their tissues showed changes due to cancerous destruction, and contained substantial amounts of radioactive material, evidently absorbed because of the girls' practice of licking the tips of their paint brushes to achieve a finer point. Other varieties of cancer associated with radiation were discovered. All too slowly, reasons were established for treating X-rays and radioactive materials with considerable caution and for initiating proper control over the use of all sources of radiation.[18,19]

The interactions of X-rays with living tissues were studied in very few laboratories, perhaps because physicists working with the new rays were not usually well informed about the biological problems which might be worth approaching, and biologists lacked the background and training in physics necessary to use X-rays in their work. Not until after the development and use of atom bombs did 'radio-biology' become a widely recognized science and its contribution to cancer research properly appreciated. Cancer research was moving forward on other fronts too.

Hormones and cancer

At about the time when Becquerel was working on the radiation from uranium, G. T. Beatson, a surgeon in Glasgow, observed that the inoperable breast cancers of three patients underwent regression after he had removed their ovaries.[20] He experimented on the effects of removing the ovaries of rabbits, and speculated extensively, but lacked too many important facts to make much sense of his discoveries. Other surgeons and radiotherapists obtained benefit for some patients when they were sterilized by irradiation. However, the advantages of either treatment were too limited and too infrequent for them to be widely adopted.

We should remember that very little was known at that time about the ways in which the ovaries influence distant organs. The extensive studies on removal and transplantation of the ovaries by Knauer in Vienna (see Chapter 5) appeared in 1900 and the more general concept of hormones or blood-borne chemical messengers was elaborated for the first time by Starling in his Croonian lectures a decade after Beatson's observations (see Chapter 5). It was hardly possible to understand the effects of removing the ovaries until ovarian hormones had been purified and shown to produce effects comparable to those seen at puberty. The isolation and chemical identification of several sex hormones in the late 1920s and the 1930s had many practical consequences. Direct evidence was obtained that certain ovarian hormones, the oestrogens, were responsible for the growth of the breasts, both in humans and other mammals. In mice, in particular circumstances, the growth was uncontrolled and resembled cancer. But it was a far cry to advance from recognizing that some breast cancers diminished when deprived of oestrogens to achieving an effective cure of a common and usually fatal condition.

Investigation of the physiology of the male prostate yielded more immediate progress. This organ has received much attention from surgeons, who are confronted with the problems which arise when the prostate enlarges late in life and obstructs the urinary channel which passes through it. In the 1930s the Chicago surgeon, C. A. Huggins, made extensive studies of the prostate gland of humans and of dogs. He observed that the tumours of the prostate which arise spontaneously in aged dogs and are sometimes but not usually cancerous, got smaller after the dogs were castrated. He achieved the same result by administering female sex hormones, which indirectly cause the testes to become inactive.[21,22]

Huggins and his colleagues showed that, either removal of the testes or treatment with hormones benefited the majority of patients with carcinoma of the prostate. The cancer was not eliminated, but its growth ceased, and for many years could be prevented from spreading to distant tissues as long as treatment with hormones continued. The synthetic oestrogen, stilboestrol (see Chapter 5), was as effective as naturally occurring oestrogens. The treatment was investigated in a co-operative study involving many surgeons: by 1950 some 1800 cases had been reported, leaving little doubt that treatment had been found which at least delayed the progress of an otherwise intractable cancer.[23] Huggins shared a Nobel prize in 1966 with Peyton Rous. As so often happens, the passage of time showed an increasing number of disadvantages of the treatment, including a greater frequency of heart disease and strokes in men receiving oestrogens.[24] Subtler methods of modifying endocrine balance continued to be explored.

Control of cancer of the breast was to prove more difficult. The activity of cells in the breast is influenced by oestrogens and progestogens (cf. Chapter 5) and by androgens from the adrenal glands. Some tumours grow if oestrogens are present, and others are checked by their administration. Such differences in the behaviour of different cancers of the same organ does not make it easy to identify the right hormones for treatment. As late as 1975, an authoritative source stated, 'The first cardinal principle' (in treating cancer of the breast) 'is that hormonal therapy should be reserved for patients for whom surgical treatment or radiotherapy has been considered and deemed no longer of value'.[25] One advance, still perhaps not fully evaluated, arises from the principle of competitive antagonism and the discovery that certain compounds compete with oestrogens and block the tissue receptors on which the oestrogen acts. The drug tamoxifen, first described in the late 1960s, has some value as a palliative in patients with oestrogen-dependent tumours.[26]

In spite of many attempts to extend the use of hormones to the treatment of other cancers, it appears that their important effects are on the organs which they normally control. Hormones of the adrenal cortex influence the behaviour of lymphocytes, the white blood cells involved in the development of immunity: the adrenal hormones, and synthetic related substances have been found useful in the control of those leukaemias which are, in effect, tumours of lymphocytes.

Hormones modify the cells which may become cancerous, rendering them less, or more, likely to do so, but the hormones do not act primarily on the processes which turn any quiescent cell into one that divides without control. The effects of hormones may themselves take a very long time to appear. About 1970 a number of young adult women were found to have a rare cancer of the vagina: the appearance of several cases close together prompted enquiry about possible causes. It was discovered that the mothers of all the patients had taken the drug diethylstilboestrol during the first 3 months of their pregnancies. The connection has been amply confirmed and leaves no doubt that indirect exposure of the fetus to abnormal amounts of oestrogen is a potential cause of cancer.[27,28] The discovery of an effect with such a long latent period was no mean feat of epidemiology.

Swords into ploughshares: war gases and the arrest of cancer

In 1938, just before Huggins reported on the treatment of prostatic cancer, Goodman and Gilman's celebrated textbook of pharmacology made its first appearance. It is interesting to note that no section of the book was devoted to cancer and the word cancer did not even appear in the index.[29] Such was the state of worthwhile knowledge at that time. It is interesting that Gilman and Goodman themselves were the first to discover an effective anticancer drug.

Many valuable drugs have been recognized first as poisons. Atropine from deadly nightshade, eserine from the ordeal bean of old Calabar, the alkaloids from the arrow poison curare all yielded useful medicines when their properties were studied carefully and the right dose was used for beneficial instead of destructive ends. Other noxious substances may be very good drugs too if they are understood and properly used. The first effective anticancer drug was discovered in this way. Its story is all the more piquant, because the starting point, the harmful substance, was one almost universally regarded as an exceptionally evil human 'achievement'. The volatile oily liquid beta-chloro-beta-ethyl sulphide was first synthesized in 1854, and in 1887 it was reported to produce blisters if it touched the skin. It was called mustard gas and was used at Ypres in 1917, when it caused many thousands of casualties.

Chlorine was used at the same time, and one may suppose that confusion arose in many minds about their separate effects. Research was essential to identify the ill-effects attributable to each substance and to devise means of protecting against them and of overcoming them. By 1919, mustard gas was known to injure not only the skin but also the bone marrow, preventing the formation of white blood cells and other essential elements of the blood.[30] But the war was over, and the pressure to investigate chemical warfare agents disappeared. Mustard gas was disagreeable and hazardous to work with. Chemists maintained a sporadic interest in related compounds, including 'nitrogen mustard', synthesized in 1935, in which a nitrogen atom replaces the sulphur.

The outbreak of a new war made defence against chemical warfare agents once again an urgent problem. In Britain, and later in the United States, government agencies arranged to supplement the activity of arsenals by sponsoring work in university laboratories. An academic approach to these problems, not usually studied in university laboratories, proved fruitful, both in advancing basic ideas and in identifying new drugs. Most of them did not escape from wartime secrecy and were not properly developed until the war was over.

At Yale, under a contract between the University and the Office of Scientific Research and Development, Goodman and Gilman supplemented our understanding of mustard gas by exploring the properties of the nitrogen mustards.[31] Like mustard gas, the nitrogen mustards caused blisters when they came into contact with skin, and they damaged many other tissues when they were absorbed or injected into the circulating blood. The most vulnerable cells were those which the body renews most frequently; especially the white blood cells, including the lymphocytes. Working with another scientist, Thomas Dougherty, who was particularly interested in tumours of white cells in mice, Goodman and Gilman used a nitrogen mustard to treat a mouse with a lymphoma, a large solid lymph cell tumour. The tumour shrank dramatically, in a way which had not been seen before. But the shrinkage did not last, and further courses of treatment were less effective. Finally the mouse died, but its life had been extended. More experiments on more lymphoma-bearing mice followed, and the doses likely to achieve the best effect were assessed with great care.[32]

A surgical colleague, G. E. Linskog, agreed to try the treatment in human patients. It was a hazardous undertaking: everyone thought of the nitrogen mustards as dangerous compounds which could destroy vital cells in many organs. One of Linskog's patients had a lymphatic tumour which had become resistant to radiotherapy; situated in the jaw, it was making eating and breathing difficult. The drug, called 'compound X' for reasons of wartime secrecy, was given in doses which Gilman later described as 'a most fortunate guess'. In 2 days some benefit was detectable: in 4 days it was obvious. Later, the patient was in severe, but expected, danger from a depletion of his own healthy white blood cells. But the 'guessed' dose had been well chosen, and the bone marrow, which makes the white cells, recovered. The tumour had shrunk, and the patient was, temporarily, rescued from his miserable state. However, the tumour was not entirely destroyed, and it too recovered. They saw for the first time what was to become a regular and tragic sequence of events: a dramatic response to the first treatment, a lesser one to a second, and in the end delayed death from a condition which had become as resistant to drugs as it was to radiation therapy.[32]

Nevertheless, no drug had previously achieved such success against any cancer-like proliferation of cells. It was enough to warrant an investigation of related compounds and a closer exploration of how they worked. Several laboratories and clinics were involved, in Salt Lake City, in Portland, Oregon, and in Chicago, where an independent study of another nitrogen mustard gave results like those found in Yale.[33-35] Under wartime agreements for the exchange of secret information, the news reached England, and investigations were pursued, particularly at the Chester Beatty Institute[7] in London. When publication was permitted, the principal findings were summarized by Gilman and Philips.[31,36] Nitrogen mustards in suitable doses damage only cells and tissues which normally exhibit relatively high rates of proliferation and growth. They stop the process known as mitosis, in which the chromatin of the nucleus is divided into precisely matching portions for each daughter cell. After larger doses, chromosomes are seen to break and cell nuclei to fragment. Once again there was an association between action against cancer and changes in chromatin. Indeed, the changes were very like the effects of irradiation with X- or ultra-violet rays, but this time they were being produced by a chemical instead of a physical agent.

At that time, the inhibition of enzymes was a particular focus of interest as a mechanism of action for drugs, and so nitrogen mustards were tested for such properties, without very striking results. A certain formal similarity of the compounds to acetylcholine (see Chapter 4) was explored but this too led nowhere. One of their chemical properties which attracted attention was their ability to alkylate other molecules, i.e. attach a simple organic radicle such as $-CH_3$ to them. Such an addition could readily cause (and explain) big changes in the biological properties of the alkylated compound. At the Chester Beatty Institute, it was suggested that nitrogen mustards might alkylate two adjacent molecules at once and so link them together. If the molecules were some part of the chromosomes, this might account for the many nuclear changes which had been observed.[37]

Several lines of research developed. The most obvious was to seek better compounds than the original nitrogen mustards, and numerous alkylating agents emerged during the following 30 years. They were active against leukaemias of various kinds and against some related solid tumours. These new compounds were often specific for a particular kind of leukaemia, and sometimes they were so effective that remissions appeared to be permanent. But for other cancers they were of little use.

A more fundamental approach was to investigate the reactions between nitrogen mustards and the chromatin which filled the cell nucleus and which they were suspected of alkylating. The time was ripe for such an approach, because progress in chemistry was just beginning to make possible the effective study of very large molecules such as those that formed the main constituents of chromatin.

First whisperings of DNA

The cell nucleus, so prominent in stained microscopical sections of tissues, often attracted chemical investigation but presented many difficulties. The first serious attempt was made as early as 1869 by a young Swiss doctor, Johannes Friedrich Miescher (1844–1894), whose uncle was Wilhelm His (1831–1904), a most unusual anatomist who maintained the forward-looking proposition 'La solution finale du problème du développement tissulaire se trouve dans la chimie' (The final solution of the problem of development of organs is to be found in chemistry). Inspired by his uncle, Miescher went, at the age of 25, to Tübingen as the first pupil of Felix Hoppe-Seyler (1825–1895), the young professor of the then new science of physiological chemistry. Miescher studied lymphocytes, which consist of large nuclei and little else. He obtained these cells from an unattractive source, the pus on discarded surgical dressings, which had the merit, in pre-chemotherapy days, of being very readily available. He achieved the very considerable feat of isolating a material, demonstrating its purity and getting an analysis of the elements present. He named it 'nuclein', and noted that it was rich in phosphorus. Later, after his return to his native city of Basle, he obtained similar material from other cells, mostly the spermatozoa (again cells containing little else except nuclei) of salmon caught in the Rhine outside the window of his laboratory. From salmon sperm he also isolated a protein, named protamine, and observed that nuclein was closely associated with proteins if it was not actually a protein itself.[38]

In the next 30 years considerable progress was made in purifying and crystallizing proteins. As crystallization occurs only when a substantial fraction of the molecules present are alike in size, composition, and shape, this achievement implied that the protein crystallized had a definite molecular structure. Attempts to determine their molecular weight suggested that proteins were behaving as though they were substances with a fixed and definite composition and, probably, structure. Analysis of purified proteins, notably by Gowland Hopkins (see Chapter 7), also showed that the variations apparent in crude preparations disappeared with purification. But the molecular weights were enormous, implying molecules which contained thousands of atoms, much too complicated for any details of their structure to be worked out. With an instinct for what was later called the 'Art of the Soluble',[39] most biochemists confined themselves to more manageable topics.

It was fashionable at the time to study the colloidal state, a condition in which small molecules are associated in loose, somewhat indefinite aggregates. Ostwald in Germany and Bancroft in the United States, who held dominant positions in the world of physical chemistry, especially as editors of the principal journals, were influential in promoting such ideas.[40] The idea extended from chemistry into biology, so that 'protoplasm' was envisaged, not as a giant molecule such as Ehrlich suggested, but as a colloidal system of indefinite composition. The notion was sufficiently vague to be easy to understand and difficult to disprove. At the other end of the scientific spectrum, organic chemists were reluctant to extend their studies beyond compounds consisting of small molecules with not more than perhaps 50 or 100 atoms. Their closely defined concepts of the structure of such molecules had brought enormous advances in knowledge, but extending these rigid ideas to larger molecules was not easy. Even Emil Fischer, in propounding the idea that proteins were built up by a standard linkage (-CO-NH-) between different amino acids,[41] apparently did not envisage molecules containing more than 30 or so such units, corresponding to no more than 300 atoms which is much too small.

What were needed, and gradually emerged between 1900 and 1940, were technical methods for investigating large molecules. Even such an elementary requirement as the accurate control of acidity and the alkalinity was difficult until, in about 1912, Sørensen devised the scale of measurement known as pH, and pioneered the use of solutions which 'buffered' changes of pH. The crystallization of enzymes, the invention of the electrical method called electrophoresis which separates molecules according to their size and their electric charge, the development of very high speed centrifuges in which the larger proteins are separated from smaller ones as they spin, and later the application of X-ray crystallography to large molecules have all contributed to progress. None of these methods seemed at the time to have any relevance to therapeutics, and the scientists who developed them would have found it very difficult to persuade the distributors of funds for medical research to support them, but without them the practical advances which came later would have been impossible. Tedious though this point may be, it is particularly in need of emphasis today, when the value of research is so often judged only by expected and short-term practical applications.

The main constituents of chromatin were known as 'nucleins'. An early paper[42] (1909) by McCollum, of vitamin fame (see Chapter 7), describes nucleins as combinations of protein and nucleic acids, the latter having a high content of phosphorus in combination with certain organic molecules, known as purines and pyrimidines, and with a sugar. This description was entirely correct, but it took nearly 50 years to work out exactly how the different parts fitted together. In New York Phoebus Levene (1869–1940) produced a definite theory of their structure, based on units which contained one each of the four bases adenine and guanine (purines) and cytidine and uridine (pyrimidines). For a long time, too long in fact, this theory held the field, until improved analytical methods showed that various samples of nucleic acid did not have the exact constant proportions of the different bases required by Levene's tetra-structure. The sugar was identified as either ribose or deoxyribose: for a time it was believed that plant nucleic acids always contained ribose and animal nucleic acids deoxyribose. This belief was dispelled in the 1940s when advances in technique showed that both kinds of nucleic acid were present in all cells, and that the deoxyribonucleic acid (DNA) was relatively stable while the ribonucleic acid (RNA) was being continually synthesized and decomposed.[43] The spiral structure of DNA and RNA molecules, the 'double helix', was not proposed until late in the 1950s.

None of this had any direct relevance to the discovery of drugs, and one may assume that many pharmacologists did not follow these developments. Students of cancer had more reason to be interested in chromatin and its role in cell division, but chemical approaches to cancer were predominantly concerned with molecules which could be identified at the time, i.e. with molecular weights under 1000. Purines and pyrimidines, ribose and deoxyribose were all well in this range, and we shall discuss the essential role they play. But, with that unpredictability which makes biology such a fascinating science, it was some biochemists interested in nutrition who laid foundations for the next major advance in the treatment of cancer.

Converging discoveries

By the 1940s, physiological chemistry had become biochemistry and biochemistry was a flourishing science which spread into almost every aspect of animal, plant, and microbial life. The processes which keep alive a microbe, a spinach plant, and a human turned out to have more in common than might have been expected. Ideas spread quickly from one field to another. Scientists noticed discoveries, apparently irrelevant to their work, which contained an inspiration for solving some current problems of their own. Here we can mention only a few events which led to major therapeutic developments, and pass by much else which gave the clues and spurs for progress.

In Chapter 7, we saw how several nutritional factors were detected— Will's principle, found in yeast extracts and active against a tropical anaemia; a similar material to that which prevented anaemia in monkeys; the substance derived from spinach which was necessary for the growth of various microbes; and substances from yeast and from liver also needed by microbes. Gradually it became clear that all these factors were chemically identical, and the single name, folic acid, was coined.

The exact structure of folic acid was established[44] in 1946. It had three parts; the amino acid glutamic acid, the nitrogenous base pteridine, and a substance we have met previously, *p*-aminobenzoic acid (PABA), which linked the two other components together. PABA is essential for those microbes which are attacked by sulphonamides (see Chapter 8) and the discovery of folic acid threw fresh light on the way sulphonamides worked. PABA, glutamic acid, and a source of pteridine, are the building blocks from which the microbe constructs its own folic acid. We have already seen that a sulphonamide acts as a false building block; it fits into the construction but is the wrong shape for further blocks to be added. Now the identity of the further blocks was known, and it was evident that sulphonamides prevented folic acid from being put together. Theoretically, the sulphonamide block could be by-passed if ready-made folic acid were available, but in practice microbes which make their own folic acid are unable to take it up from outside. Most organisms, ranging from other microbes to man, can absorb folic acid and rely on it to provide their needs, and so they are not affected by sulphonamides.

Folic acid itself is of great interest to anyone concerned with the formation of red and white blood cells, and with the abnormal and excessive formation of white cells in leukaemia. Was folic acid necessary for abnormal proliferation? Might proliferation occur because the cells in question lacked the essential material and got out of control? This point was rapidly settled when folic acid was shown not to check but to *accelerate* the development of certain leukaemias. This discovery suggested that an analogue of folic acid would block its use and prevent the multiplication of blood cells, just as an analogue of PABA prevented the multiplication of microbes which used PABA.

Such an analogue might block the normal use of folic acid and cause intractable and fatal anaemia, so it would be a hazardous substance to try. But also it might have a greater effect on leukaemic cells, which divide more frequently, probably use more folic acid, and are thus more vulnerable to a competitive antagonist. If this was the case, the proliferation of cells in leukaemia might be brought under control.

Analogues were made, in particular by the group at the Lederle laboratories which had identified the structure of folic acid. The substance aminopterin, made by replacing an -OH group by an $-NH_2$ group on the pteridine part of folic acid, was a milestone in the history of chemotherapy. With it, the paediatrician S. A. Farber and his colleagues in Boston obtained striking results in the treatment of acute leukaemia in children. Apparent recovery took place, though it was only temporary: it was followed by a fresh proliferation of leukaemic cells which no longer responded to treatment.[45] Nevertheless, a considerable advance had been made. Aminopterin had disadvantages, and the normal process of making and testing a series of closely related compounds led to a better drug, amethopterin, which differed by a single additional substituent but proved to be one of the most effective drugs for the treatment of blood cancers.

At the time there was little understanding of the role of folic acid in the normal working of the body. Observations on microbes appeared 'further to relate [folic acid] with nucleic acid metabolism and to emphasize the probability of a role for this vitamin in cellular proliferation.'[46] Much had yet to be done in the study of nucleic acid chemistry.

A search with many facets

In 1940, no one knew how the components of nucleic acids fitted together, but there was no doubt that they included certain pyrimidines (cytosine, thymine and uracil) and purines (adenine and guanine). Very little was known about the origin of the pyrimidines and purines. Since the amount of nuclear material increases as cells grow and divide, new pyrimidines and purines had to come from somewhere. Food was unlikely to provide exactly the right nucleic acids ready-made, and it was more likely that they were being built up from their components. So it was time to attend to the synthetic pathways, about which very little was known. As folic acid was evidently important in some way for the formation of new cells, it was natural to seek a role for it in the synthesis of pyrimidines and purines.

The advantages of using microbes for the early stages of such studies were well known and their biochemistry provided a convenient starting

point for several investigators, especially in America. Among them was
G. H. Hitchings, who had worked at Harvard on the quantitative
estimation of purines and on the purification of the anti-anaemia principle
in liver, and at Western Reserve on studies which led later to the discovery
of folic acid. In 1942 he came to the Research Laboratories which
Burroughs Wellcome had established at Tuckahoe on the outskirts of New
York. There he embarked on a study of nucleic acid synthesis with the twin
objectives of seeking fundamental knowledge about the roles of pyrimidine
and purine bases in growth, and of discovering new chemotherapeutic
agents.[47] As parasitic tissues generally depend for survival on a more rapid
synthesis of nucleic acid than the tissues on which they prey, anything
which stopped the synthesis of nucleic acids had a chance of harming
parasites more than their hosts.

On the basis of its dietary habits and its particular synthetic capacities,
Hitchings and his colleagues chose to work with a milk-fermenting
organism, *Lactobacillus casei*. Even in such a simple organism as *L. casei*, the
interactions turned out to be very complex, far beyond those of a single
enzyme sytem and full of 'inviting pitfalls for unwary investigators'.[48] Their
search led, by the middle of the 1940s, to a compound, 2:6-diaminopurine,
which had effects in mammals and chicks 'about what one would expect of
a substance which interferes with nucleic acid metabolism in some way'.[49]
It prolonged the life of mice with leukaemia, and it produced dramatic
remissions in some leukaemia patients. But it made the patients very sick,
and the remissions did not last. The leukaemias resisted further treatments
and were eventually fatal.

Hitchings and his colleagues, most notably Gertrude Elion, continued to
investigate purines and pyrimidines, by no means confining their attention
to the treatment of cancer.[50] In other hands (see Chapter 10), investigations
based on the pyrimidines led to the antimalarial drug chloroguanide.
Chloroguanide was not itself a pyrimidine, but it was sufficiently close for it
to be tested in *L. casei*, where it showed interesting effects, resembling
some of Hitchings' compounds. Turning the argument round, Hitchings
suggested to colleagues in the Wellcome Tropical Medicine Laboratories
in London that they might test some of the compounds being made in New
York against experimental malarial infections. The results were immediately encouraging. After much transatlantic exchange of information, the
synthesis and study of more than 300 compounds, and collaboration with
physicians in many parts of Africa and Asia, the drug christened
'Daraprim' and later known as pyrimethamine emerged in 1952 as an
antimalarial of considerable importance.[51]

This most productive diversion was followed by further work on anti-
cancer drugs. In 1951, another purine, 6-mercaptopurine (6-MP), was
found which caused remissions in some leukaemias. It was tolerated well

enough to become a useful drug, but it had a very brief period of action; 6-MP is attacked by the enzyme xanthine oxidase, a normal constituent of the human body (and of widespread occurrence elsewhere) and is converted to a substance, thio-uric acid, which is not active against tumours. Moreover, the leukaemia cells of patients treated with 6-MP adapted to it, so that the disease became resistant to further treatment. But these limitations contained the seeds of fresh advances.

Drugs which are rapidly inactivated have advantages, because the risk of overdosage is minimized and there are no cumulative effects. But they are also inconvenient, because frequent doses must be given to maintain a continuous effect. So a search began for ways of overcoming the difficulty. One way was to synthesize compounds which had all the useful properties of 6-MP but were not attacked by xanthine oxidase: this was achieved with the synthesis of a substance later named azathioprine. The second way was to find a compound which would compete for or block the inactivating enzyme. This way needed no search: the substance later known as allopurinol had already been synthesized by Hitchings and his colleagues, and it inhibited the enzyme xanthine oxidase (which converts xanthine to uric acid and mercaptopurine to thio-uric acid). As allopurinol did not inhibit the growth of *L. casei* nor interfere with the growth of experimental tumours, nor did it appear to have any other properties which made it unsuitable, it received further investigation in man.

Another possibility, at that time very speculative, was to see whether allopurinol would prevent uric acid from being formed from xanthine in man. If it did, it might be valuable in patients with gout, when abnormal deposits of uric acid are responsible for the development of intensely painful joints. Whether preventing the formation of uric acid would be beneficial, or whether the accumulation of unconverted xanthine would do more harm than good was open to question, and only to be settled by cautious clinical trial. In fact, fears were unfounded and the hoped-for benefit was achieved. Allopurinol has withstood substantial trials, and remains a standard drug for the treatment of gout.[52]

The use of 6-MP for the treatment of leukaemia had an unexpected consequence. Early trials of the new drug took place at a time when the transplantation of kidneys and other organs was being vigorously investigated. The principal obstacle to transplantation lay in the subtle differences between the tissues of one individual and another, differences which provoked immune reactions between the recipient and the transplanted organ. Such reactions are the normal response to any foreign material and an essential protection against invasive microbes. The importance of immune reactions in preventing successful organ grafting was shown conclusively when a kidney was transplanted from one identical twin to another and was not rejected. (Identical twins do not differ genetically, so their tissues are immunologically indistinguishable and there is no stimulus to an immune reaction.)

The principal problem for would-be organ grafters was therefore that of overcoming normal immune reactions to alien tissues. Robert Schwartz, a young physician working under William Dameshek, the chief of haematology at the New England Medical Centre, was assigned the task of solving this problem. His success[53] depended, as so often happens in research, on entirely fortuitous events, of the kind which sometimes contribute as much as careful planning to the attainment of desired objectives. In Schwartz's words:

The concept that formed the basis of this search was that immunologically competent cells were stimulated to proliferate on contact with antigen. At that time, this was a relatively novel idea without much basis in fact. Nevertheless, it seemed reasonable that chemotherapeutic agents known to block the proliferation of cells might also be immunosuppressive. We were then using two agents for the treatment of leukaemia: methotrexate and 6-mercaptopurine (6-MP). It was therefore decided to test their effects on the immune responses of rabbits. Now for the lucky part. Since neither drug was available commercially, I wrote to the Lederle Laboratories and to the Burroughs Wellcome Company for a supply of methotrexate and 6-MP, respectively. I never received an answer from Lederle. By contrast Burroughs Wellcome sent me a generous supply of 6-MP along with some easy-to-follow directions for its administration. I believe the directions were written by Trudy Elion. The experiments in rabbits were begun and within a month it was apparent that 6-MP was a potent suppressor of the immune response. Subsequently, a sample of methotrexate was finally obtained and tested in the same manner in the rabbit. It was found to be without effect. Later metabolic studies of methotrexate revealed that it is virtually without effect in the rabbit because of a metabolic peculiarity. In retrospect, it is highly likely that if Lederle had responded to my letter and Burroughs Wellcome had not, the whole idea of immunosuppressive chemical therapy would have been dropped as an interesting, but unsubstantiated, speculation.[54]

So a lucky accident provided a means of overcoming the outstanding barrier to organ transplantation. An English surgeon, (Sir) Roy Calne, then in Boston, investigated the use of a derivative of 6-MP, azathioprine, and after repeated experiments showed that it prevented rejection of kidney grafts in dogs.[55] In the next few years many clinical experiments established the effective use of drugs to suppress immune responses. The obvious hazard, that immunosuppression would render patients vulnerable to all the infections against which immune mechanisms normally work, was found, somewhat surprisingly, not to be a major problem. The longer-acting drug azathioprine was more satisfactory than 6-MP, and the adrenal cortical steroids also contributed to the prevention of graft rejection and to establishing organ transplantation as a regular therapeutic practice.

The reasoning which led to the discovery of the antimalarial pyrimethamine was applied equally fruitfully 15 years later, to the drug trimethoprim, which was as specific for certain bacteria as pyrimethamine was for malaria parasites.[56] All the drugs mentioned represent an astonishing range of achievements. History is still being made with antiviral agents from the same laboratories.[57] All this arose from the wise and fortunate decision to study the biosynthesis of nucleic acids. As Hitchings himself has pointed out, 'we were uncommitted with respect to specific disease targets, but we were bound to follow wherever our thoughts and antimetabolites led us'.[58] At the end of 1988, Hitchings and Elion and Sir James Black (see Chapter 12) shared a Nobel prize for 'discoveries of important principles of drug treatment'.

Progress in cancer research

Would-be developers of anti-cancer drugs have continued to seek competitive antagonists which might block one or another stage in the synthesis of pyrimidines and purines. At the same time, great advances have been made in finding out how best to use the many drugs now available. The margin between doses which destroy only malignant cells and those which damage healthy cells is very small, and the cure of a cancer depends on removing or killing all the malignant cells which elude the body's own control. Many years of research in many places have been devoted to working out the best ways of using surgery, radiation, and drugs to treat each kind of solid tumour and of using radiation and drugs for those disseminated cancers, such as the leukaemias, in which no central growth can be removed. Often, more can be achieved by a combination of surgery, radiation, and drugs, or by combinations of drugs than can be done by any single means.

Among the very many antimetabolites which have been investigated, only a small number have shown the combination of properties necessary to be therapeutically useful. More or less empirical screening is very unreliable for showing drugs active against cancers, but more refined methods have evolved. Several useful agents have been found in unexpected places. They include antibiotics produced by fungi (dactinomycin, daunorubicin), alkaloids derived from some plants of the Vinca species (vinblastine, vincristine), synthetic organic compounds containing platinum (cisplatin), the antiviral agent interferon (see Chapter 10), and even a vaccine made from an unfamiliar bacterium, *Corynebacterium parvum*. The activity of corticosteroids (see Chapter 5) against lymphocytes has offered a means of controlling lymphocyte tumours. A variety of other agents have shown specific, useful activities, with a fairly limited range of application. Often their mode of action provides fresh problems and, at least potentially, new insights into the processes which lead to the growth of cancers.

Overall, the search for 'a' cure for cancer has resolved itself into a large number of separate questions, many of which have still to be solved. It is no disparagement of the successes which have been achieved to say that cancer remains, largely, unconquered. For many patients with cancer, the prospect of long-term survival without recurrence is much more uncertain than it is, say, for infections, such as tuberculosis or cholera. Neither the researches funded by charities specifically directed towards cancer nor the great drives sponsored by governments and planned by committees have been a fruitful source of new treatments. Research into cancer has been well funded. As early as 1913, when the British Medical Research Committee (later Council) was established (see Chapter 4), cancer was the major topic to which it did *not* direct attention, because a charitable fund, the Imperial Cancer Research Fund, was already making substantial provision.[59] As has been recorded in this chapter, the principal anti-cancer drugs have been discovered either as a by-product of military research financed by government or by research in the pharmaceutical industry, much of which had its roots in research not primarily concerned with cancer.

The laboratories supported by the various cancer research agencies have, of course, made many valuable discoveries. Spectacular scientific advances have come from them just as they have from other institutions. But the problems of curing cancer have not been solved. At present, more can be achieved by avoiding exposure to recognized carcinogens[60] than by relying on the discovery of new remedies. Such prevention is in the province of every citizen and every politician, and it is sad that the political appeal of 'getting cancer licked' can divert extensive resources to activities of doubtful value. Conferences, committees, grandiose planning of operations divorced from day-to-day contact with laboratories and clinics, and the blind screening of endless substances have not been notably efficacious. We described, at the beginning of this chapter, a society set up in 1801 to investigate the nature and cure of cancer. Much was written but no cures were found. In 1971 the United States legislature passed a National Cancer Act, requiring the Director of the National Cancer Institute to produce a five year plan of his intentions. By 1973 a plan had been created from a basic structure set up with the aid of a 'systems management specialist'. Administrators of the institute and some 250 cancer specialists met in groups at a retreat in Virginia between October 1971 and March 1972, to list every conceivable way of approaching cancer research.[61] With all the possibilities set out in great detail, action could be taken, but the results are not striking. Sixteen years later, it was claimed that 'nobody in biomedical research wanted the War on Cancer'.[62] This exercise in planning research from outside has done little good to patients with cancer, and the politicians have given place to others. Apart from the dollars spent, one may wonder at the concealed costs of taking so many consultants away from their patients and research workers from their laboratories and clinics. The limited range of drugs which are effective against cancer have been discovered by uninterrupted research in laboratories, often on subjects which have had little very obvious relevance to cancer, but a lot to do with the chemistry of living cells. That field is in no way exhausted.

Notes

1. Schoenberg, B. S. (1975). A programme for the conquest of cancer: 1802. *Journal of the History of Medicine* 30, 3–22.
2. [No author stated] (1909). *Secret remedies. What they cost and what they contain*. British Medical Association, London.
3. Fishbein, M. (1964). History of cancer quackery. *Perspectives in Biology and Medicine* 8, 139–66.
4. Horsfall, F. L. (1964). Some facts and fancies about cancer. *Perspectives in Biology and Medicine* 8, 167–79.
5. Ackernecht, E. H. (1953). *Rudolph Virchow. Doctor, Statesman, Anthropologist*. University of Wisconsin Press, Madison.
6. Haddow, A. and Kon, G. A. R. (1947). Chemistry of carcinogenic compounds. *British Medical Bulletin* 4, 314–25.
7. Brunning, D. A. and Dukes, C. E. (1965). The origin and early history of the Institute of Cancer Research of the Royal Cancer Hospital. *Proceedings of the Royal Society of Medicine* 58, 33–6.
8. Kennaway, E. L. and Hieger, I. (1930). Carcinogenic substances and their fluorescent spectra. *British Medical Journal* i, 1044–6.
9. Rous, P. (1911). A sarcoma of the fowl transmissible by an agent separable from the tumor cells. *Journal of Experimental Medicine* 13, 397–411.
10. Roentgen, W. C. (1896). On a new kind of rays. *Nature* 53, 274–6.
11. Martindale, W. H. (1938). *The extra pharmacopoeia* 22nd edn, vol. 2, p. 966. The Pharmaceutical Press, London.
12. The measurement of radiation and the absorption of radiation by materials through which it passes, and the effects which result from such absorption are too complicated for any simple unit of measurement to be satisfactory. The roentgen was an early attempt at such a unit but is now superseded.
13. Deeley, T. J. and Hale, B. T. (1973). The first seventy-five years in radiotherapy. *British Journal of Radiology* 46, 906–10.
14. Spear, F. G. (1973). Early days of experimental radiology. *British Journal of Radiology* 46, 762–5.
15. Allen, G. E. (1978) *Thomas Hunt Morgan. The man and his science*, p. 148, footnote 113. Princeton University Press, Princeton, N.J.
16. Carlson, E. A. (1981). *Genes, radiation and society. The life and work of H. J. Muller*, Cornell University Press, Ithaca and London.
17. Burrows, E. H. (1986). *Pioneers and early years. A history of British radiology*. Colophon, Alderney.

18. Martland, H. S., Condon, P. and Knef, J. P. (1925). Some unrecognized dangers in the use and handling of radioactive substances. *Journal of the American Medical Association* 85, 1769–76.

19. Colwell, H. A. and Russ, S. (1932). Radium as a pharmaceutical poison. *Lancet* ii, 221–3.

20. Beatson, G. T. (1896). On the treatment of inoperable cases of carcinoma of the mamma: suggestions for a new method of treatment, with illustrative cases. *Lancet* ii, 104–7; 162–5.

21. Huggins, C. and Hodges, C. V. (1941). Studies on prostatic cancer. 1. The effect of castration, of estrogen and of androgen injection on serum phosphatases in metastatic carcinoma of the prostate. *Cancer Research* 1, 293–7.

22. Huggins, C., Stevens, R. E. and Hodges, C. V. (1941). Studies on prostatic cancer. II. The effects of castration on advanced carcinoma of the prostate gland. *Archives of Surgery* 43, 209–23.

23. Nesbitt, R. M. and Baum, W. C. (1950). Endocrine control of prostatic carcinoma. *Journal of the American Medical Association* 143, 1317–20.

24. Veterans Administration Cooperative Research Group. (1967). Treatment and survival of patients with cancer of the prostate. *Surgery, Gynecology and Obstetrics* 124, 1011–17.

25. Calabresi, P. and Parks, R. E., Jr. (1975). In *The Pharmacological Basis of Therapeutics* 6th edn, eds A. G. Gilman, L. S. Goodman and A. Gilman, p. 1303, col. 1, Macmillan, New York.

26. Klopper, A. and Hall, M. (1971). A new synthetic agent for the induction of ovulation: primary trials in women. *British Medical Journal* i, 152–4.

27. Greenwald, P., Barlow, J. J., Nasca, P. C. and Burnett, W. S. (1971). Vaginal cancer after maternal treatment with synthetic estrogens. *New England Journal of Medicine* 285, 390–2.

28. Herbst, A. L., Ulfelder, H. and Poskanzer, D. C. (1971). Adenocarcinoma of the vagina. *New England Journal of Medicine* 284, 878–81.

29. Goodman, L. S. and Gilman, A. (1938). *The Pharmacological basis of therapeutics*. Macmillan, New York.

30. Krumbhaar, E. B. and Krumbhaar, H. D. (1919). The blood and bone marrow in yellow cross gas (mustard gas) poisoning. *Journal of Medical Research* 40, 497–507.

31. Gilman, A. and Philips, F. S. (1946). The biological actions and therapeutic applications of the β-chlorethylamines and sulfides. *Science* 103, 409–15.

32. Gilman, A. (1963). The initial clinical trial of nitrogen mustard. *American Journal of Surgery* 105, 574–8.

33. Rhoads, C. P. (1946). Nitrogen mustards in the treatment of neoplastic disease. *Journal of the American Medical Association* 131, 656–8.

34. Goodman, L. S., Wintrobe, M. M., Dameshek, W., Goodman, M. J., Gilman, A., and McLennan, M. T. (1946). Nitrogen mustard therapy. *Journal of the American Medical Association* 132, 126–32.

35. Jacobson, L. A., Spurr, C. L., Barron, E. S. G., Smith, T., Lushbaugh, C., and Dick, G. F. (1946). Nitrogen mustard therapy. *Journal of the American Medical Association* 132, 263–71.

36. Philips, F. S. (1950). Recent contributions to the pharmacology of bis(2-haloethyl) amines and sulfides. *Pharmacological Reviews* 2, 281–323.

37. Goldacre, R. J., Loveless, A., and Ross, W. C. J. (1949). Mode of production of

chromosome abnormalities by the nitrogen mustards. The possible role of cross-linking. *Nature* 163, 667–9.

38. Meuron-Landholt, M de (1970). Johannes Friedrich Miescher: sa personalité et l'importance de son oeuvre. *Bulletin Schweiz Akademie Medizinsche Wissenschaft* 25, 9–24 (and following papers).

39. Medawar, P. B. (1982). *Pluto's republic*, pp. 2–3. Oxford University Press, Oxford.

40. Hess, E. L. (1970). Origins of molecular biology. *Science* 168, 664–9.

41. Fischer, E. (1907). Synthetical chemistry in its relations to biology. Faraday memorial lecture. *Journal of the Chemical Society* 91, 1749–65.

42. McCollum, E. V. (1909). Nuclein synthesis in the animal body. *American Journal of Physiology* 25, 120–41.

43. Davidson, J. N. (1968). Nucleic acids. The first hundred years. *Progress in Nucleic Acid Research* 8, 1–6.

44. Jukes, T. H. and Stokstad, E. L. R. (1948). Pteroylglutamic acid and related compounds. *Physiological Reviews* 28, 51–106.

45. Farber, S. A. *et al.* (1948). Temporary remissions in acute leukemia in children produced by folic acid antagonist, 4-aminopteroylglutamic acid (aminopterin). *New England Journal of Medicine* 238, 787–93.

46. Jukes, T. H., Franklin, A. L., and Stokstad, E. L. R. (1950). Pteroylglutamic acid antagonists. *Annals of the New York Academy of Science* 52, 1336–41.

47. Hitchings, G. H. (1969). Chemotherapy and comparative biochemistry. G. H. A. Clowes memorial lecture. *Cancer Research* 29, 1895–903.

48. Hitchings, G. H., Elion, G. B., Falco, E. A., Russell, P. B., and Vander Werff, H. (1950). Studies on analogs of purines and pyrimidines. *Annals of the New York Academy of Science* 52, 1318–35; at p. 1321.

49. *Ibid.*, at p. 1333.

50. Hitchings, note 47.

51. Symposium on Daraprim (1952). *Transactions of the Royal Society for Tropical Medicine and Hygiene* 46, 465–508.

52. Elion, G. B. (1978). Allopurinol and other inhibitors of urate synthesis. In *Uric acid. Handbook of Experimental Pharmacology. Vol. 51*, eds W. N. Kelly and I. M. Weiner, pp. 485–514. Springer-Verlag, Berlin.

53. Schwartz, R. and Dameshek, W. (1959). Drug-induced immunological tolerance. *Nature* 183, 1682–3.

54. Schwartz, R. (1976). Perspectives on immunosuppression. *Design and achievements in chemotherapy. A symposium in honour of George H. Hitchings*, pp. 39–41. Science and Medicine Publishing Co.

55. Calne, R. Y. (1976). Mechanisms in the acceptance of organ grafts. *British Medical Bulletin* 32, 107–12.

56. Garrod, L. P., James, D. G. and Lewis, A. A. G. (eds) (1969). The synergy of trimethoprim and sulphonamides. *Postgraduate Medical Journal* 45, Symposium Suppl., 1–104.

57. Elion, G. B. *et al.* (1977). Selectivity of action of an antiherpetic agent, 9-(2-hydroxyethoxymethyl)guanine. *Proceedings of the National Academy of Sciences, USA* 74, 5716–20.

58. Hitchings, G. H. (1969). Chemotherapy and comparative biochemistry, G. H. A. Clowes memorial lecture. *Cancer Research* 29, 1895–903 at p. 1895.

59. Thomson, A. L. (1975). *Half a Century of Medical Research. Volume 2. The Programme of the Medical Research Council (UK)*, pp. 10–12. HMSO, London.
60. Doll, R. and Peto, R. (1981). *The causes of cancer*. Oxford University Press, Oxford.
61. Culliton, B. J. (1973). Biomedical research (II): will the 'wars' ever get started? *Science* 181, 921–5.
62. Report (1987). Recollections on the war on cancer. *Science* 237, 843.

Chronology of Chapter 12

Date

1750	1800	1850	1900	1950

Blood pressure

Transmitters and mediators

Rauwolfia

Adrenaline receptors

Salt and water

Evalu-ation

1750	1800	1850	1900	1950

NB The blocks show periods discussed in the text, and do not mean that there was no activity about the subject at other times.

12

Cause unknown

Internal medicine

When the cause of an illness is unknown, the search for a cure (in the strict sense of the word) is inseparable from the search for a cause. But the progress of a disease can be mitigated and the disability and suffering of patients can be relieved without finding a cure. Diseases of the heart and blood vessels, asthma and other allergies, gastric and duodenal ulcers, and kinds of arthritis all fall into this category. They are not the result of any known lack of food or hormones, and no microbes have been proved to cause them. Nor does knowledge of contributing causes give an easy way out. The familiar enemies—cigarette smoking, too much alcohol, too much fat and sugar, and lack of physical exercise—do no good to health, but the cure by a frugal and temperate life is unattractive. Most people prefer to enjoy themselves and forget the cost until it is too late. Preventing or curing disease by a change in lifestyle raises problems to do with education and politics that are somewhat beyond our scope.

In most common diseases of western society, we know more or less *what* has gone wrong, even if we do not know *why* it has gone wrong. Also we often know of drugs which will reverse some of the changes; drugs which will steady an irregular heartbeat, drugs which will lower a high blood pressure, and so on. And if we do not have the right drugs for a particular adjustment, research on orthodox lines will often find one. So numerous drugs have become available, but the adjustments they achieve are sometimes disappointing. The body is full of compensatory mechanisms, and corrective treatments which are adopted for apparently sound reasons have often turned out to do more harm than good.

Blood pressure

Reasoning about diseases of the heart and blood vessels began to have a sound basis when, in the seventeenth century, William Harvey established that blood circulates from the heart, through arteries to the minute

capillary vessels in every organ, and back by the veins to the heart.[1]
Surgeons, and anyone concerned with the seriously wounded, knew that
blood was under considerable pressure in the arteries, because it spurted
powerfully if an artery was severed, but the arterial pressure was not
measured until Stephen Hales (1677–1761), 'Rector of Farringdon,
Hampshire, and Minister of Teddington, Middlesex',[2] undertook a
fearsome experiment while exsanguinating a horse.[2,3] Time passed and
more humane, non-invasive methods of measuring blood pressure were
devised. The familiar instrument in common clinical use depends on
measuring the pressure which checks the flow of blood in an artery, and
was introduced at the end of the nineteenth century.[4] Other instruments
were invented which recorded the actions of the heart and the flow in the
vessels, and it became possible to investigate how the whole system was
regulated throughout all the different activities of life. Clinical research
showed how far the human circulation behaved like that of other animals,
and how it changed as diseases advanced. Without this progress,
understanding the existing drugs and inventing new ones would have been
impossible.

The circulation depends on having enough blood in the system to fill the
vessels, on the small arteries acting as taps which can be adjusted to send
the blood where it is needed, and on the heart beating hard enough to push
the blood through the resistance given by the taps. The system is regulated
by the controlling, involuntary or autonomic nerves (chapter 4), by
hormones, and by the centres in the brain which co-ordinate all these
component parts. A defect in any of them can disturb the balance of the
whole system, and the use of the right drug to restore balance had evident
possibilities, if the right drugs could be discovered or invented.

The causes of high blood pressure, or hypertension, and of clotting, or
thrombosis, are still obscure, in spite of the mountains of evidence
accumulated about the mechanisms by which the blood presssure may be
raised or the blood may clot. Plenty of drugs lower the blood pressure, and
several prevent clotting, and all look as though they should be useful
treatments. However, it is not quite so simple. In the nineteenth century
the name 'essential hypertension' was invented, meaning that in some
patients a rise in blood pressure is a protective mechanism, essential in
order to overcome narrowing of the arteries and force blood through them
to reach vital organs. The high blood pressure was, in this view, essential
for the survival of the patient, and, if it were so, drugs which reduced the
blood pressure were more likely to do harm than good. On the other hand,
many patients with high blood pressure ultimately had strokes, apparently
because their vessels burst under the excessive pressure, or they died
because their hearts failed in the continuous task of driving blood through
their constricted arteries. Controversy about essential hypertension has

waxed and waned for a century, and what is orthodox today may change with time. And what can the cure-seekers do about a disease which is so difficult to understand?

Between 1945 and 1956, many research findings identified other factors which affected the blood pressure. They included the amount of salt in the body, the activity of the brain, the sympathetic nerves, and the adrenal glands, and the role of the kidneys in secreting a hormone-like substance which would indirectly raise the blood pressure.[5] Low-salt diets and sedative drugs contributed to the methods of treatment. These and many other factors are of great importance to practising physicians, but to do justice to them and to the controversies which they provoked[6] is beyond the purpose of this book. We shall focus on the search for drugs which lower blood pressure, while noting that they do not cure, that their role is essentially palliative, and that their use is not always appropriate. Before 1940, few such drugs were available. Most acted directly on the blood vessels. Those which caused a fall in pressure by weakening the heart appeared unsuitable, and some of those which, like nitrates (see chapter 2), cause all vessels to dilate extensively were beneficial only in special circumstances. More effective drugs had their origin in the basic research on chemical transmission of nerve impulses described in chapter 4.

Transmitters and mediators

The transmitter substances which we described in chapters 4 and 6 provided a valuable basis from which to start searching for new drugs. Dale and his colleagues in the 1930s showed that acetylcholine, in addition to its other properties, was a local hormone relaying messages at autonomic ganglia, that is at critical points in the nerve pathways which control the heart and blood vessels, and at the sites where nerves activate voluntary muscles. The first drugs which could be relied on to reduce the dangerously high arterial pressure which precedes strokes and heart failure were substances which blocked the actions of acetylcholine at these ganglia, and so prevented the passage of nervous messages which put up the blood pressure.

A series of synthetic compounds, with a long pedigree going back to the work of Crum Brown and Fraser in the 1860s (see chapter 2), were produced simultaneously by R. B. Barlow and H. R. Ing in the Pharmacology Laboratory at Oxford[7] and by W. D. M. (later Sir William) Paton and Eleanor Zaimis at the National Institute for Medical Research.[8] Some of them are potent ganglion blocking agents[9] and were introduced into clinical medicine, but they had grave disadvantages. Drugs of this sort interfere with the moment-to-moment control of blood pressure. All is well as long as the subject is lying down, but when he stands up, the reflexes which normally make arteries all over the body contract to meet the new hydrostatic conditions no longer operate effectively, and a sharp and dangerous fall of blood pressure follows. If the dose is adjusted carefully, and changes of posture are made slowly, it is possible to get about, and so the compounds are not unuseable. When they were given to patients with high blood pressure, the immediate results were encouraging. During the 1950s, several such compounds were introduced, and they were used extensively. But the search for better ganglion-blocking drugs was unfruitful. All the drugs weakened the fine control of blood pressure, so that patients were liable to faint if they stood up too suddenly. Control of other organs was also impaired, and led to difficulties in focusing the eyes, to abdominal distension when the muscles of the gut failed, and to impotence. They were not drugs which made patients happy.

More selective agents were sought, although this was theoretically unpromising. If the receptors for acetylcholine were all the same, agents which blocked any of them were expected to block all of them. However, detailed analysis of the ways in which receptors are blocked revealed many unforeseen complications. New substances were synthesized which acted rather like the ganglion-blockers and lowered blood pressure, but showed distinct and important differences. More selective compounds did emerge, and some of them continue to play an important role.[10–12]

Acetylcholine, adrenaline, and other hormones with both general and local effects, were originally discovered (see chapters 4 and 5) by examining extracts of tissues of the body for substances which showed physiological effects. Many scientists were interested in other tissue extracts, several of which were known to affect blood vessels[13] but were not identified until after 1945. It was often difficult to guess which ones had important physiological roles and which were by-products or artefacts formed when tissues were damaged during extraction. The pursuit of their properties was intellectually exciting but something of a gamble if the aim was to discover a drug for a specific purpose.

For example, a substance which lowered the blood pressure was first discovered in 1934 in extracts of the prostate gland and in semen.[14] It was called prostaglandin, but in time was found to be a mixture of many related substances with very complex chemical and biological properties. In the 1960s the chemical structure of one of these substances and subsequently of many other members of the family were established.[15] More people became interested and the field expanded rapidly once the exact composition of the prostaglandins was known. They seemed to have many roles, in reproduction, in blood vessels, in certain enzyme systems, and in the activity of nerves.[16] An unexpected but intriguing discovery was that aspirin and related pain-relieving drugs stop the formation of some prostaglandins.[17] This discovery gave fresh impetus to research aimed at developing new drugs which, like aspirin, would relieve pain and control inflammation, and also it provided a new basis for testing candidate compounds. It became clear that prostaglandins have a prominent role in tissue inflammation, as well as in such diverse functions as blood clotting and contraction of the uterus. Thus the original study spread into wider fields.[18] The original observation that an extract containing prostaglandins lowered blood pressure was a good reason for further investigation, but did not mean that it would lead to an agent for treating high blood pressure.

Indian snake-root

It is perhaps surprising that plant tissues contain substances which affect animals, but plants have always been a useful source of drugs, and indigenous remedies from all parts of the world were studied when research interests expanded after 1945. Most of them proved disappointing because no activity could be demonstrated, but one Indian medicine proved to be important. *Rauwolfia serpentina*, or snake-root, is a climbing shrub long used by the inhabitants of India and neighbouring countries as a sedative, and for other medicinal purposes. During the 1930s several alkaloids were prepared from the plant and investigated,[19] but their properties did not correspond to the sedative and hypotensive properties of the crude drug.[20] Another alkaloid, named reserpine, was isolated in 1952 in the Ciba laboratories in Basle.[21] It fascinated pharmacologists, mainly because it displaced stores of adrenaline-like transmitter substances both from the brain and from other tissues (see chapter 13). Among the consequences was a fall in blood pressure, particularly when the drug was given to patients with a high pressure. Reserpine became the focus of further clinical studies.[22] Its sedative effects were valued, but sometimes progressed to pathological depression with suicidal tendencies, so its use was limited.

Rauwolfia is often quoted as an example by those who think that more attention should be paid to indigenous plant remedies. However, one must remember that the number of plants reputed to relieve one condition or another is enormous, and that most of them have not stood up to critical investigation. Substantial sums have been spent in the deliberate search for medicinal plants, and they have seldom been rewarded. It must be remembered, too, that plants which contain medicinally active substances often do not contain them all the time in all their parts. Substances may accumulate and disappear according to the functions which they serve and the stages in the plant's life cycle. Cultivating such plants on a sufficient scale for worldwide needs may be difficult in primitive surroundings where the plant grows naturally and impossible elsewhere. Aesthetic and romantic ideas of the beneficent properties of 'natural' remedies do not always stand up to the realities of practical life.

Receptor for adrenaline

Although the nerve pathways from the brain to the blood vessels and heart use acetylcholine as their transmitter at ganglia, they end in terminals which release an adrenaline-like substance, so a possible way of controlling blood pressure is to block the action of this substance. No suitable drugs for this purpose were known before 1950, and the exact identity of the transmitter itself presented difficulties. It had been established as a close relative of adrenaline,[23] named noradrenaline, late in the 1940s. But a rather elaborate hypothesis emerged, particularly popular in the USA, which involved excitatory and inhibitory 'Sympathins' which were thought to combine with the transmitter to produce the active mediators (see chapter 4).[24]

This theory dampened interest in new evidence and new interpretations.[25] In 1948 Raymond Ahlquist, then at the University of Georgia, made careful measurements of the effects of adrenaline and closely related compounds on several functions of the sympathetic nerves.[26] His work suggested that there were two kinds of receptor for adrenaline and adrenaline-like substances. His conclusions were far reaching and, although the experimental evidence was sound, it was not thought to be sufficiently convincing to challenge the ideas about sympathin. As Ahlquist wrote later,[27] 'The original paper was rejected by the *Journal of Pharmacology and Experimental Therapeutics*, was loser in the Abel Award competition, and finally was published in the *American Journal of Physiology* due to my personal friendship with a great physiologist, W. F. Hamilton'. Little attention was paid to it, and 10 years passed before the existence, let alone the exact functions of these receptors, non-committally named alpha and beta, was recognized.

In common with many other groups, researchers at the Eli Lilly Laboratories in Indianapolis were ringing the changes on the structure of adrenaline,[28] with the initial objective of developing a compound which, like adrenaline, would relax the bronchial muscle and so be valuable in treating asthma. A good compound would not have the unwanted actions of adrenaline on the heart and blood pressure, and would go on acting for longer. One drug with these properties, isoprenaline, was already well known.[29] Among the substances produced at Eli Lilly, some not only relaxed the airways but, unexpectedly, made the tissue less sensitive to adrenaline. It appeared that these substances not only stimulated the adrenaline receptor but remained attached to it and blocked it. A blocker could be useful in several ways, and a compound (dichlorisoprenaline) was developed accordingly. As it turned out, it blocked just those actions of adrenaline which Ahlquist had grouped with the beta receptors, and it had no appreciable effect on the alpha receptors. Unfortunately it could not be used clinically. However, this discovery greatly strengthened the hypothesis that there *were* alpha and beta receptors, and the idea at least became widely accepted.

Among the actions which Ahlquist had attributed to beta receptors was the stimulation of the rate and force of the heartbeat. It was known that stimulation caused the heart to increase its oxygen consumption. J. W. (later Sir James) Black, who was at that time an academic pharmacologist in Glasgow, had the idea that blocking the action of adrenaline on the heart might reduce its need for oxygen and so be beneficial in angina pectoris (a condition in which pain arises because the oxygen supply to heart muscle is insufficient). When Black moved to the Imperial Chemical Industries Pharmaceutical Laboratories near Manchester he began a 4-year search for suitable compounds. Eventually a compound was identified which met the specification. Named Nethalide or pronethalol, it was the first effective clinical beta-blocking agent.[30]

Pronethalol had only just come into clinical use when it was found to produce tumours in mice. As this suggested that it might also be a cause of cancer in man, it was promptly withdrawn, to be succeeded by a more acceptable agent, propranolol. Many other beta-blockers were subsequently developed by different companies, because the market was very large. Clinical evidence began to accumulate, suggesting that the new drugs had a wider range of useful activities than had been predicted from experiments in animals. They lowered blood pressure in patients with hypertension,[31,32] prevented irregularities of the heartbeat which might be caused by adrenaline, reduced mortality after coronary thrombosis,[33,34] and, perhaps most importantly of all, they revealed the complexity of the factors which control cardiac activity in man, and the difficulties of predicting the effects drugs will have without first carrying out extensive and detailed experiments, both in the laboratory and in the clinic.

Histamine, another active substance first extracted from tissues (see chapter 4) is not involved in the treatment of high blood pressure, but the study of receptors has been extremely fruitful in exploring its properties and discovering other drugs. Substances that blocked the actions of histamine were discovered from the late 1930s[35] onwards, and by the end of the 1940s were well known as 'antihistamines'. They were useful for treating allergic disorders and also as sedatives and remedies for motion sickness. However, these antihistamines did not block all the actions of histamine, and, in particular, they did not affect its ability to stimulate the stomach to secrete acid. When substances were discovered which did block these actions[36] a second histamine receptor was postulated and the search began for clinically acceptable drugs. Black, by this time at the laboratories of Smith, Kline, and French, led a team which achieved results no less interesting than the discovery of beta-blockers, and of equal practical importance. Their work culminated in the drug Tagamet or cimetidine, soon widely adopted as the most effective remedy yet discovered for gastric overacidity and gastric and duodenal ulcers. For his achievements in the design and discovery of drugs, Sir James Black was to share a Nobel prize with Doctors Hitchings and Elion (see chapter 11) in 1988.

Salt and water drugs

After Ahlquist made his first observations on the alpha and beta receptors, and before they had any practical development, another series of drugs appeared which, somewhat unexpectedly, contributed to the treatment of hypertension. They were discovered in a search for diuretics, i.e. drugs which increase the volume of urine, and were not the first diuretics to have additional valuable properties. Withering's interest in digitalis was aroused primarily because it promoted a flow of urine and relieved swollen, dropsical tissues of their load of water (see chapter 1). But here the main action of digitalis is on the heart, and the removal of excess water is a consequence of improving the circulation of blood and the transport of water.

Other diuretics have been known for a long time, especially arsenical compounds of mercury. Early in the twentieth century, when the success of Ehrlich's organic compounds had been recognized, organic mercurial compounds were tested as treatments for syphilis. The rationale was simple: if arsenic could be made into a better drug by incorporating it into an organic compound, why not mercury? Although the results were not striking, organic mercurial drugs began to be widely used. Clinical measurements were becoming popular about this time, and many interesting things were being discovered as a result of this more quantitative approach. A. Vogl[37] wrote a charming account of how, as a young doctor on the wards of a Vienna hospital, he discovered that patients passed very large volumes of urine after they had received injections of one of the new organic mercurial compounds. He described how difficult it was to interest his seniors in this useful discovery. However, it did eventually become well known, and for 20 years organic mercurials were the most potent diuretics in clinical use. However, they were effective only when injected, and something better was desirable.

In 1957 a better diuretic was found, but only after a logical but circuitous investigation, described in detail by K. H. Beyer of Merck, Sharpe, and Dohme.[38] On the way his team investigated an undesirable property of some sulphonamides, which were liable to crystallize in the kidneys and cause serious damage in consequence; the enzymes inhibited by sulphonamides and by compounds containing mercury; and the normal working of the kidney. Luckily they chose experimental animals which happened to include a species sensitive to one of the new agents under trial. If rats, traditionally used for studies on urine flow, had not been replaced by dogs, just because larger animals were needed, the new drug, chlorothiazide, would have been missed, since its activity in rats is insignificant. Chlorothiazide is the forerunner of a large class of similar drugs, which have become known as the thiazide diuretics. They provide an effective way of eliminating excess water and salt, and are very safe when properly used. Once the research costs had been met, they were not particularly expensive. Organic mercurials were completely displaced.

Heart failure is common in patients with high blood pressure, and so these patients were often given thiazides. It soon became evident that the thiazides had an antihypertensive effect of their own and that they augmented the effect of most other antihypertensive drugs. The later thiazides have comparable activities, and have come to be a mainstay of therapy in mild hypertension, and important contributors in more severe conditions. Like the beta-blocking drugs which came into clinical use later, their effects were not predicted but were undoubtedly useful. So the circuitous route to the new diuretics had a final twist: the drugs were not only diuretics but antihypertensives as well.

Evaluation

The drugs introduced between 1950 and 1965 came to dominate the treatment of hypertension. Before 1950 there were no generally recommended drugs to lower the blood pressure. Twenty years later at least four major families of drugs were widely used, as well as a considerable number of supporting agents which were believed to help some patients in particular circumstances. What was achieved by the great expansion of research which produced these drugs and the clinical innovation which adopted them so freely?

There is no doubt that these potent drugs caused many problems. Once it became possible to lower a patient's blood pressure substantially, it was not merely of academic interest to know whether the raised blood pressure was really 'essential', that is necessary to maintain a sufficient flow of blood through narrowed arteries to vital organs, and, if it was, whether modest treatment (to ward off the obvious hazards of strokes or heart failure) could be tolerated. It became important to distinguish a raised blood pressure caused by an unknown progressive disease from one associated with anxiety, stress, and fear, particularly fear of medical treatment and its implications ('white coat hypertension').[40] And above all, hard facts were needed about all the consequences of any of the regimes of treatment which could be adopted. If the use of a drug seemed appropriate, what dose should be used, and for how long should it be continued? Although there is no fundamental objection to life-long medication—many diabetics rely on daily injections of insulin with no prospect of stopping—no drug is without hazard, and the longer the treatment continues, the stronger the case should be for it. All this made evaluation the new drugs very difficult. It had to take place over a long period, in the face of a prodigious number of uncontrolled variables which were likely to confuse the outcome.

For patients with severe hypertension and an expectation of life of months or at most a year or two, the benefits of lowering the blood pressure were soon seen. Clinical trials in which comparable patients received either one kind of treatment or another confirmed this belief. But when studies were extended to the less seriously ill, uncertainty prevailed. Patients lived for many years, and some died from causes unrelated to their blood pressure. The effect of the drugs on their length of life was very difficult to judge, and small differences could be attributed to many incidental causes, such as differences in smoking, dietary habits, the amount of exercise they took, or to changes in all these habits over the years. Not least among the 'incidental' factors was the probability that a particular treatment might be so disagreeable that it was abandoned, or not used often enough. 'Compliance' received increasing attention, and the phrase 'quality of life' began to appear. Therapeutic trials were undertaken to investigate as many of these factors as was feasible, and necessarily involved many patients and many doctors, nurses, secretaries, and controllers.

Some doctors became rather sceptical, and some began to protest that the multiplication of therapeutic trials was diverting resources away from more basic research.[41] However, valiant attempts were made to conduct massive trials in which as much information as possible could be included.[42] The results, obtained at considerable cost, were full of interesting facts but they were not always very conclusive. Beliefs which had become established almost as folk-law, such as the universally harmful effects of too much salt, had less support than might have been expected.[43] But the practical necessities of treating patients remained, so the majority of physicians continued to use as many of the available remedies as they thought necessary, relying on their personal experience, the opinions of colleagues, and the reports of such trials as they had time to study. Most, although certainly not all, of the remedies of 1840 were innocuous, and most had certainly not been tested in any formal way. Those of today have always had some evaluation, but seldom enough to cover all the circumstances in which they may be used, and they include many of such potency that their misuse can have dire consequences. Thus although much has been achieved, it would be a mistake to believe that most of the problems have been solved.

Hypertension is but one among many diseases, and progress has been made similarly in developing new drugs for regulating or adapting other parts of the body: drugs to strengthen the heart, to promote formation of blood, to help blood to clot or to prevent blood from clotting, to aid respiration, to increase the flow of urine, or selectively to increase or to diminish the amount of some selected component, to prevent conception or to promote fertility, drugs to stimulate the production of hormones or to block their actions, drugs to influence some particular aspect of the metabolism, and so on.[44–46] To describe all these discoveries is beyond our scope.

Taking a less detailed view, the patterns of discovery do not differ greatly. New knowledge of the basic physiology and biochemistry of the body provides new ideas for developing drugs. Sometimes chance discovery directs research onto new and profitable lines. Often there are discoveries which lead nowhere and do not attract much, or indeed any attention. Some may be seen 20 or more years later to have been seminal but were ignored because they did not fit into the theories current at the time. Many are forgotten, buried in old scientific journals or the archives of individual pharmaceutical companies. Attempts to rescue and re-evaluate them are apt to be less rewarding than making a fresh start.

Drugs which act on the brain are comparable in many ways to any other drugs. Some of them modify mental as well as bodily functions and have effects beyond the repertoire of conventional laboratory experiments in pharmacology. They have provided remedies for the treatment of some kinds of insanity, and some of them alter conscious experience in ways which are of great philosophical interest. They are discussed in the next chapter.

Notes

1. Whitteridge, G. (1971). *William Harvey and the circulation of the blood*. Macdonald, London.
2. Hales, S. (1733). *Statical Essays: containing haemastaticks; or, an account of some hydraulick and hydrostatical experiments made on the blood and blood-vessels of animals. Vol. II*. Royal Society, London.
3. Clark-Kennedy, A. E. (1929). *Stephen Hales, D.D., F.R.S., An eighteenth century biography*. Cambridge University Press, Cambridge.
4. Booth, J. (1977). A short history of blood pressure measurement. *Proceedings of the Royal Society of Medicine* 77, 793–9.
5. Fishberg, A. M. (1954). *Hypertension and nephritis*, 5th edn. Ballière, Tindall and Cox, London.

6. Moyer, J. H. (ed) (1959).*Hypertension. The first Hahnemann symposium on hypertensive disease*. Saunders, Philadelphia.
7. Barlow, R. B. and Ing, H. R. (1948). Curare-like action of polymethylene bisquaternary salts. *Nature* 161, 718.
8. Paton, W. D. M. and Zaimis, E. J. (1948). Curare-like action of polymethylene bisquaternary salts. *Nature* 161, 719.
9. Paton, W. D. M. and Zaimis, E. J. (1952). The methonium compounds. *Pharmacological Reviews* 4, 219–53.
10. Green, A. F. (1982). The discovery of bretylium and bethanidine. *British Journal of Clinical Pharmacology* 13, 25–34.
11. Maxwell, R. A. and Wastila, W. B. (1977). Adrenergic neuron blocking agents. In Antihypertensive agents. *Handbook of Experimental Pharmacology, Vol. 39*, ed. F. Gross, pp. 161–261, Springer-Verlag, Berlin.
12. Boura. A. L. A. and Green, A. F. (1981). Historical perspective. Noradrenergic neurone blocking agents. *Journal of Autonomic Pharmacology* 1, 255–67.
13. Gaddum, J. H. (1936). *Vasodilator Substances of the Tissues*. Fiftieth Anniversary edition with notes by F. C. McIntosh. (1986). Cambridge University Press, Cambridge.
14. Euler, U. S. von (1934). Zur Kenntnis der pharmakologischen Wirkungen von Nativsekreten und Extrakten männlicher accessorischer Geschlechtsdrüsen. *Archiv für Experimentelle Pathologie und Pharmakologie* 175, 78–84.
15. Bergstrøm, S., Ryhage, R., Samuelson, B., and Sjøvall, J. (1963). Prostaglandins and related factors. 15. The structure of prostaglandin E_1, $F_{1\alpha}$ and $F_{1\beta}$. *Journal of Biological Chemistry* 238, 3555–3564.
16. Horton, E. W. (1969). Hypotheses on physiological roles of prostaglandins. *Physiological Reviews* 49, 122–161.
17. Ferreira, S. H., Moncada, S., and Vane, J. R. (1973). Prostaglandins and the mechanism of analgesia produced by aspirin-like drugs. *British Journal of Pharmacology* 49, 86–97.
18. Vane, J. R. (1982). Prostacyclin: a hormone with a therapeutic potential. The Sir Henry Dale Lecture for 1981. *Journal of Endocrinology* 95, 3P–43P.
19. Chopra, R. N., Gupta, J. C., and Mukherjee, B. (1933). The pharmacological action of an alkaloid obtained from *Rauwolfia serpentina* Benth. A preliminary note. *Indian Journal of Medical Research* 21, 261–71.
20. Chopra, R. N., Gupta, J. C., Bose, B. C., and Chopra, I. C. (1943). Hypnotic effect of *Rauwolfia serpentina*: the principle underlying this action, its probable nature. *Indian Journal of Medical Reearch* 31, 71–4.
21. Müller, J. M., Schlittler, E., and Bein, H. J. (1952). Reserpin, der sedative Wirkstoff aus *Rauwolfia serpentina* Benth. *Experientia* 8, 338.
22. Woodson, R. E., Youngken, W. H., Schlittler, E. and Schneider, J. A. (1957). *Rauwolfia: botany, pharmacognosy, chemistry and pharmacology*. Little, Brown, Boston and Toronto.
23. Euler, U. S. von. (1946). A specific sympathomimetic ergone in adrenergic nerve fibres (sympathin) and its relations to adrenaline and nor-adrenaline. *Acta Physiologica Scandinavica* 12, 73–97.
24. Cannon, W. B. and Rosenblueth, A. (1933). Studies on conditions of activity in endocrine organs. XXIX. Sympathin E and Sympathin I. *American Journal of Physiology* 104, 557–74.

25. Bacq, Z. M. (1975). *Chemical transmission of nerve impulses. A historical sketch*, pp. 44–6. Pergamon Press, Oxford.
26. Ahlquist, R. P. (1948). A study of the adrenotropic receptors. *American Journal of Physiology* 153, 586–99. For an appreciation of Ahlquist's work see Black, J. W. (1976). Ahlquist and the development of beta-adrenoceptor antagonists. *Postgraduate Medical Journal* 52, suppl. 4, 11–13.
27. Ahlquist, R. P. (1973). Adrenergic receptors: a personal and practical view. *Perspectives in Biology and Medicine* 17, 119–22. Ahlquist himself here referred to the alpha and beta receptors as an 'abstract concept conceived to explain observed responses of tissues produced by chemicals of various structure' and added 'It is now my opinion that there is only *one* adrenergic receptor. Designed to respond to epinephrine, the alpha-ness or beta-ness of this receptor is determined by its physiological environment.'
28. Powell, C. E. and Slater, I. H. (1958). Blocking of inhibitory adrenergic receptors by a dichloro analog of isoproterenol. *Journal of Pharmacology and Experimental Therapeutics* 122, 480–8.
29. Konzett, H. (1940). Neue broncholytisch hochwirksame Körper der Adrenalinreibe. *Archiv für experimentelle Pathologie* 197, 27–40.
30. Black, J. W. & Stephenson, J. S. (1962). Pharmacology of a new adrenergic beta blocking compound (Nethalide). *Lancet* ii, 311–13.
31. Prichard, B. N. C. and Gillam, P. M. S. (1964). Use of propranolol (Inderal) in the treatment of hypertension. *British Medical Journal* ii, 725–7.
32. Veterans Administration Cooperative Study Group on antihypertensive agents (1977). Propranolol in the treatment of essential hypertension. *Journal of American Medical Association* 237, 2303–10.
33. Green, K. G. (co-ordinating secretary) (1975). Improvement in prognosis of myocardial infarction by long term beta-adrenotropic blockade using practolol. A multicentre international study. *British Medical Journal* iii, 735–40.
34. Julian, D. G., Chamberlain, D. and Baber, N. S. (eds) (1982). Symposium. Beta blockade and myocardial infarction. *British Journal of Clinical Pharmacology* 14, 1S–64S.
35. Tréfouel, J., Tréfouel, Mme. J., Bovet, D. and Nitti, F. (1946–47). The contribution of the Institut Pasteur, Paris, to recent advances in microbial and functional chemotherapy. *British Medical Bulletin* 4, 284–9.
36. Ash, A. S. F. and Schild, H. O. (1966). Receptors mediating some actions of histamine. *British Journal of Pharmacology* 27, 427–39.
37. Vogl, A. (1950). The discovery of the organic mercurial diuretics. *American Heart Journal* 39, 881–3.
38. Beyer, K. H. (1977). Discovery of the thiazides: where biology and chemistry meet. *Perspectives in Biology and Medicine* 20, 410–20.
39. O'Brien, E. and O'Malley, K. (1988). Overdiagnosing hypertension. *British Medical Journal* 297, 1211–12.
40. Pickering, T. G. *et al.* (1988). How common is white coat hypertension? *Journal of the American Medical Association* 259, 225–8.
41. Goldring, W. and Chasis, H. (1966). Antihypertensive drug therapy: an appraisal. In *Controversy in internal medicine* (eds F. J. Ingelfinger, A. S. Relman, and M. Finland), pp. 83–91. Saunders, Philadelphia.
42. Levy, R. I. and Sondik, E. J. (1978). Initiating large-scale clinical trials. In *Issues in research with human subjects*. U.S. Department of Health, Education, and

Welfare, NIH publication no. 80-1858. (A useful account of the difficulties of such trials).

43. Swales, J. D. (1988). Salt saga continued. *British Medical Journal* 297, 307–8.
44. Binden, J. S. and Ledniger, D. (eds) (1982). *Chronicles of drug discovery*, Wiley, New York.
45. Mann, R. D. (1984). *Modern drug use: an enquiry on historical principles*. MTP Press, Lancaster.
46. Sneader, W. (1985). *Drug discovery: the evolution of modern medicines*. Wiley, Chichester.

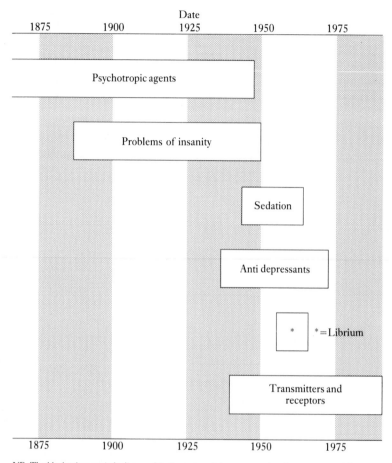

Chronology of Chapter 13

NB The blocks show periods discussed in the text, and do not mean that there was no activity about the subject at other times.

13

Drugs and the mind

Poppy or mandragora

The earliest human records contain evidence of the use of substances which alter mood or behaviour. At least four have been in use for so long that we have no idea when they were discovered. Alcohol, opium, cannabis, and tobacco have been known for centuries and the history of each fills many books. These drugs are still widely used, and for many people one or more of them is a regular part of everyday life. None of them is harmless, and all can present problems when they are abused. Strong social pressures often support or repudiate their use, and sometimes the pressures lead to control or prohibition by governments.[1,2]

Alcohol is the most widespread drug in use and was probably the first to be discovered. Written records attributed to the Hittites of Asia Minor of 1500BC contain references to wine, and its use is familiar from Homeric legends onwards. Opium was probably known in Roman if not in Greek times, and was used to promote sleep as well as to relieve pain. Cannabis comes from hemp, which was among the first plants to be domesticated, especially in parts of South-East Asia, where it is a valuable source of fibre for making rope and twine. The plant, *Cannabis indica*, yields a resin known by various names including marihuana and hashish. Its use is recorded by the Greeks and Romans, and by the Chinese.[3,4] One can only guess how these drugs were originally recognized, though it is easy enough to imagine how the fermented juice of squashed grapes began to be enjoyed. The first use of tobacco is also unrecorded, but the plant was known to American Indians and was brought to Europe in 1558 by Spanish explorers. The practice of smoking the prepared leaves of the plant was popularized in England by Sir Walter Raleigh (1552–1618).

It is widely agreed that such psychoactive plants or preparations from them were much used in religious ceremonies, where their effects were taken as evidence for religious reality; visions of Paradise which could be reached only by the faithful. At the festivals of Dionysus, or Bacchus, wine was central to the proceedings. The experience of a mind-moving drug offered, and still offers, an agreeable escape from the monotony of life. Medicinal properties, whether real or supposed, were gradually attributed to these substances. But the ill-effects, especially those that developed slowly, took much longer to be identified. Cirrhosis of the liver was recognized in the eighteenth century as a consequence of too much drinking. Lung cancer was not seen as a sequel to smoking until the middle of the twentieth. Dependence on each of these drugs was also recognized rather slowly and perhaps reluctantly. Even in the late nineteenth century opium could be purchased readily in some pharmacies in England.[5]

Explorers of distant countries found other plants which produced curious mental effects. They included the soothing mandragora, the energizing coca leaves (see Chapter 6), and the givers of disordered visions, the Mexican cactus *Peyote* and certain mushrooms of the *Amanita* and *Psilocybe* families. On the other hand, a lot of folk-lore about plants was investigated to no avail.[6]

An understanding of the substances which produced these curious effects came with the advance of chemistry in the eighteenth and nineteenth centuries. Distillation liberated the spirits from fermented grains and fruit juices, and in time ethyl alcohol was purified. Lavoisier established that carbon, hydrogen, and oxygen were the sole constituent elements of this alcohol, and his successors determined its exact structure as $CH_3.CH_2OH$. Much was learnt, and much remains to be learnt, about the innumerable other constituents which give particular qualities to particular beers, wines, and spirits. But identifying the principal active ingredient of alcoholic drinks was a vital step forward in that it allowed the quantity of intoxicant to be measured accurately, always the first step in any proper study of drugs and their actions.

Doctors and philosophers were beginning serious studies, first of the effects of the crude mood-and-mind drugs, and later of purer principles. In 1883 the German physician Emil Kraepelin (1856–1926), who was very influential in psychiatric circles, suggested that studying the intoxicating effects of hashish might aid the understanding of other kinds of mental disorder. The active principles of cannabis were not purified until well into the twentieth century, but one of Kraeplin's pupils was among those who made early studies of mescaline,[7] isolated from the Mexican *Peyote* cactus in 1896. The chemical structure of mescaline was established in 1918 and was found to be closely related to that of adrenaline, which leads to some interesting speculations. The drug is still occasionally used experimentally by scientists, psychiatrists, and philosophers, as well as by dilettante drug takers.[8]

New chemicals synthesized by man were found sometimes to have effects on the mind. The effects ranged from the hilarity and confusion produced by laughing gas (nitrous oxide in less than anaesthetic doses) to a simple diminution of the sense of pain by fever-reducing drugs of the coal-tar dyestuffs industry (see Chapter 2). The origin of these substances did not suggest that they would have magical or mystical properties, and so began to dispel ideas that psychotropic drugs had any metaphysical significance. William James, in his study *Varieties of Religious Experience*, discussed the effects on consciousness of alcohol and of anaesthetics, including nitrous oxide, without distinguishing sharply between 'naturally occurring' and man-made substances.[9] Pharmacologists also adopted a down-to-earth approach. Buchheim, professor at Dorpat in the 1860s, is said to have rated the botanical characteristics of drug-containing plants as no more important to medicine than the case in which a surgeon carried his scalpels was to surgery. But exclusion of the mystical did not advance knowledge very far. The human brain was very inaccessible to any sort of experimental investigation. Little progress was made in this field before the 1950s, but in the 1940s a particularly notable drug came to light unexpectedly.

Once again ergot (see Chapters 4 and 6) was involved. One day in April 1943 Albert Hoffmann, a chemist who was working at Sandoz on the development of ergot alkaloids, felt unwell and went home early, where for some hours he experienced a variety of disordered visions. He guessed that these effects might be due to intoxication by one of the compounds he was working with, and confirmed his idea by deliberate experiment. The substance was identified and named lysergic acid diethylamide (in German, Lysergische Säure-Diäthylamid, or LSD). It became one of the best known 'psychotogenic' drugs.[10] Like its predecessors, it attracted attention from many sources, and in the liberal or liberated social climate of the post-war years it probably created more problems than it solved. Some of its pharmacological properties are of great interest to scientists trying to understand how the brain works.

Problems of insanity

In spite of the interest which Kraepelin and others had taken in them, strange mental states produced by drugs did not help much towards understanding the causes of or finding treatments for insanity. The classical approach, through post-mortem examinations, also gave little benefit. Indeed, the search for anatomical abnormalities seemed to create a division between conditions, such as haemorrhage into the brain or a cerebral tumour, where an organic cause of disease was obvious, and conditions where a patient was all too evidently suffering from a mental disturbance, but for which no physical basis could be found.

Attempts to classify mental disorders and give them names ran into many difficulties. No two patients were quite alike and objective criteria were scanty. Some 'patients' were simply of very low intelligence or 'feeble minded'. Some, often but not necessarily in early adult life, behaved peculiarly and described visions; they were said to have dementia praecox or, in more recent language, schizophrenia. Some had excessive swings of mood, and were either excessively energetic or elated, or, more often, inert and plagued by beliefs of their own worthlessness and guilt; the manic-depressive psychosis. Some had delusions of persecution or of grandeur. Some gradually lost their memory and all their skills as they aged, and were classified as suffering from senile dementia. Alzheimer described a similar condition beginning in middle life.[11] Most of these patients died for reasons unrelated to their strange behaviour, and such microscopic abnormalities as might be found in the brain seemed too obscure to account for the disturbance of the mind. The doctrine that mental disease did not have a physical cause became widely established, so much so that a study published in 1915 which reported definite changes in the left cerebral hemispheres of schizophrenic patients was largely ignored until it was rediscovered 70 years later.[12]

However, the borderline between those disorders which were classified as physical and those which were seemingly mental began to shift as knowledge advanced. One mental condition after another turned out to have a physical basis and, sometimes, to be alleviated when the physical defect was corrected. The apathy of patients with myxoedema was associated with failure of the thyroid gland and was overcome by treatment with the whole gland or one of its active principles (see Chapter 5). The 'feeble minded' children labelled cretins also suffered from thyroid deficiency, or, usually, from a deficiency of the iodine which is necessary for the gland to function effectively. They could be cured if they were treated in time with iodine, whole gland, or pure hormone.

Once the spirochaete which caused syphilis had been identified and Wassermann had invented the diagnostic reaction which perpetuates his name, the condition called general paralysis of the insane was recognized to be the result of a long-lasting infection. Means of relief were sought, and it was found that fever, produced artificially or by deliberate infection with malaria,[13] produced remissions. Effective anti-syphilitic drugs, the arsenicals and later penicillin, made prevention possible, though they could not effect a cure if damage to the brain had gone too far.

Another form of mental disorder, pellagra, was associated clinically with diarrhoea and dermatitis. It occurred mainly in communities with poor diets and was eventually identified as a deficiency disease which could be promptly and completely cured by supplying adequate food. Later the missing dietary ingredient was identified as nicotinic acid.[14]

Each advance transferred one more kind of 'mental' disorder to the category 'physical'. But until about 1950, the hard core of insanity, especially the schizophrenic and the manic-depressive psychoses, appeared to have no physical basis. The special nature of their illness resulted in patients being shunned socially and isolated in special institutions. As there was no effective treatment, the asylums were more custodial than therapeutic and did not usually cater for investigation and research. The great change which followed can be attributed partly to changes in social beliefs and attitudes in the 1950s and 1960s, and partly to the success of three strands of investigation, which began more or less independently but soon converged.

In university laboratories and research institutes, studies of nerve cells, and particularly of the transmission of impulses from one nerve to another, developed widely from the 1920s onward (see Chapter 4). The transmission at sites outside the brain by acetylcholine or by noradrenaline was established unequivocally (see Chapters 6 and 12) and later it was established that both these substances were present at many sites inside the brain and spinal cord. Whether they functioned as transmitters was difficult to prove: the technical problems of access to particular sites inside the skull, located in the middle of the dense networks of nerve cells and fibres which make up the brain, are still far from being completely solved. But the *idea* that this is how messages would pass from one nerve cell to another presented no difficulty. It implied that there would be ways of modifying it, just as peripheral transmission at ganglia could be blocked by methonium compounds and transmission at certain nerve endings by beta-blockers (see Chapter 12).

For those working in industrial pharmaceutical laboratories, the technical problems of designing drugs to act on the 'mind' were formidable. There were no good models in animals of human mental disorders, and no obvious way of devising them. If a new substance produced unusual behaviour in a mouse or a rat, it was worth further study, but this approach required skilful observation and took time. Batteries of tests were devised, and very large numbers of compounds produced by innovative chemists were studied. Like most screening operations, the yield was poor and the significance of the positive findings was limited. Occasionally, very occasionally, an oustanding discovery rewarded this speculative labour.

In hospitals a number of new drugs were being used. These drugs were intended to have specific effects and on the whole it was hoped that they would have as few other effects as possible. But sometimes unexpected things happened. Alert clinicians noticed surprising events and related them (not always correctly) to some new drug which was being used. The general ethos in western Europe and America favoured the publication of any novel findings, often far in advance of facts being properly established. So ideas passed rapidly from one institution to another and promising discoveries stimulated the development of new advances in many different quarters, sometimes with remarkably fruitful results.

The devious paths to tranquillizers

The traditional sedatives, laudanum, mandragora, bromide, paraldehyde, chloral, barbiturates, all had disadvantages and were gradually displaced. The antihistamines, which became well known late in the 1940s (see Chapter 12), turned out to have several properties which would not have been predicted from what was known about histamine. They reduced nausea and prevented vomiting, and many of them made people sleepy. Antihistamines became popular remedies for motion sickness, and a search, only marginally successful, began for effective compounds which did not cause drowsiness. Their sedative action naturally raised the possibility that histamine was a neurotransmitter. Little evidence supported this theory, and there was rather more reason to think that anti-histamines might be blocking noradrenaline receptors. As the anti-histamines were developed from early attempts to find adrenaline blockers, this was not entirely surprising.

One of the most powerful antihistamines, Phenergan (promethazine) was developed in the laboratories of the French firm Rhone-Poulenc. Many related compounds were made in the same laboratories and were investigated in more or less detail according to the promise they showed. The French psychiatrist Pierre Deniker, records that a substance closely related to promethazine and later known as chlorpromazine was syn-thesized in 1950 and 'would have remained on the shelves had the surgeon and physiologist Henri Laborit not asked the manufacturer for a drug with central effects stronger than those of promethazine'.[15] Laborit was investigating the potential of controlled hypothermia, that is, chilling patients during surgical operations so that they would be less reactive to any disturbance, and the sedative properties of promethazine helped to achieve the effects he sought. Several other groups, including some psychiatrists, began to investigate the drug chlorpromazine about the same time. Deniker lists fourteen publications in French journals between February and October 1952. It emerged that, in conscious and especially in excited or agitated patients, the drug produced a remarkable state of inactivity or indifference. Some authorities maintained that it was the first time a single drug had been shown to be useful in controlling psychotic patients. At the very least it was more selective than various predecessors which by that time had fallen into disrepute. Further investigation showed that it had a large number of actions: so large that the trade name Largactil was coined. The basic chemical structure common to promethazine and chlorpromazine provided a good basis from which to begin the search for compounds which were better than chlorpromazine, and in the next few years many came into clinical use.[15]

At about the same time, the alkaloid reserpine, extracted from the Indian plant *Rauwolfia serpentina*, was attracting interest for the treatment of high blood pressure (see Chapter 12). It had been used in India to control over-excitement and mania. As was soon discovered, it acts rather like chlorpromazine, but it is slower to take effect. Patients sedated by reserpine are liable to become depressed and have suicidal urges, and repeated large doses of the drug produce tremor and rigidity resembling Parkinson's disease, so its clinical usefulness is limited. But the discovery that it depleted the brain of certain transmitters made it a valuable tool in investigating brain mechanisms.

Quite by chance, and while he was working on a different problem, an Australian psychiatrist, J. F. Cade, discovered that, in guinea pigs, lithium salts made the brain less excitable. He tried using lithium salts to quieten manic patients.[16,17] As lithium salts were not amenable to being patented, they did not attract the attention of the drug industry: also they had a bad reputation for their toxic effects, and so Cade's findings were neglected for a long time. It was nearly 20 years before Cade's discovery was re-investigated and found to be a useful treatment for maniacal patients. It is not known how lithium salts work. They are mentioned here to illustrate that lines of thought and research do not always converge. A chance finding, outside the mainstream of scientific research, can sometimes be very valuable, but it is often difficult to see its potential.

Curare and its effects have been mentioned previously (see Chapter 6). Psychiatrists envisaged that curare might be useful in relieving muscular tension and so diminish the feedback from over active muscles which perpetuates a sense of anxiety in tense patients. Curare is inactive when taken into the body by mouth (which is why it is a good arrow poison for hunting game: any poison in the meat of the paralysed animal is not absorbed and does no harm to the eater). So curare itself was unsuitable, but a substance named mephenesin achieved the desired relaxation, by acting on the spinal cord rather than on the junction between nerve and muscle. Mephenesin was not a good drug: its effects were too short lived, and it had other unwanted effects. A derivative, meprobamate, appeared in 1953. Under the trade names Miltown and Equanil, it was widely advertised, especially in the United States, and became very generally known. Perhaps, more than any other drug, it promoted to the general public ideas about new 'tranquillizers', and did more to establish the questionable notion that anxiety could be banished with pills.[18]

So there were a large number of new drugs, all purporting to quieten disturbed patients, and very little means by which to evaluate them properly. In practice the phenothiazines and lithium proved to be of great importance in managing schizophrenia and mania, respectively, and in helping to restore many patients to everyday life. The others were more difficult to assess. How, indeed, were mental states to be measured, and how was a mental state to be judged to be lessened or heightened if it could not be measured, even crudely? How were clinical trials, so convincing for antituberculous drugs, to be brought into the sensitive area of subjective experience? Could a course of treatment be judged without reference to the accompanying activities of the doctors who arranged it? Could one, indeed, evaluate the effects of the doctors themselves, and, if not, what was the good of having them?

The new drugs were widely acclaimed, as so many other medical and surgical procedures had been. Critical examination by the most reliable of tests and measurements, usually showed that they had little if any action greater than that of placebos, or that their effects on performance were deleterious rather than beneficial.[19,20] But were the tests sensitive or relevant enough not to miss therapeutically important benefits? The evaluation of drugs had become more complex.

Relief of depression

Whatever benefits were conferred by the phenothiazines, they gave little help to patients who were suffering from excessive mood swings or who had entered into long periods of helpless depression. The most effective treatment before 1960 for severe and disabling depression was electro-convulsion therapy. This practice had a curiously devious origin, in a supposed antagonism between epilepsy and schizophrenia. It was inferred, by what seems now a rather wild speculation, that artificially produced fits might arrest the progress of schizophrenia. The idea was tested by using drugs which produced convulsions. In some schizophrenic patients, particularly those who were unresponsive to human approaches and showed other features of depression, the treatment appeared to work. When electrical stimulation was used instead of drugs, and was properly managed, regulation of the 'dose' was easier. The improved treatment was investigated for patients whose only disorder was severe depression, and found to give much benefit. The occurrence of convulsions carried risks of bone fractures and of other injuries, but ingenious psychiatrists encouraged the development of standardized preparations of curare which could be used clinically to prevent the muscular component of the electrically-induced seizures. 'Modified E.C.T.' became an established and very valuable practice by which countless depressed patients were restored to health. Until the late 1950s it was practically the only effective remedy for a most distressing disorder.[21,22]

The first effective antidepressants evolved from discoveries in a quite different field. When the new antituberculous drug isoniazid (see Chapter 10) was used, there was a striking improvement in the well-being of the patients. Their appetites improved; they became cheerful and they gained weight remarkably quickly. Another product of the same line of research, Marsilid or iproniazid, had similar effects. Initially there appeared to be little to choose between them, but careful comparative trials[23] showed that Marsilid made more contribution than isoniazid to the improvements in appetite and the weight gain. However, Marsilid also had more unwanted effects than its companion drug, and some patients experienced psychotic episodes. The interest of psychiatrists was aroused, and so was that of biochemists.[24]

The chemical structures of isoniazid and Marsilid were, somewhat distantly, related to that of several of the amines known to be present in living tissues. These amines are metabolized by an enzyme now named monoamine oxidase. Some of the amines which are decomposed by monoamine oxidase occur in the brain and are thought to function as neurotransmitters. When it was shown that both isoniazid and Marsilid inhibited the activity of monoamine oxidase and that Marsilid was considerably more potent, the possible implications aroused a great deal of interest. Did Marsilid act on the enzyme in the brain and protect a transmitter amine from premature destruction? If the amine was essential to nervous processes which made people alert and responsive, it could explain the more energetic behaviour of tuberculous patients and also the abnormal excitement which they sometimes showed. Marsilid was tried as a treatment for depressed patients, and, given over a period of weeks rather than days, but it had disadvantages as well as advantages.[25]

The next step was to devise more potent inhibitors of amine oxidase in the laboratory. This was a relatively straightforward exercise, because the chemical structures which served as the starting point were simple and the enzyme, which was easy to prepare, provided a much simpler test system than experiments involving whole animals. Of course, such potent compounds had still to be assessed in animals, before they could be used in patients. The most promising compounds were selected and studied in many laboratories, and in the next few years a substantial number of monoamine oxidase inhibitors went into clinical trials, and some into widespread clinical use.[26]

Opinions about the value of the new drugs varied, as they had about tranquillizers. The more scientifically rigorous the clinical trials, the more some observers judged that the trial itself had created an environment unfavourable to the effectiveness of the drug. Others commented that the subjective experience of patients was outside the range of what could be recorded and judged objectively. Apart from these difficulties it soon became only too obvious that the drugs could be unexpectedly toxic. Jaundice associated with liver failure appeared without warning and was sometimes fatal. Also, it was discovered that monoamine oxidase removed certain amines from the blood stream—amines derived from food, chemically related to adrenaline, and thus liable to raise the blood pressure. If patients being treated with monoamine oxidase inhibitors ate quantities of cheese or other foods rich in these amines they were prone to attacks of high blood pressure, sometimes sufficient to precipitate strokes or heart failure.

Fortunately another quite unrelated series of compounds was found to alleviate depression. The discovery arose from the investigation of substances related to chlorpromazine. They were all expected to have similar properties, but, prudently, in some clinics the new substances were given to patients with conditions other than those in which chlorpromazine and its allies were usually effective. So it was that R. Kuhn, in Switzerland, administered compounds which had been prepared in the laboratories of J. R. Geigy S. A. to patients who were withdrawn, inactive, and depressed. He observed that they improved.[27] According to Kuhn, '. . . thoroughness was not the only reason for doing this. There was also our conviction that it must be possible to find a drug effective in endogenous depression. This conviction arose from the literature study as well as from the great deal of experience we had acquired in the shock treatment of these depressions'.[28] The faith of the psychiatrist in whatever remedies he uses is an important contribution to, and perhaps sometimes the main reason for its effectiveness or its reputation, but this drug, later named Tofranil or imipramine, was found to be effective in many other clinics and came to be used on a wide scale. It was the first of a series of compounds which are known as the 'tricyclic antidepressants', because their chemical structure contains three rings and distinguishes them from the monoamine oxidase inhibitors.[29]

Both the monoamine oxidase inhibitors and the tricyclics were discovered empirically, but they aroused great hopes that the chemical disturbances in the brains of depressed people might at last be understood. Monoamine oxidase inhibitors would prevent the inactivation of noradrenaline and other amines thought to act as neurotransmitters: might they be making good a lack of one of these amines in the tissues? Quite quickly it was discovered that the tricyclics prevented the re-uptake of amines liberated at nerve endings, so they too were capable of increasing local concentrations of amines. It was easy to imagine depression as a condition in which the stores of certain amines became exhausted: then the monoamine oxidase inhibitors would prevent destruction of noradrenaline and conserve what was available, while the tricyclics prevented losses from the stores. Alas, it was not nearly so simple, and many more complex problems remained to be solved. But it was now clear that depression had a biochemical basis, even if its original causes lay in heredity and in the experiences of life.

Rescued from the remnants: a new range of tranquillizers

The search continued for drugs which would relieve anxiety, although in the absence of good models, approaches were bound to be somewhat speculative, and good luck also played its part. One can enjoy its contribution to the story of Librium or chlordiazepoxide, the first of the benzodiazepines.

A development programme was undertaken by chemists at the Roche laboratories in New Jersey during the 1950s.[30,31] With commercial considerations about patenting strongly in mind, they pursued unfamiliar compounds, not related to the barbiturates or other sedatives or tranquillizers then known. The wide range of existing tranquillizers had no common chemical features by which they could be identified, so this approach had some merit, particularly in inspiring chemists and setting them free to pursue interesting chemical problems with no biological constraints. To pick out compounds which might be potential drugs, a number of animal experiments were adopted, including the assessment of muscular relaxation and reduction of aggressive behaviour. The experiments gave positive results with meprobamate, at that time the most popular agent against anxiety in use in the USA.

One of the chemists concerned, Leo H. Sternbach, has described what happened. He chose as his novel substances the class of compounds on which he had worked 20 years earlier when he was a Ph.D. student at the University of Cracow, Poland. The compounds had not led, as he had then hoped, to new dyestuffs, and it appeared that no one had yet investigated their biological properties. At Roche he made some new compounds of the same class, but they were no more rewarding. The chemical problems were interesting, but the biological tests were all negative, and there were more urgent jobs to be done. The programme was therefore abandoned. In April 1957, when what remained of the work was being cleared up, some specimens which had been laid by to crystallize were found to have done so, and so they were sent for pharmacological investigation. it was supposed that, like their predecessors, they would be inactive and so would not waste much of the tester's time. Sternbach promised that no more of the series would be sent.[30]

However, one last compound was active, more so than meprobamate in quietening mice, preventing electrically-induced convulsions, relaxing muscles, and reducing aggressive behaviour in monkeys. The compound did not make animals sleep and very large doses could be given without causing death. So it received a full pharmacological and toxicological examination. No faults were found, and, as it behaved well in early clinical trials, it was duly marketed, under the name Librium. It was an astonishing success, in terms of the extent to which it was used, the absence of immediately obvious ill-effects, and the financial returns on its production. As usual it was the forerunner of a series of related drugs, and we may suppose Sternbach was allowed to withdraw his promise not to burden the pharmacologists.

The benzodiazepines rapidly became very popular. All the problems of evaluation remained, but the people who took them felt better, and the doctors who prescribed them were relieved to find a means by which unhappy patients could be helped. Depressed and anxious patients who sought refuge from their troubles by taking an overdose remained unconscious for a time but did not die of respiratory failure or asphyxiation, as happened only too readily after an overdose of barbiturate. The subtler aspects of their effects on behaviour were widely investigated but remained extremely difficult to evaluate. A decade passed before there was any general awareness that it is more difficult to stop taking benzodiazepines than to start, and that weaning dependent patients from their drugs could be a painful process.

Brain mechanisms and drugs

In spite of the continuing scepticism and regardless of the clinical faults of the new drugs, many attempts were made to find out how and where in the brain they worked. Progress was easier because many new techniques were available. By the 1950s and 1960s, delicate equipment for reaching and investigating single nerve cells could be built with new materials. Specialized apparatus, which formerly had been made for specific purposes in the laboratory where it was required, could be obtained from manufacturers of scientific instruments, saving much time and labour. Novel methods of analysis made it possible to do chemical separations in minutes instead of days or weeks. Physical methods of estimating substances, for instance by measuring the absorption or emission of light at specific wavelengths, all increased the power, speed, and sensitivity of investigations. Automatic equipment meant that many of the relatively routine tasks could be done more quickly and more accurately. Technically, the methods which Dale and his colleagues used had become archiac, as Dale (who lived to the age of 93 in full intellectual vigour) well knew. However, in principle, they were unchanged. Messages were passed from one cell to another by chemical messengers, and many drugs imitated, obstructed, or prolonged the activity of the messengers. The problem was to identify the messengers and find out what messages each one carried.

A few of the salient discoveries are shown in table 13.1.

Table 13.1 Drugs affecting neurotransmitter mechanisms

Drug	Transmitter affected	Result
Hyoscine	Acetylcholine receptors blocked	Drowsiness
Phenothiazines	Dopamine receptors blocked	Anxiety relieved, voluntary movement impaired overdose
Reserpine	Dopamine and other amine stores depleted	Depression Parkinsonian symptoms
Levodopa	Replenishes dopamine stores	Relief of Parkinson's disease
Monoamine oxidase inhibitors	Inactivation of amines diminished	Relief of depression
Tricyclic antidepressants	Re-uptake of amines diminished	Relief of depression
LSD	Serotonin receptors blocked	Hallucinations
Benzodiazepines	Gamma-aminobutyric acid re-uptake blocked	Anxiety relieved
Morphine	Opiate receptor activated	Pain suppressed

The transmitter, gamma-amino butyric acid, often referred to as GABA, first discovered from studies of inhibitory factors in the spinal cord, was found to be basic to the actions of the benzodiazepines. Another, 5-hydroxytryptamine or serotonin, presents some tantalizing questions. Its actions are blocked by the hallucinogen LSD. Does this mean that our mental images depend on this amine in some way? Are the disordered visions we call hallucinations a result of blocking serotonin? What experiments could be done to find out? The first transmitters discovered, acetylcholine and noradrenaline, are still being investigated. A precursor of noradrenaline, named dopamine, has attracted a lot of attention since it is involved in the actions of the phenothiazines, in Parkinson's disease, and quite probably in schizophrenia.

So basic research and the discovery of new drugs for mental disorder are almost inseparable. The academic scientist who seeks and identifies a new chemical messenger is providing his industrial colleagues with the basis for a whole new range of drugs, and the industrial pharmacologist who screens a range of new chemical entities and finds one which alters the behaviour of the brain is offering a new tool for his academic brethren. It is all a very long way from the days of the preposterous proposition, when the American Society for Pharmacology and Experimental Therapeutics forbade its members to accept jobs in the pharmaceutical industry (see Chapter 6).

New light on old drugs

None of these newly-discovered substances explained the effects of one of the oldest of all the drugs used to relieve anxiety and pain, morphine. Most of the potent drugs obtained from plants are known to act on specific physiological systems. Indeed, the very specificity of their actions has provided a crucial tool in understanding these systems, as Langley showed with nicotine (see Chapter 4). But morphine! Morphine relieves fear and dismisses pain. It has some effects, not obviously related, on coughing (for which it is a useful suppressant) and on the bowels (where it causes inconvenient constipation). It creates a sense of well-being so attractive that few can resist using it if they have been exposed to it too often. Can it be related to a system with a biological purpose?

However disagreeable pain may be, it is essential because it draws attention to bodily damage. Pain is usually allied to behaviour which will protect injured parts and provide better conditions for local recovery. Many thinkers have been troubled by the problems implicit in the occurrence of pain and suffering. But biologically, pain can be seen to have survival value, and the benefit of a mechanism which suppresses pain and cancels an important warning mechanism, is not immediately obvious.

But pain often persists after all steps have been taken to relieve the cause, and practical investigators looked for drugs which gave the benefits of morphine without causing addiction or other inconvenient effects. Codeine, which occurs with morphine in opium, relieves pain, but more feebly, and rarely causes dependence. Heroin, derived from morphine by a modest chemical change, is more potent and more strongly addictive than morphine itself. Many compounds, more or less related to morphine, have been prepared and tested, and a substantial list of alternative drugs provides possible substitutes with merits for particular purposes. Chemically, they differ considerably from morphine, but they can usually be seen to correspond to part of the morphine molecule, and are thus envisaged as acting on the same receptors.

Advances in understanding other receptors and in devising methods for isolating them gave a more fundamental approach to the morphine problem. In 1973 experiments were reported which described the isolation of portions of tissue containing something which combined specifically with morphine and other potent analgesics. [32-34] Plainly these were opiate receptors: but what were they there for? It went against all biological sense to suggest that the receptors and the elaborate physiological adaptations which they set in motion had no purpose, or had evolved in response to taking opium. There must be a naturally occurring substance which acts on the receptors, and, if so, it was likely that it was made and, when appropriate, let loose close to the opiate receptors, especially in the brain. It might be any sort of compound, but a lot of evidence was accumulating that peptides, i.e. compounds containing a chain or ring of amino acids, frequently occurred in specific parts of the brain and had specific functions as local or systemic hormones. Search for a peptide was rewarded by the discovery of a compound containing five amino acids, which combined powerfully with the opiate receptor and had all the necessary properties for it to be recognized as a new transmitter substance. It was named enkephalin [35,36] and its discovery was a major advance in the search for new analgesics. Later opium-like activity was found in a group of larger peptides, named endorphins, and of which enkephalin was shown to be a key fragment.

The developments which followed are too recent and perhaps too elaborate to belong in this essentially historical account of the discovery of drugs. The fact that a transmitter substance has been found does not mean that a new drug has been discovered. Indeed, transmitters, by their very nature, are seldom suitable as therapeutic agents. It is essential that a transmitter can be inactivated as soon as it has conveyed its message; if it were not so, the system would be clogged or would receive needless repetitions of the same message. A drug, on the other hand, must be stable enough to reach its target, and there it must last long enough to achieve the desired effect. Acetylcholine was identified as a transmitter in the 1920s and 1930s, and what was known of its biology and chemistry guided the search for new drugs in the 1940s and 1950s. So much greater was the pace of science in the 1970s, that, at a very crude estimate, what took 20 years for acetylcholine took 5 for the much less tractible opiate receptor and enkephalin. At the time of writing we have yet to see what new drugs will come from this foundation.

What we now know as a result of studies on the enkephalin–endorphin system has brought a new understanding of pain and of bodily responses to injury. Endorphins are released in response to severely painful stimuli. They over-ride the normal protective mechanisms associated with pain. Their survival value is clear: when severe injuries first happen it may be more important to get away at all costs than to stop and attend to them. The experience of soldiers in battle is well documented, and there are many accounts of other circumstances in which horrible wounds are accepted with numbness but not pain. It is also interesting to note that endorphins are released when acupuncture is practised [37] and perhaps make a significant contribution to the therapeutic effects which may be achieved.

Body and mind

The discoveries made in the last 30 years about the action of drugs on the brain are important medically and philosophically. It is beyond the scope of this book, and its author, to pursue the philosophical implications. But one may note the steady accumulation of evidence that all kinds of mental disorder have a physical basis, and that biochemical adjustment brought about by drugs often provides more effective help to a disturbed person than any other process. This is not to say that methods of biochemical adjustment are sufficient in themselves, or that a psychiatrist, however biochemically oriented, can neglect the pastoral aspects of caring for his patients. [38] It also seems inescapable that, however much conscious experience depends on the orderly biochemical working of the body, consciousness is itself outside the ordinary material concepts of matter and energy. The disturbances of consciousness that are produced by drugs are profoundly interesting, and they may help us to understand its nature. Or they may be no more than symptoms of a chemical fault in the consciousness-generator, and of no greater significance than that.

Notes

1. Parnham, M. J. and Bruinvels, J. (eds) (1983). *Discoveries in pharmacology. Vol. 1; Psycho- and neuro-pharmacology*. Elsevier, Amsterdam.
2. Leake, C. D. (1975). *An historical account of pharmacology to the twentieth century*. Charles C. Thomas. Springfield, Illinois.

3. Brunner, T. F. (1973). Marijuana use in ancient Greece and Rome. *Bulletin of the History of Medicine* 47, 344–55.
4. Kramer, J. C. and Merlin, M. D. (1983). The use of psychoactive drugs in the ancient old world. In *Discoveries in pharmacology. Vol. 1: Psycho- and neuro-pharmacology*, eds M. J. Parnham and J. Bruinvels, pp. 23–47. Elsevier, Amsterdam.
5. Lomax, E. (1973). Use and Abuses of Opiates in 19th Century England. *Bulletin of the History of Medicine* 47, 167–76.
6. Ayd, J. F. and Blackwell, B. (eds) (1970). *Discoveries in biological psychiatry*. Lippincott, Philadelphia.
7. Knauer, A. and Maloney, W. J. M. A. (1913). A preliminary note on the psychic action of mescalin, with special reference to the mechanism of visual hallucinations. *Journal of Nervous and Mental Diseases* 40, 425–36.
8. Mayer-Gross, W. (1951). Experimental psychoses and other mental abnormalities produced by drugs. *British Medical Journal* ii, 317–21.
9. James, W. (1904). *The varieties of religious experience. A study in human nature*. Longmans Green, London.
10. Stoll, W. A. (1947). Lysergsäure-diäthylamid, ein Phantastikum aus der Mutterkorngruppe. *Schweiz Archiv fur Neurologie und Psychiatrie* 60, 279–323.
11. McMenemy, W. H. (1970). Alois Alzheimer and his disease. In *Alzheimer's disease and related conditions*, Ciba Foundation Symposium, eds. G. E. W. Wolstenholme and M. O'Connor. Churchill, London.
12. Southard, E. E. (1915). On the topographical distribution of cortex lesions and anomalies in dementia praecox, with some account of their functional significance. *American Journal of Insanity* 71, 603–71. Quoted by T. J. Crow (1986). Left brain, retrotransposons, and schizophrenia. *British Medical Journal* 293, 3–4.
13. Wagner-Jauregg, J. (1927). The treatment of dementia parlytica by malaria inoculation. *Nobel lectures physiology or medicine, 1922–1941*, pp. 159–69. Published in 1964 for the Nobel Foundation, Elsevier, Amsterdam.
14. Goldberger, J. and Tanner, W. F. (1924). A study of the treatment and prevention of pellagra. *United States Public Health Reports* 39, 87–107.
15. Deniker, P. (1983). Discovery of the clinical use of neuroleptics. In *Discoveries in pharmacology. Vol. 1: Psycho- and neuro-pharmacology* eds M. J. Parnham and J. Bruinvels, pp. 163–80. Elsevier, Amsterdam.
16. Cade, J. F. J. (1949). Lithium salts in the treatment of psychotic excitement. *Medical Journal of Australia* 2, 349–52.
17. Cade, J. F. J. (1970). The story of lithium. In *Discoveries in biological psychiatry* eds J. F. Ayd and B. Blackwell, pp. 218–29. Lippincott, Philadelphia.
18. Domino, E. F. (1962). Human pharmacology of tranquillizing drugs. *Clinical Pharmacology and Therapeutics* 3, 599–664.
19. Foulds, G. A. (1958). Clinical research in psychiatry. *Journal of Mental Science* 104, 259–65.
20. Fox, B. (1961). The investigation of the effects of psychiatric treatment. *Journal of Mental Science* 107, 493–502.
21. Hamilton, M. (1986). Electroconvulsive therapy. Indications and contra-indications. *Annals of the New York Academy of Sciences* 462, 5–11.
22. Kalinowsky, L. B. (1986). History of convulsive therapy. *Annals of the New York Academy of Sciences* 462, 1–4.

23. Ogilvie, C. M. (1955). The treatment of pulmonary tuberculosis with iproniazid (1-isonicotinyl-2-isopropyl hydrazine) and isoniazid (isonicotinyl hydrazine). *Quarterly Journal of Medicine* 24, 175–89.
24. Davis, W. A. (1958). The History of Marsilid. *Journal of Clinical and Experimental Psychopathology* 19, Suppl., 1–10.
25. Kline, N. S. (1958). Clinical experience with iproniazid (Marsilid). *Journal of Clinical and Experimental Psychopathology* 19, Suppl., 72–8.
26. Zbinden, G., Randall, L. O. and Moe, R. A. (1960). MAO inhibitors for depression. *Diseases of the Nervous System* 21, sec. 2, no. 3, 89–100.
27. Kuhn, R. (1957). Über die Behandlung depressiver Zustände mit einem Iminodibenzylderivat. *Schweizerische Medizinische Wochenschrift* 87, 1135–40.
28. Kuhn, R. (1970). The imipramine story. In *Discoveries in biological psychiatry*, eds J. F. Ayd and B. Blackwell, pp. 205–17. Lippincott, Philadelphia.
29. Sulser, F. and Mishra, R. (1983). The discovery of tricyclic antidepressants and their mode of action. In *Discoveries in pharmacology. Vol. 1: Psycho- and neuro-pharmacology*, eds M. F. Parnham and J. Bruinvels, pp. 233–47. Elsevier, Amsterdam.
30. Sternbach, L. H. (1978). The benzodiazepine story. *Progress in Drug Research* 22, 229–66.
31. Haefely, W. (1983). Alleviation of anxiety—the benzodiazepine saga. In *Discoveries in pharmacology. Vol. 1: Psycho- and neuro-pharmacology*, eds M. F. Parnham and J. Bruinvels, Pp. 269–306 Elsevier, Amsterdam.
32. Pert, C. B. and Snyder, S. H. (1973). Opiate receptor: demonstration in nervous tissue. *Science* 179, 1011–14.
33. Simon, E. J., Hiller, J. M. and Edelman, I. (1973). Stereospecific binding of the potent narcotic analgesic ^{3}H-etorphine to rat-brain homogenate. *Proceedings of the National Academy of Sciences, USA* 70, 1947–9.
34. Terenius, L. (1973). Stereospecific interaction between narcotic analgesics and a synaptic plasma membrane fraction of rat cerebral cortex. *Acta Pharmacologica Toxicologica* 32, 317–20.
35. Hughes, J. (1975). Isolation of an endogenous compound from the brain with pharmacological properties similar to morphine. *Brain Research* 88, 295–308.
36. Hughes, J. *et al*. (1975). Identification of two related pentapeptides from the brain with potent opiant agonist activity. *Nature* 258, 577–9.
37. Chung, S. H. and Dickenson, A. (1980). Pain, enkephalin and acupuncture. *Nature* 283, 243–4.
38. Gellner, E. (1987). Psychiatry and salvation: discussion paper. *Journal of the Royal Society of Medicine* 80, 759–61.

14

The present and the prospects

Some lessons of history

We have come a long way from Paracelsus and the earliest applications of science to medicine. Historians are rightly cautious about extending a historical study to very recent events, because of the lack of perspective needed for making balanced judgements. But the growth of science and the impact it has had on medicine in the last 40 years cannot be ignored. Of all the man-hours which have been devoted to the search for medicines, at a rough guess 90, perhaps 95, per cent can be accounted for since 1940. Much of the work has been complex, repetitious, and without spectacular results, so it has been treated here in less detail than many of the earlier discoveries. But the general change in the amount and in the style of research is obvious.

The previous chapters have shown that discovery cannot be reduced to a simple formula or a standard intellectual process, nor is the logical route likely to be straight. The discovery of cancer drugs began with work on radioactive elements, investigation of chemical warfare agents, and studies of the diet of microbes. Many of the drugs used in preventing and relieving heart attacks and high blood pressure evolved from very academic studies of the transmission of messages from one cell to another. Discovery of psychopharmacological drugs was often prompted by astute clinical observers who recognized unforeseen properties in drugs intended for other purposes.

Progress has usually come in a series of jumps, separated by long periods with little change in ideas or in methods of medical treatment. Lister knew about the antibacterial power of a *Penicillium* in the 1870s, but could make nothing of it. Fleming rediscovered it by chance 50 years later, but could make little of it. Another decade passed before Florey and Chain embarked on a more systematic study of antibiosis, and found a practical application which suddenly and completely changed the treatment of many diseases. The periods of inactivity are not surprising: progress with a special problem of this sort often awaits technical advances in more basic sciences and then depends on someone seeing the possibilities created and having the necessary zeal and determination to apply the new techniques to good effect.

Much thought has been given to designing rather than discovering drugs. Enough is known about some of the receptors on which drugs act and about the shapes of molecules, to make a more precise approach than is implied, for instance, by a half-intuitive search for competitive antagonists. The game is made more feasible and more exciting by the power of computers, both in performing complex calculations and for the visual display and manipulation of shapes. As in so many other fields, this allows years of mental and manual work to be reduced to hours and to be repeatable within seconds.[1]

It would be nice to think that the growth of new techniques combined with the enormous expansion of laboratories in old and new universities, in industry, and in government institutions has given so much power to research that all the necessary basic advances can now be made quickly and the remaining medical problems rapidly solved. However, expansion has brought problems as well as power, and the old and provenly effective style of imaginative thought and work is under threat. The new insistence of accounting for expenditure, much of which has always been spent on correcting unlucky guesses rather than on advancing knowledge, can be seen to waste a lot of money. Formal procedures replace what was done intuitively, so that scientists spend time in meetings to plan research and in writing applications for grants instead of thinking about their scientific problems and doing experiments at the bench. It has become essential for professors, as heads of departments, to be good managers and fund-raisers. Scientists who have achieved a sound reputation are at grave risk of spending their lives reading and reviewing other grant applications and contributing to policy matters. Not all of this is new: the search for funds has always been difficult and frustrating, and the leaders of science have had to lead as well as continuing to practice their art. But research becomes more expensive, both in the price of more elaborate apparatus and in the cost of skilled persons to run and maintain it. Planning and justification are the order of the day, and threaten to destroy the climate in which original ideas flourish.

As the funds spent on research in the western world have increased perhaps twenty-fold between 1949 and the early 1970s, expenditure from government sources has naturally attracted more and more scrutiny from parliaments and administrators. It has become essential to convince many people, who have no idea of the ways in which research progresses, that the enormous funds involved are being at least properly, if not always wisely, used. The problem becomes more severe now that funds from government sources are shrinking. So bureaucracy has grown. Applicants for research grants are expected to have a clearly defined programme of what they would be doing 3 or 5 years hence, presumably based on the assumption that they would discover nothing unforeseen in the meantime, or that if they did they would not let it distract them from the approved line of work. Other bodies, especially charities which support medical research,[2] can take a more relaxed view, with great advantage, but they too have responsibilities to their supporters, and naturally tend to follow a similar pattern of careful control of the money they provide.

It has always been difficult to decide what research will be worth doing, or funding. The policy of 'Back the man, not the project', often attributed to Dale and stated explicitly by Mellanby,[3] was perhaps the greatest strength of the Medical Research Council until the freedom which Sir Walter Fletcher did so much to achieve (see Chapter 4) became increasingly curtailed by the growing responsibilities of the Council. To go back to Pasteur and 'Chance falling on the prepared mind', it is not difficult to identify some of the minds which are well prepared, but hardly possible for grant-awarding bodies to predict where chance will fall. And chance still plays an important role in discoveries about living matter. The more rigorously research is controlled and costed, the less opportunity there is for innovation.

It is sometimes argued that controls are essential to prevent wasteful duplication of research. Certainly there are many examples of duplicated discovery. Several have appeared in these pages—ergometrine four times in 1933–1934 (see Chapter 6), polymyxin three times about 1946 (see Chapter 9), isoniazid twice in 1952 (see Chapter 10). Would not much unnecessary work have been avoided by good planning? As usual it is easy to be wise after the event and to forget how much scientists communicate with each other and how often different teams working on the same problem have been supported by each other's activities. Before results have been reached, what planner, confronted with four or five parallel schemes would have the foresight to pick the winner reliably? What committee can even recognize a winner when it has it under its nose? The wartime story of chloroquine should not be forgotten (see Chapter 10). Independence means much to any worthwhile research worker and competition is a spur to some.

Apart from problems of funding, scientists in the laboratory have become more restricted in other ways. Research in biology and medicine has profited, and indeed simply could not have achieved what it has without the introduction of methods based on physical and chemical discoveries. The training necessary to handle such techniques both effectively and safely is an additional burden on doctors and biologists who are tempted towards research. The maintenance of safety in a laboratory has become more difficult, and more regimented. No one likes to be diverted from the primary object of a successful experiment, so conflicts between independent-minded investigators and good management are sometimes unavoidable. In industry, where discipline is usually rated above independence, the conflicts have been familiar for a long time. Research chemists were quarreling with management in the Bayer laboratories in the 1890s,[4] and the problems are not very different today.

Nevertheless, the opportunities for a career, or part of a career, in medical research have changed. Both the desire for fame and the influence of grant-giving bodies has encouraged some to excessive publication, often premature and sometimes misleading. Scientific literature, already far beyond the capacity of any human study, is swollen by multiple publications of the same results, and even fraudulent papers containing bogus results are not unknown.[5,6] One could continue a pessimistic catalogue, but any temptation to see the incipient collapse of scientific and medical research must be resisted. The far-seeing inventor of early calculating 'engines', Charles Babbage (1792–1871), published his *Reflexions on the decline of science in England* in 1830,[7] so there is nothing original in the idea that science is going downhill. The important question, which it is too early to answer, is whether medical science is evolving into an even more fruitful period of advance, or whether innovation is gradually being strangled by its own overgrowth.

The safety of medicines

No one expects a knife to be safe all the time. It depends how sharp it is, who wields it, and what it cuts. But the half-magical faith in medicines, and ignorance about how they work, blinds most people to the fact that drugs are like knives, dangerous when mishandled, and that only very feeble drugs, like very blunt knives, can be used with impunity in unskilled hands. Not all the risks of new drugs can be predicted, and it is inevitable that they come to light gradually. Society is still not good at managing the problems, but much can be learnt from past mistakes.

A number of drugs have, for one reason or another, become infamous, especially when tragic episodes have received great publicity. The names of thalidomide, practolol, and benoxaprofen spring to mind as drugs which have done harm and have been discarded, and there are many others. A little while ago a careful review, based on the combined opinions of a number of expert physicians, listed the numerous ill-effects of new drugs which had been reported between the time of thalidomide (about 1962) and 1983.[8] A dozen or more were regarded as particularly serious. Most of the troubles were novel: that is to say, no drug already known had exactly that effect. The details were complex; sometimes they depended on the doses involved, sometimes on the state of the patients being treated. It was not always possible to distinguish what was caused by the drug from new symptoms arising in the course of a disease, a disease perhaps prolonged by the drug so that new manifestations were possible. Larger and more extensive reviews of the 'side-effects' of drugs are available and show the complex ramifications.[9,10]

The phrase 'side-effects' became popular in the 1950s, and soon aroused protests.[11,12] What were 'side-effects', it was asked? What were they at the side of? The only clear answer was that they were beside the main effects, and it was implied that, by providence, the main effects of drugs were those we wanted and the side-effects were all unwanted. Somehow the phrase 'side-effect' was reassuring, and made such reactions seem to be irregular in appearance and relatively insignificant.

Thus do we all delude ourselves with words. Drugs react with particular bodily components. They are chosen for their purposes because some of these reactions are beneficial, but it is asking for trouble to dismiss the other, unwanted, reactions as 'side-effects', as though they were in some way regrettable mistakes which ought not to attract attention. No drug now comes into general use without a great deal being known about its regularly observable actions. However, it is impossible for there to be *complete* knowledge of *every* action it will have in unusual circumstances. The history of the discovery of drugs shows how different kinds of failure have had tragic consequences and have each fostered the growth of more and more precautions directed to that impossible ideal, a completely safe yet potent drug.

Seventy years ago, adverse reactions were accepted more readily than they are today. In the first flush of chemotherapy, drugs related to quinine were tested against lobar pneumonia (see Chapter 3). A few patients found that their vision was impaired, sometimes permanently. The chances of surviving an attack of lobar pneumonia were not good and the likelihood of long-lasting, chronic infection was substantial. In the medical papers of the time, the risk of partial blindness was mentioned as a factor to be considered when administering the drug. However, it did not prohibit the use of the drug. The chance of death from pneumonia was about 30 per cent anyway, so the alterative of possible loss of sight was cautiously accepted. Presently these drugs disappeared, as much because they were ineffective as because they were toxic (see Chapter 8).

In the second flush of chemotherapy, when the use of sulphanilamide was spreading round the world, many local endeavours were made to dispense drugs in forms which would be attractive to patients and remunerative to the dispenser. To avoid using tablets, which were large and tiresome to swallow, liquid preparations with a sweet taste, were sometimes devised. One such preparation was made by a small enterprise in the USA. It was named Elixir of Sulfanilamide-Massengill, and was duly prescribed. During the months of September and October 1937, at least 76 human beings in various places in the USA died after consuming the elixir. The investigations which inevitably followed showed that the elixir was made by suspending the drug in 72 per cent ethylene glycol, a sweet-tasting substance often used as an anti-freeze and sometimes, even in very recent years, as an adulterant of wine. It is well known that ethylene glycol is converted, after it is swallowed, to the poisonous substance oxalic acid. Apparently the manufacturers of the elixir were, like others since, unaware that ethylene glycol was harmful, and the elixir was distributed without its safety being tested either on animals or on man.[13] The repercussions of the incident were extensive and had a great influence on the development of the policies of the United States Food and Drug Administration.

This tragedy reflected the primitive attitudes to medicines at that time, when their manufacture and marketing was lawful even by persons who were quite ignorant of simple and dangerous properties of the substances which they were handling. But professional knowledge alone is not a guarantee against disaster. In the German town of Lübeck in 1930, about 250 infants were treated with a supposed vaccine (BCG) against tuberculosis. Three-quarters of the children developed clinical tuberculosis, and of these about one-third died. It was widely alleged that the organisms in the vaccine must have regained their virulence, and the reputation of the vaccine and its inventors, Calmette and Guérin, suffered accordingly. A commission of enquiry, however, decided that a virulent human culture had been substituted in error for the vaccine. By mismanagement, of a kind that any experienced laboratory administrator would instantly condemn, both materials had been kept in the same room and a situation arose in which it was all too easy to select the wrong material.[14] The vaccine was acquitted, but the unmerited stain on its character lasted for many years. Only after effective antituberculous drugs (see Chapter 10) brought reassurance that *if* another such disaster occurred, the resulting infection could be controlled, did BCG vaccine take its place as a useful means of protecting against tuberculous infection.

Of all the drug tragedies, the thalidomide disaster is the most familiar. The drug was introduced by a German chemical business during the 1950s as a sleeping tablet with a very wide margin of safety between the dose which caused sleep and that which brought danger of death. The right to market it was licensed to other firms in different countries. In the United States, scientific assessors in the Food and Drug Administration had doubts about its safety and had not recommended its acceptance by the time its dangers became known. In other countries, use of thalidomide became popular because of concern about the number of suicides attributed to barbiturate poisoning and the need for a safer hypnotic. Later, the benzodiazepines (see Chapter 13) would fill this role, but they had not yet been discovered. Thalidomide seemed to be helpful for expectant mothers troubled by the discomforts of early pregnancy. After the drug had been on the market for 2 years, clinical concern about its safety came from reports that peripheral nerves deteriorated after its use.[15] Also obstetricians noted that within their area of practice babies with a particular and rare kind of deformity were being born much more often than had ever previously been observed.[16] It took another 2 years to connect the deformities with thalidomide. As the drug was injurious only when taken at a particular and brief period early in pregnancy, a proportion of mothers who had taken thalidomide at some time during their pregnancy did not have ill-effects, and so tracing the connection was not as easy as it might seem.[17] The position was aggravated by attempts on the part of the German company to suppress information prejudicial to their product.

Much debate ensued on the extent to which suitable tests in pregnant animals could have identified the hazard before humans were involved. Careful reviews of the effects of drugs on the fetus showed that such effects *could* be forecast,[18] but many drugs were still being introduced into clinical practice without being tested in this way. Later, experiments with thalidomide, and with many other substances, showed that damaging effects on the fetus often varied from species to species, and that negative results in a particular species were of limited predictive value for others. From the point of view of ensuring the safety of drugs, the accumulated evidence on this point was very discouraging.

The thalidomide affair caused a great increase in concern about the safety of all drugs, especially any which were 'new', and especially when taken during pregnancy. The USA had already progressed further than any other nation in developing a regulatory system, which provided a model for other countries with sufficient resources to undertake controls. The pharmaceutical industry, anxious to show its concern, co-operated, and procedures intended to demonstrate the safety of new substances were elaborated. These procedures depended particularly on determination of the size of lethal doses in small animals, on the administration of massive doses for long periods to at least two species of mammal, and on certain specialized tests such as dosing pregnant animals and examining the fetuses. The procedures were in no way scientific experiments, in the sense that precise hypotheses were to be tested by appropriate experiments, nor did they seek novel kinds of toxic effect.[19] The fallibility of making predictions from animal studies to man was glossed over and the history of good drugs, recovered after they had been rejected on the basis of sketchy toxicity studies (e.g. mapharside and chloroquine), was known to few people. The methods adopted for toxicity testing were laborious and very expensive in terms of the animals used and the staff involved. Consequently, the great yardstick by which scientists judge experiments, the fact that the results can be reproduced in other laboratories by other scientists, was not applied. As one far-sighted independent toxicologist wrote in 1963:

Testing procedures outlined by authority whether national or international will either be so vague as not to be worth dissemination or become by necessity more precise. The recommended tests would then be carried out by scores of unthinking technicians who will supply a mass of data eventually to be pushed under official noses. The scientific study of toxicology will atrophy and the hazards from new drugs remain as much of a problem in the future as it is to-day.[20]

These procedures did not, of course, reduce the amount of investigation to which a drug was submitted, sometimes in human volunteers and always in patients. The requirements for such studies increased, and the detailed submission of papers, and their inspection and assessment by staff of the regulatory authority consumed much time. It greatly delayed, if it did not in the end prevent, the drug from being licensed for human use in the countries which proceeded so thoroughly. The 'drug lag', particularly in the USA,[21,22] became notorious, and calculations began to appear of the number of American citizens who had died prematurely because the drugs available in other countries which would have saved them were awaiting approval on the desks of the Food and Drug Administration. In view of the pressures to which US officials can be exposed in Congressional enquiries,[23] their caution was only to be expected.

The regulatory approach had a considerable effect on the discovery of new drugs. The costs involved promptly led to pressures to pursue only lines leading to drugs for which there was a large and rewarding market. Any disease which was confined to the third world consequently had a very low priority. Drugs for rare diseases brought an income to their manufacturers so small that in 20 or 30 years it would not have paid for the tests needed to meet regulatory requirements. So they were not commercially viable, and physicians treating such patients had great difficulties[24] in obtaining the drugs which were known to be effective but were not licensed for marketing. New compounds which showed unpromising features in experiments in animals were abandoned, although plenty of precedents existed which showed that such features in other substances had been entirely compatible with acceptable safety in man.

Nor did the regulatory approach prevent further adverse reactions. The beta-blocking drug practolol (see Chapter 12) seemed to be better than its predecessors pronethalol and propranolol. It was submitted to all the tests that the rules required, and more besides,[25] passed them and was accepted accordingly. As thorough clinical trials showed, and experience of wider use confirmed, it was a useful and successful drug. But in 1974, a unique constellation of symptoms and signs were observed in some patients treated with practolol, including rashes, changes in the eyes and ears, and formation of abnormal fibrous tissue round internal organs. If injury had not gone too far, the condition was resolved when the patients were transferred to other beta-blocking drugs. The reason why this small minority of patients reacted remained obscure. Attempts to reproduce the conditions in animals failed, and no method was found by which potential reactors could be identified. The condition was too serious to be regarded as an acceptable risk, and so the drug was withdrawn. The manufacturer, Imperial Chemical Industries, though advised that no legal liability existed, set up a compensation system for patients who had suffered permanent injury.[26]

Practolol was the most striking but not the only example of a drug which produced unacceptable ill-effects in a small minority of patients. It became obvious that it was such minority effects which were the principal source of difficulty and that it was logically impossible to discover such effects in the ordinary studies which preceded marketing.[27] Such investigations vary in scale, but might reasonably be confined to, say, 5000 patients. A serious unwanted effect occurring in one patient in 10 000, which is enough to make most drugs unacceptable, could hardly be recognized when only 5000 patients have been treated. To discover the minority effects, it was necessary to know what happened to all the patients who received the drug for a time after it was marketed, and schemes were devised accordingly.[28] Such surveillance is now widespread. Its advantages, both in giving an early warning of impending trouble and in providing information about other factors associated with the development of unwanted reactions, have yet to be fully evaluated.[29]

Just as drugs have often been credited with useful properties because patients get better after taking them, any recently consumed drug (or food, especially if it contains some chemically-identified ingredient) is apt to be blamed for observed ill-effects. Sometimes the blame is entirely unjust. Congenital abnormalities, like those caused by thalidomide, occurred before thalidomide was invented, and continue to occur although thalidomide is no longer used in pregnancy.[30] A medicine containing more than one ingredient, Debendox or Bendectin, was devised to relieve the nausea common in early pregnancy and was consumed by millions of women for this purpose without ill-effect. Inevitably, these millions included mothers whose babies developed abnormalities as a matter of course. The association is fortuitous, but in individual cases uninformed observers have naturally sought to blame the drug, and its reputation has been so damaged that it is no longer expedient to produce it.[31,32] It appears to have been a good drug for its purpose, so the losers are the patients who need it.

Even when a drug clearly has some ill-effects, it does not follow that it is worthless. Thalidomide has been shown (in male patients) to relieve the skin lesions and pain of leprosy,[33] but, because it is a 'banned' drug, supplies are difficult to obtain and beyond the means of the (usually very poor) patients.

Problems of safety are not absolute. The simplest of substances are dangerous in particular circumstances. For patients with congestive heart failure, an ordinarily acceptable amount of common salt may be lethal. For patients with diabetes mellitus, sugar is a serious hazard. Effective drugs are more dangerous than sugar and salt, and their safety depends on using them only in the right conditions, at the right dose for the circumstances, and with a sense of the relative risks of using or not using the drug.

Alternatives

One way of dealing with difficult matters is to take refuge in a simpler (and perhaps imaginary) past in which the difficulties did not arise. The last 20 years have seen the revival of interest in various forms of alternative or 'complementary' medicine, including herbalism, homeopathy, acupuncture, osteopathy, and various forms of faith healing. Of these, herbalism overlaps with and homeopathy disagrees with the physiochemical approach to discovering remedies for diseases. How do they compare with the orthodox scientific approach to medical treatment? Various attempts have been made to assess the effects of different systems of therapy, either by deliberately designed trials or by comparing the progress of patients who adopted such treatments with the progress of others who did not.[34,35] The latter approach involves a considerable bias because patients who choose a particular treatment are different from those who do not. Nor are controlled trials easy to arrange, or to design in such a way that the trial does not conflict with the doctrines of the alternative treatment. On the whole it seems as though most forms of alternative medicine are equally effective (or ineffective) and that they appear to work best in conditions which have unexplained ups and downs, or which usually get better with the passing of time. It is difficult to judge how much improvement should be attributed to the treatment and how much is coincidental. It is more comforting to do something in the face of difficulties than to accept them passively, so alternative treatments, like much conventional medicine, have provided a good deal of comfort and happiness, at least in the short term.

Studies of alternative methods of treatment ought to include an assessment of their safety. Herbalism comes closest to orthodox medicine. It might be regarded as persisting in the crude use of plants, with all the disadvantages of impurity, uncertain and variable composition, and the absence of any assessment of toxicity. Skilled herbalists avoid the more potent plants, such as monkshood, nightshade, hemlock, and foxglove, but dangerous ill-effects occur from the use of popular herbs including comfrey (liver damage and cancer), ginseng (nervous tension and enlargment of male breasts), liquorice (high blood pressure due to salt retention), laetrile (cyanide poisoning), and pennyroyal (liver damage).[36,37] Identification of plants is not easy and contamination of herbal remedies with poisonous species is not unknown, especially when herbs are collected by untrained people. At least one recent popular account of medicinal herbs is refreshingly cautious about the hazards of potent plants.[38] There are sound reasons for continuing to regard herbs and other plants as a potential source of new medicines, but the pathway by which such discoveries are made has been trodden many times and leads through classical pharmacology to fundamental chemistry, the isolation of active principles of known composition and purity, and quite possibly their manufacture thereafter by chemical synthesis.

Homeopathy is in a different position, because some of its basic tenets are incompatible with elementary scientific knowledge. In as far as materials are diluted to such a degree that a given dose probably does not contain any of the original material at all, the procedures enter the realm of magic and deserve study as such.[39] When the homeopathist studies his individual patient and adjusts his procedures accordingly, he is following a form of psychotherapy which may have admirable results and should be judged by suitable criteria. But the process is not related to drugs and medicines as they are defined in this book, unless, as is sometimes reported to have happened, the homeopathic remedy has been diluted, adulterated, or fortified with some potent substance.[40,41]

The release of endorphins by painful stimuli such as acupuncture has been mentioned (see Chapter 13), and provides a possible physical mechanism by which benefits may be obtained. The theoretical basis of acupuncture rests on a system of physiology different from that which has emerged from occidental laboratories, and we shall not attempt to reconcile the two systems.[42] Psychological mechanisms may be expected to contribute to the results of acupuncture as much as to any other form of therapy. The hazards of conveying infection, for instance hepatitis or AIDS, by inserting an inadequately sterilized needle arises in any operation in which the intact skin is penetrated.

The therapeutic value of prayer and other forms of faith healing is also beyond the scope of this book. It does not seem impossible to evaluate their merits by the standards which we have been discussing, but attempts to do so are not well known.[43] Again the problem arises of reconciling an enquiring approach with a practice which depends upon trust,[44] and we will not speculate here on how such reconciliation can be achieved.

The problems of success

To revert to the orthodox scientific approach to medicine, one may ask how much further can it go? Many diseases, which create suffering and usually shorten life, present unresolved problems. Cancer, heart disease, arthritis, rheumatism, mental disorders, and numerous other conditions await fuller understanding, which in turn may provide a means of arresting or reversing their progress. The widespread deterioration of physical and mental powers with advancing years clearly has a biochemical basis and perhaps may become controllable. The scientific problems are fascinating and there are good reasons for hoping that some at least will be solved.

However, the social problems created by medical success are formidable. Belatedly, methods of preventing conception have been added to the many ways of delaying the arrival of death that had already been discovered. But the contraceptives which exist are being used by too few people to prevent the world's population from continuing to increase at an alarming rate, while the world itself gets no larger. Mental and social adaptations are needed to cope with the great increase in children surviving in poorer countries, and in the aged members of affluent nations. The implications of having an ever growing number of centenarians are disconcerting, especially if they are handicapped by whatever diseases remain uncontrolled and if food production does not continue to increase in the striking way which it has over the last 40 years.[45]

Such anxieties may or may not be more serious than other portents of doom, such as the overheating of the earth, the gap in the ozone layer, viruses from outer space, and so on. Wherever they may come from, new parasites should be expected to be able to survive existing modes of chemotherapy and benefit from the reduced competition for particular ecological niches. The recent appearance of human immunodeficiency virus, facilitated perhaps by widespread changes in human sexual habits, is a reminder that man continues to be in competition with other species besides his own, and that new problems are inevitable. It is desirable that we should be well equipped to solve them.

One infectious disease appears to have been totally conquered. This major achievement of preventive medicine began with Edward Jenner in the 1790s and the start of vaccination against smallpox. In 1977 the last naturally occurring case of smallpox in the world was reported in Merka, Somalia.[46] After 10 years during which no fresh cases have been reported, the disease can be regarded as eradicated. Because there was no specific moment at which one could say '*Now* it has gone', the achievement has received disgracefully little publicity, and many people are still unaware that smallpox is no longer with us. To achieve its demise took many years of devoted work by the staff of the World Health Organization. Much ingenuity went into improving the vaccine so that it survived tropical conditions in primitive surroundings, and much also into finding ways of administering it which were within the scope of persons available as vaccinators. The problems of curing and eradicating diseases do not end with the discovery of effective remedies. But it does help to have them, and finding them needs every single advance which can possibly be made in understanding the *basic* science of the molecules of life.

Notes

1. Dean, P. M. (1987). *Molecular foundations of drug—receptor interaction.* Cambridge University Press, Cambridge.
2. *The Association of Medical Research Charities handbook 1988—89.* Association of Medical Research Charities, London.
3. Mellanby, E. (1938). The state and medical research. *Lancet* ii, 929–36.
4. Meyer-Thurow, G. (1982). The industrialization of invention: a case study from the German chemical industry. *Isis* 73, 363–81.
5. Lock, S. (1988). Scientific misconduct. *British Medical Journal* 297, 753–4.
6. Lock, S. (1988). Misconduct in medical research. Does it exist in Britain? *British Medical Journal* 297, 1531–5.
7. Howarth, O. J. R. (1931). *The British Association for the Advancement of Science: a Retrospect 1831—1931.* British Association, London.
8. Venning, G. R. (1983). Identification of adverse reactions to new drugs. I: What have been the important adverse reactions since thalidomide? *British Medical Journal* 286, 199–202.
9. Meyler, L. (1952). *Side effects of drugs.* Elsevier, Amsterdam.
10. Davies, D. M. (1985). *Textbook of adverse reactions.* Oxford University Press, Oxford.
11. Modell, W. (1964). The extraordinary side effects of drugs. *Clinical Pharmacology and Therapeutics* 5, 265–72.
12. Weatherall, M. (1965). Side effects. *British Medical Journal* 1, 1174–6.
13. Geiling, E. M. K. and Cannon, P. R. (1938). Pathologic effects of elixir of sulfanilamide (diethylene glycol) poisoning. A clinical and experimental correlation: final report. *Journal of the American Medical Association* 111, 919–26.
14. Parish, H. J. (1965). *A history of immunization*, p. 101. Livingstone, Edinburgh.
15. Fullerton, P. M. and Kremer, M. (1961). Neuropathy after intake of Distaval (thalidomide). *British Medical Journal* ii, 855–8.
16. Lenz, W. (1961). Kindliche Missbildungen nach Medikament Einnahme während der Gravidität? *Deutsche Medizinische Wochenschrift* 86, 2555–6.
17. Report (1964). Deformities caused by thalidomide. *Ministry of Health Reports on Public Health and Medical Subjects no. 112.* Her Majesty's Stationery Office, London.
18. Baker, J. B. E. (1960). The effects of drugs on the foetus. *Pharmacological Reviews* 12, 37–90.
19. Zbinden, G. (1982). Current trends in safety testing and toxicological research. *Naturwissenschaften* 69, 255–9.
20. Barnes, J. M. (1963). Are official recommendations for the testing of drugs for

toxicity dangerous? *Proceedings of the European Society for the Study of Drug Toxicity* 2, 57–63.

21. Wardell, W. M. (1974). Therapeutic implications of the drug lag. *Clinical Pharmacology and Therapeutics* 15, 73–96.

22. Wardell, W. M., DiRaddo, J., and Trimble, A. G. (1980). Development of new drugs originated and acquired by United States-owned pharmaceutical firms, 1963–1976. *Clinical Pharmacology and Therapeutics* 28, 270–7.

23. Shubin, S. (1979). Triazure and public drug policies. *Perspectives in Biology and Medicine* 22, 185–204.

24. Walshe, J. M. (1975). Drugs for rare diseases. *British Medical Journal* ii, 701–2.

25. Cruickshank, J. M., Fitzgerald, J. D., and Tucker, M. (1984). Beta-adrenoceptor blocking drugs: pronethalol, propranolol and practolol. In *Safety testing of new drugs. Laboratory predictions and clinical performance*, eds D. R. Laurence, A. E. M. McLean and M. Weatherall. Academic Press, London.

26. *Ibid.*, p. 123.

27. European Workshop (1977). Towards a more rational regulation of the development of new medicines. *European Journal of Clinical Pharmacology* 11, 233–8.

28. Inman, W. H. W. (1981). Postmarketing surveillance of adverse drug reactions in general practice. I: Search for new methods. *British Medical Journal* 282, 1131–2.

29. Inman, W. H. W., Rawson, N. S. B., Wilton, L. V., Pearce, G. L., and Spiers, C. J. (1988). Postmarketing surveillance of enalapril. I: results of prescription-event monitoring. *British Medical Journal* 297, 826–9.

30. Weatherall, J. A. C. (1988). Surveillance of congenital malformations and birth defects. In *Surveillance in Health and Disease*, eds W. J. Eylenbosch and N. D. Noah, pp. 101–14. Oxford University Press, Oxford.

31. Holmes, L. B. (1983). Teratogen Update: Bendectin. *Teratology* 27, 277–81.

32. Walters, W. A. W. (1987). The management of nausea and vomiting during pregnancy. *Medical Journal of Australia* 147, 290–1.

33. Iyer, C. G. S. *et al.* (1971). WHO co-ordinated short-term double blind trial with thalidomide in the treatment of acute lepra reactions in male lepromatous patients. *Bulletin of the World Health Organization* 45, 719–32.

34. Struthers, G. R., Scott, D. L., and Scott, D. G. I. (1983). The use of alternative treatments by patients with rheumatoid arthritis. *Rheumatology International* 3, 151–2.

35. Reilly, D. T., Taylor, M. A., McSherry, C., and Aitchison, T. (1986). Is homeopathy a placebo response? Controlled trial of homeopathic potency, with pollen in hayfever as model. *Lancet* ii, 881–5.

36. Penn, R. G. (1983). Adverse reactions to herbal medicines. *Adverse Drug Reaction Bulletin* 102, 376–9.

37. Phillipson, J. D. and Anderson, L. A. (1984). Pharmacologically active compounds in herbal remedies. *Pharmaceutical Journal* 232, 41–4; 111–14.

38. Stodola, J. and Volak, J. (1984). *The illustrated book of herbs*, ed. S. Bunney. Octopus Books, London.

39. Maddox, J., Randi, J., and Stewart, W. W. (1988). 'High-dilution' experiments a delusion. *Nature* 334, 287–90.

40. Morice, A. (1986). Adulterated 'homeopathic' cure for asthma. *Lancet* i, 862–3.

41. Morice, A. (1987). Adulteration of homeopathic remedies. *Lancet* i, 635.

42. Chung, S. H. and Dickenson, A. (1980). Pain, enkephalin and acupuncture. *Nature* 283, 243–4.
43. Bartholomew, D. J. (1988). Probability, statistics and theology. *Journal of the Royal Statistical Society, Ser A* 151, 137–78.
44. Joyce, C. R. B. and Welldon, R. M. C. (1965). The objective efficacy of prayer. A double-blind clinical trial. *Journal of Chronic Diseases* 18, 367–77.
45. Blaxter, K. (1988). Future problems—presidential address 1988. *Biologist* 35, 173–8.
46. Hopkins, D. R. (1983). *Princes and peasants. Smallpox in history*, p. 310. University of Chicago Press, Chicago. (Last case of smallpox.)

Index

Abel, John, J. 86, 104
Abraham, Sir Edward 167
Académie des Sciences, Paris 27, 53
accessory food factors 123, 125, 129
acetanilide 36,37
acetylcholine 74, 76, 218, 237, 263
acetylsalicylic acid (aspirin) 37, 151, 195,
    238–9
acridine 149
Act, Cruelty to Animals, 1876 45, 54
    Medical, 1858 12
    National Insurance, 1911 75
    National Cancer, U.S.A. 1971 228
actinomycetes 180
acupuncture 264, 279
acyclovir 200
addiction 20 (table)
Addison, Thomas 83, 132
admiralty 7
adrenal gland 83, 86–7, 94
adrenaline 71, 73, 86, 94, 107, 239, 240–1,
    252
Aesculapius 3
Agfa 36
aggression 259
ague 7
Ahlquist, Raymond 240–1
AIDS 141, 281
Aikin, J. 8
air, hospital 8, 15, 41
albumin 119
alchemist 5
alcohol 145, 235, 251
alkaloid 17, 19, 21, 23, 239, 257
alkylating agents 219
Allen & Hanbury 122
allergy 73, 74
allopurinol 225
alpha receptor 240
alternative medicine 280
Alzheimer, Alois 254
Amanita 71, 252
American Medical Association 112
amethopterin 223
amidopyrine 36,37
amine 125, 259, 263 (table)

amine oxidase 259, 263 (table)
amino acid 119, 147, 220 265
aminopenicillanic acid 177
aminopterin 223
ammonia 34, 70
amyl nitrite 30–1
amoeba 151
anabolism 118
anaemia 9, 19, 132–5
anaemia
    Addisonian or pernicious 132–5
    monkeys 134
    Wills's 133–4
anaesthesia
    general 30, 37, 253
    local 108–10
    relaxation in 108
anaesthetic 18, 29–30, 253
    local 105–10
analgesic 37
anatomy, morbid 7–8, 42
Andersag, H. 191
angina pectoris 31, 241
animals, experiments 17, 21, 44–5, 54
antagonism between drugs 23, 34, 107, 240–2
antagonism, competitive 153, 195, 223–7
anthrax 40
antibiotic 162–82
    synthetic 182
antibodies 142
antidepressant 258–61, 263 (table)
antifebrine 36, 37
antigens 142
antihistamine 241–2, 256
antimony 5
antipyrine 37
antiseptic 41, 149
antitoxin 52–5, 85, 141–4
antiviral agents 197–200, 227
anxiety 256, 261, 263 (table)
Apothecaries, Society of 12
Aristotle 3
army 143, 163, 191, 192
arsenic 5
arsenicals 59–64, 147, 254
arsphenamine 60–4

arteries 31, 235
arthritis 235
Asklepios 3
aspirin 37, 151, 195,238–9
assay
   biological 56, 76–7, 107–8
   insulin 92–3
   thyroid hormone 88
asthma 235, 240
atebrin 150, 190–1
atoms, definition 27
atoxyl 59
Atropa belladonna 4 (table)
atropine 20, 23, 71, 107
attenuation 40
autumn crocus 20
Avicenna 4
Avogadro, Amadeo 27
azathioprine 225

Babbage, Charles 274
Bacchus 251
Bacon, Francis 117
bacteriology 42–3, 51–7, 141–8, 163, 168
bacteriophage 162–3
Bacillus influenzae 168
Balard, Antoine 30–1, 78
Baltimore 104, 125
Banting, Sir Frederick 90–2, plate 17
barbiturate 256, 262, 276
Barger, George 73–5, 89, 110–11
bark, cinchona or Peruvian 9, 28
bark, willow 10, 37
barley 121
Barlow, R. B. 237
Bassi, Agostino 14, 40
Bayer 36, 149, 191, 193, 274
Bayer 205 (Germanin) 149
BCG vaccine 276
Beatson, G. T. 214
Becquerel, A. H. 212
Beecham Laboratories 177
Behring, Emil von 52, 56, plate 12
belladonna 4 (table), 9
Bendectin 279
benzoic acid 34–5
benzodiazepine 261–2, 263 (table)
beri-beri 120–2, 126
Berkeley, Gloucestershire 13
Berlin 53, 56
Bernard, Claude 18, 108, plate 8
Bernheim, F. 195
Berzelius, Jons 28
Best, Charles 90–2, plate 18
beta-blocker 241
beta receptor 240–1
Beyer, K. H. 242

Bichat, Xavier 16, 17
bio-assay, *see* assay; standardization
biochemistry 34–5, 122–6, 146–8, 219–21
biotransformation 34–5
bismuth 148
Black, Sir James 241, 242
Bland-Sutton, Sir John 126
bleeding 9, 16
blindness (atoxyl) 60
blindness (optochin) 148–9
block, competitive 153
block, ganglion 237–8
blood cells, white 217–19
blood pressure 73, 95, 235–45
blood vessels 31–2, 235–45
Boccacio 6
Boerhaave, Hermann 8
bone 126–9
bone marrow 132–5
Boston (USA) 15, 30, 77, 133, 223, 226
Bourne, Alec 111
Boyle, Robert 27
Braddon, Leonard 124
Bradley, Richard 14
brain 245, 253–66
breast cancer 214–16
British Association 31, 130
British Pharmacological Society 106
Brodie, Sir Benjamin 108
bromide 32–3
bromine 30, 32
Brown, A. Crum 23, 33–4, 237
Brown-Séquard, E. C. 83, 85
Browning, Carl 149
brucine 17
Brunton, Sir Thomas Lauder 31
Buchheim, Rudolph 21, 35, 103, 253
Buchner, E. 145
buffer 220
bullet, magic 45, 61–3
Bulloch, William 61
Burn, J. H. 111
Burroughs, Silas 72
Burroughs Wellcome 72, 224
butter 123, 128

Cabanis, Pierre 16
Cade, J. F. 257
caffeine 17
Calabar beans 22
calcium 126–9
Calne, Sir Roy 226
Cambridge (UK) 70, 78, 110, 122, 146, 165
cancer 95, 199, 209–29, 252
cannabis 252
Cannon, Walter B. 77
carbohydrate 118

carbolic acid (phenol) 41, 148, 162
carcinogen 211
Carleton, H. M. 166
casein 123
Castle, W. B. 133
catabolism 118
cattle 74, 194
causes, knowledge of 209, 235
Caventou, Joseph 17, 20, plate 7
cell, chemical activities 130–1, 145, 147
cells 39, 45, 145
centrifuge 221
Cephaelis ipecacuanha 4 (table)
cephalosporin 177
Chain, Sir Ernst 167, 173–7, 200, 271,
    plate 22
chance 39
charbon (anthrax) 40
charity 201, 228, 273
charlatan 11
charts, temperature 37
chemicals, novel 29
chemistry 27
chemistry of food 117–20
chemistry, organic 27, 34
chemotherapy of cancer 216–27
chemotherapy of infections 58–64, 148–54,
    190–200
chicken 121
chickenpox 197
childbirth 8, 14–15, 72–3, 110–12, 141, 162
chimney sweep 211
chloral 35, 37
chloramphenicol 182
chlordiazepoxide 261
chloroform 30, 35
chloroguanide 191, 224
chloroquine 191–3
chlorothiazide 243
chlorpromazine 256
cholera 15, 51, 209
cholera, fowl 40
cholesterol 95, 129
cholic acid 95
cholinesterase 77
Christison, Robert 22
chromatin 218, 221
chromosomes 218, 219
CIBA 239
cigarette 235
cimetidine 242
Cinchon, Countess of 9
Cinchona species 4 (table) 9, 10
circulation, blood 9, 83, 235–7
cirrhosis, liver 252
cisplatin 227
Clark, A. J. 107

Clinical Society of London 84
clotting 245
Clouston, T. S. 33
Clowes, G. H. A. 91
coal tar 36, 95, 211
Coatney, R. 193
coca 4 (table), 20 (table), 109
cocaine 20 (table)
cod liver oil 126–8
codeine 20, 265
coffee 9, 17, 20
colchicine 20
Colebrook, Leonard 61, 152, 163
Collip, J. B. 90–2
colloid 220
Columbus, Christopher 5
committee 191, 209–10, 228–9, 272
Committee, Medical Research (UK) 75, 127–
    9, 146, 149, 228
competition, microbial 162, 173
composition, chemical 27
compound chemical 27
compound no. 183 (USA survey) 192
compound no. 205 (Bayer) 149
compound no. 606 (Ehrlich) 60, 63
compound no. 693 (M&B) 152
compound no. 914 (Ehrlich) 63
compound no. 6911 (USA survey) 193
compound no. 7618 (USA survey) 193
computers 270
congress, medical, London 44–5, 69, 84
congress, physiological, Edinburgh 92–3
conine 20 (table)
Conium maculatum 4 (table)
Connaught laboratories, Toronto 91
contagion 8, 14
contraceptive 96–7, 166
control of quality 55, 56, 93
convulsions 258; *see also* fits
Copernicus, Nicholas 5
copper 5
cortex, adrenal 94–6
corticosteroids 95, 227
Corynebacterium parvum 227
costs and funding 75, 272–3
cough 20 (table), 264
Council, Medical Research (UK) 75, 127,
    149, 152, 167, 173, 201, 228, 273
Council, National Research (USA) 181, 192
cowpox 13
Craddock, S. R. 168–70
crystallization 17, 38, 122, 129
crystallography, X-ray 176, 221
culture
    microbes 40–3, 144–8, 197
    tissue 197
cupping 9

curare 18, 70, 108, 166, 258
Curie, Marie 212
Curie, Pierre 212
Cushny, A. R. 106, 107

dactinomycin 227
Dakin, H. D. 86, 88
Dale, Sir Henry 71–8, 92–3, 149, 166, 201–2,
    262, 273, plates 16, 23
Dalton, John 27
Daraprim 193, 224
Darwin, Charles 69
daunorubicin 227
Davis, M. 125, 128
Davy, Sir Humphry 29, 117
Decourt, P. 192
Debendox 279
deficiency disease 117–35
dementia 254
Deniker, P. 256
deoxyribose 221
dependence 262
depression 239, 254, 257, 258–61
Derosne 20
detoxication 35
diabetes 90–3
diagnosis 51
diaminopurine 224
dibenzanthracene 211
dichlorisoprenaline 240
diet
    prison 121–2
    synthetic 119, 122–4
diethylstilboestrol 95, 216
digestion 117
digitalis 4 (table), 10, 95, 103, 242
Dionysus 251
diphtheria 52, 142, 144
    antitoxin 52, 142
    immunization 142–4
discovery, duplicated 112, 181, 273
disease
    Addison's 83, 94, 132
    Gull's 84
    legionnaire's 141
disinfectant, *see* antiseptic
distemper 197
distillation 252
Dixon, W. E. 71, 76
DNA 199, 221
DNA recombinant 93, 199
Dobell, C. 151
Dodds, Sir Edward Charles 95, plate 19
dogs 53, 91, 127, 197, 226
Domagk, G. 150
dopamine 263 (table), 264
Dorpat (Tartu) 21, 35, 107

dosage 56, 57, 61–2, 106–7
Dougherty, T. 217
Dreyer, G. 173
dropsy 10–11
drug, definition v–vi
drug lag 278
Dubos, R. 171–2
Dudley, H. W. 92–3, 112
dysentery 20 (table)

East India Company 6
Edinburgh 22, 34, 71, 89, 93, 107, 132
education, medical 104–6
egg yolk 125
egg as culture medium 197
egg white 164
Egyptians 126
Ehrlich, Paul 55–64, 70, 73, 107, 148–51,
    194, plate 15
Eijkmann, Christiaan 121
electro-convulsion therapy 108, 258
electrophoresis 220–1
element
    ancient idea 3
    chemical 27
Eli Lilly 91, 133, 240
Elion, Gertrude 224, 226, plate 24
Elliott, T. R. 71
emetine 17, 20 (table), 51
Emerich, R. 162
endocrine 87
endocrinology 97
endorphin 265, 281
energy 117
enkephalin 265
enzyme 77, 145, 165, 220, 259
epidemiology 8, 15, 196–7, 211, 216
epilepsy 32–3, 258
epinephrine 86; *see also* adrenaline
Equanil 257
ergosterol 129
ergot 72–3, 110–2
ergotamine 111, 112
ergometrine 112
ergonovine 112
erysipelas 152
Erythroxylon coca 4 (table), 109
eserine 20, 22, 71, 77
ether 29–30
ethics 13, 128
ethylene glycol 276
experiment, controlled 6, 16, 40
experiment, feeding 125

factor
    food 123
    growth 129, 146, 153

famine 119, 128
Farber, S. A. 223
Farr, William 15
fat 118
fat-soluble A 125, 128
FDA (Food and Drugs Administration) 278
Feldman, W. H. 181
fermentation 39, 42, 145, 176
Ferran, Jaime 51
fever
    artificial 254
    childbed 15, 39, 152
    Lassa 141
    marsh 189
    puerperal, *see* fever, childbed
    relapsing 60
    scarlet 152
    splenic, *see* anthrax
Field, A. G. 32
Fildes, Paul 61, 147, 153, 166
filter, bacterial 197, 212
finance 75, 272–3
fits 33; *see also* convulsions
flavine 149
Fleming, Sir Alexander 61, 163–70, 177–80,
    200, 271
Fletcher, Sir Walter 75, 127, 167, 273
Flexner, Abraham 104
Florey, Howard (Lord Florey) 165–8, 173–
    80, 200, 271, plate 22
Flourens, Pierre 30
folic acid 134, 222–3
food 117
Food and Drugs Administration (USA) 278
foot and mouth disease 197
forensic medicine 107
Foster, Michael 69, 70, 75, 122
Foundation, Rockefeller 104, 173, 175, 198
foxglove 4 (table), 10, 103, 242
fowl 40
Frankfurt-am-Main 58
Fraser, Sir Thomas 22, 33, 45, 132, 237,
    plate 14
fraud 272
freeze drying 174
Freud, Sigmund 109
frog 31
Frolich, T. 124
funding of research 75, 202, 273
Funk, C. 125

GABA (γ-aminobutyric acid) 263 (table), 264
Galen 4
Gamma-aminobutyric acid 263 (table), 264
ganglion 237, 240
Geiger 20
Geigy laboratories 260

gene 213
General Register Office 15
generation, spontaneous 39, 45
genetics 146, 163
Gerhardt, Charles 37
germs, *see* microbes
germ theory of disease 43, 141
Germanin 149
Gilman, A. 216–18
gland
    adrenal 83, 94–6
    ductless 83
    pituitary 83
    sex 94–7
    thyroid 83–9
glaucoma 20 (table)
Glaxo laboratories 134
Glenny, A. 144
glonoin 32
glyceryl trinitrate 31
glycine 34, 119
glycol, *see* ethylene glycol
goitre 84
Goldsworthy, N. E. 166
gonorrhoea 141
Goodman, L. S. 216–18
Gosio, B. 161
gout 7, 20 (table), 35, 131, 225
graft, rejection 225–6
gramicidin 172
Graunt, John 8
Greenwood, Major 143
griseofulvin 171
guanidine 191
guinea-pig 124
Gull, Sir William 44, 84
Guthrie, Frederick 30
Guy's Hospital, London 55, 83
gynaecologist 111

H_2 receptor 241–2
haemoglobin 132
Haffkine, Waldemar 51, 143
Hales, Stephen 236
hallucination 253, 263 (table), 264
Hamilton (Ontario) 144
handedness (left/right) 38, 107
Hare, Ronald 168, 179, 183 (note 13)
Harington, Charles 88–9, 165
Harvey, William 9, 235
hashish 251
Hata, Sahachiro 60
heart 11, 52, 76, 241
heart disease 235
heart failure 11, 121
Heatley, N. G. 174–6
Heidelberger, M. 151

helix, double 221
hemlock 4 (table), 20 (table)
hemp 251
Hench, P. S. 95
herbal medicines 8–11, 280
herbalist 12
d'Herelle, F. 162
Herodotus 126
heroin 265
Hinshaw, H. C. 181
Hippocrates of Cos 3
hippuric acid 34–5
histamine 74, 241, 256
history, medical 72
history, natural 69
Hitchings, G. H. 224–7, plate 24
Hoechst 36, 62, 86
Hoffmann, Albert 253
Hofman, A. von 28
Hohenheim, Theophrastus (Paracelsus) 5
Holmes, Oliver Wendell 15, 30
Holst, A. 124
Holt, L. 171
homeopathy 12, 250, 281
Homer 3
Hopkins, Sir Frederick Gowland 122–31,
    146, 220, plate 20
Hoppe-Seyler, F. 219
hormone 85–97, 214–16, 237
horse 54, 236
humours, doctrine 3, 7
Hunter, John 13
Hunt, Reid 72
Huxley, Thomas 45, 69
hydrophobia, *see* rabies
hydroxytryptamine 264
hyoscine 263 (table)
hypertension 235–45
hypnotic 37, 256, 276
hypothermia 256

idoxuridine 199
IG Farbenindustrie 59
immunity 12, 51–8, 123, 141–4, 153
    cellular 52
    humoral 52–3
immunization 12, 142–4
Imperial Cancer Research Fund 228
Imperial Chemical Industries 190, 241, 279
industrial revolution 126
industry, dyestuff 29, 36
industry, pharmaceutical 29, 36, 201, 255
inflammation 6, 239
Ing, H. R. 237
injections into veins 61–2
inoculation 12
insanity 253–5

Institute,
    Brown Sanatory 54, 146
    Chester Beatty 211, 218
    Georg Speyer 60
    Haffkine 133
    Koch 53
    Lister 54, 76, 88, 124, 128
    National Cancer (USA) 228
    National Instits. of Health (USA) 55, 192,
        201
    National, for Medical Research (UK) 74–6,
        89, 106, 111–12, 198, 202, 237
    Pasteur 53, 54, 144
    Pasteur of New York 85
    Rockefeller 88, 146, 151, 171–2, 201
    Royal Jennerian 13
    Tropical Medicine, Liverpool 190
insulin 89–93
interferon 198–9, 227
iodine 32, 87–9
ipecacuanha 4 (table), 10, 17, 20
iproniazid 259
iron 5, 132
Isaacs, A. 198
isoniazid 196, 259

James, William, 253
Jenner, Edward 13, 54, plate 4
jesuit 9
jesuit's bark (table), 9, 20 (table)
Joule, J. P. 117

Kendall, E. C. 87–9, 91, 95
Kennaway, E. 211
Kenny, Maeve 152
kidney 28, 242–3
King, T. W. 83
King's College, London 42
Kitasato, Shibasaburo 52
Klarer, J. 151
Klebs, Theodor Edwin 52
Knauer, E. 94
Koch, Robert 40, 43, 44, 53–6, 59, 74, 194,
    plate 11
Koller, Carl 109
Kraeplin, Emil 252, 253
Kuhn, Roland 260

laboratories
    academic 105–6, 255
    industrial 255, 274
Laborit, Henri 256
Lactobacillus casei 224, 225
Lancaster, James 6
Landsteiner, K. 154
Langley, J. N. 70, 108
Laplace, P.-S. de 117

lard 123
Largactil 256
La Touche, C. J. 168
laudanum 5, 7, 9
laughing gas, *see* nitrous oxide
Laveran, Alphonse 117
Lavoisier, Antoine 16, 117
lead 5
Lederle laboratories 226
leeches 9
Lehmann, J. 195
lemon 6, 120
leprosy 146, 195, 280
leukaemia 216, 219, 224
Levene, Phoebus 221
levodopa 263 (table)
Librium 262
Liebig, Justus von 35, 36, 118
Liebreich, Otto 35
life 39, 45, 197
light, polarized 38
lime 6, 120
Lind, James 7, 120, plate 2
Linskog, G. E. 218
lion 126
Lister, Joseph (Lord) 41, 44, 55, 161, plate 13
literature, scientific 274
lithium 257
liver 133–4
liver, bear's 19
liver, cirrhosis 252
Locock, Sir Charles 32
Loeffler, Friedrich 52, 197
Loeb, Jacques 213
Loewi, Otto 76–7
London School of Hygiene 170–1
London Hospital 19, 61, 145, 165
Louis, P. C. A. 16
LSD, *see* lysergide
Lübeck 276
lunacy 253–5
lungwort 9, 19
lymphocytes 216, 217, 219
lymph, vaccine 13
Lyon, D. Murray 89
lysergide (LSD) 253, 263 (table), 264
lysozyme 163, 167

M&B 693 152
McCollum, E. V. 125, 128, 221
Macfarlane, R. G. 179
McIntosh, James, 61, 147
Macleod, J. J. R. 90–3
Magendie, Francois 20, plate 5
ma-huang 4 (table)
malaria 7, 10, 20 (table), 58, 189–93
mandragora 4 (table), 252

manic-depressive 254
manufacture 36, 53, 176
MAO, *see* monoamineoxidase
mapharsen 63, 227
marihuana 251
marshes 10, 189
Marsilid 259
Martin, Benjamin 14
materia medica 9, 22, 104
Mayer, J. R. 117
measles 197
meat 117, 121
Medical Research Committee, *see* Committee
Medical Research Council, *see* Council
medication, self 132
medicine
    alternative 280
    complementary 280
    preventive 235
medicine and poisons 9, 216
medicines, definition v
medicines from animals 19; *see also* hormones
medicines from plants, *see* plants, medicinal
Meister, Lucius and Bruning 53, 86
Mellanby, Sir Edward 127–9, 167, 173, 273, plate 21
Mendel, Johann 213
meningitis 178
mental, *see* mind
mepacrine 150, 190
mephenesin 257
meprobamate 257, 262
mercaptopurine (6–MP) 224–6
Merck, E. (Darmstadt) 36, 109
Merck, G. F. 36
Merck, H. E. 36
Merck—USA 95, 134, 180, 242
mercury 5, 148, 166, 242
Mering, J. von 90
mescaline 252
messenger, chemical 83, 255
metabolism 118
metabolite, essential 135, 153
Metchnikoff, Elie 52
methisazone 199
method, numerical 16
methotrexate 226
methyl 34
microbes 8, 14, 39–43, 134, 141–200, 223
microbes, soil 171, 180, 181
microbial chemistry 146
micrococcus 164
microscope 43
Middlesex Hospital, London 95, 153
midwife 12, 72
Miescher, J. F. 219
Mietzsch, F. 151

milk 42, 119, 123, 145
milk, artificial 119
Miltown 257
mind 251, 266
minerals 5, 11, 19, 118
Minkowski, O. 90
Minot, G. R. 133
mixtures, chemical 27
Moir, J. Chassar 111–12
molecule 27
monoamine oxidase 259, 263 (table)
Montagu, Lady Mary 13
Morgagni, Battista 8
Morgan, T. H. 213
Morgenroth, J. 148
morphine 17, 20, 263 (table), 264–6
morphium (morphine) 17
Morton, William 30
mosquito 189, 190
motion sickness 256
moulds 42, 161–82
mouse 57, 150, 151, 174
mucus 166
mumps 197
Murphy, William P. 133
Murray, George 84, 85
Murrell, W. 32
mustard gas 217
mutation 213
Mycobacterium bovis 74, 194
Mycobacterium tuberculosis 74, 194
myxoedema 84, 89, 254

names, chemical vi
names, drugs vi
narcotine 20 (table)
nature 69
Naunyn, Bernhard 22
navy 7, 120
neoarsphenamine 63
Neo-Salvarsan 63
nerve-muscle junction 18, 70
Nethalide 241
neuritis 121, 131
neurotransmitter 255, 263 (table), 264
New York 55, 85, 221
nicotinamide 196
nicotinic acid 254
nightshade, deadly 9
nihilism, therapeutic 105
nitrogen mustard 217–19
nitroglycerine 31
nitrous oxide 29, 30, 253
noradrenaline 240
Novocain 109
nucleic acid 191, 221–7
nucleoprotein 220

nucleus, cell 219–20
nutrition, human 117
nutrition, microbes 134, 145–8

obstetrician (accoucheur) 72, 152
Odyssey 3
oestrogen 94, 95, 214–16
Oliver, G. 86
opium 4 (table), 5, 9, 20, 36, 251, 265
optical activity 38, 167
Optochin 148
Orfila, Joseph 18, 22, 30
organic, definition, vi *see also* chemistry
Osborne, T. B. 125
osteopath 12
ovary 94, 95, 213, 215
oxalic acid 276
Oxford, 78, 161, 165–8, 173–9, 237
oxygen uptake 195

PABA (*para*-aminobenzoic acid) 153, 190, 195, 222–3
Pacini, Filippo 15
Paget, Sir James 44
pain 18, 20 (table), 108, 263–6
Paine, C. G. 170
palatability 117
Paludrine 191
pamaquin 150
pancreas 190
papaverine 20, 36
Papaver somniferum 4 (table)
Paracelsus (= Hohenheim) 23, plate 1
paracetamol 37
Paré, Ambrose 6
Paris 16–18, 22, 27, 38, 53, 119, 150
Parke-Davis 86
Parkinson's disease 257, 263 (table)
PAS (*para*-aminosalicylic acid) 195
Pasteur, Louis 16, 38–45, 53, 141, 273, plate 10
patents 91, 133
pathology 42–3, 166, 167
Paton, Noel 127
Paton, Sir William 237
patulin 171
pellagra 254
Pelletier, Joseph 17, plate 6
penicillin 42, 64, 161–82
peptide 119, 265
Pereira, Jonathan 19–21, 118
Perkin, William 28–9, 33
Persians 126
Peruvian bark 7, 9
Petrarch 6
peyote 252
pH 220

phage typing 162
phagocytosis 52, 153
Pharmaceutical Society, *see* Royal
    Pharmaceutical Society
pharmacology 21–3, 43, 103–8, 123
pharmacy 19
phenacetin 36, 37
phenazone 37
phenergan 256
phenetidin 37
phenol (carbolic acid) 41, 162
phenothiazine 256, 263 (table)
phenylbutazone 37
philosophy 252, 266
Physicians, Royal College of 11–12
physiology 17, 69, 105, 117
physostigmine, *see* eserine
phytic acid 128
Pincus, Gregory 96
planning 272–3
plants
    medicinal 9–11, 19, 239
    metabolism of 118
Plasmaquin 150
Plasmodium 189–91
pneumococci 152, 170
pneumonia 141, 142, 152
poisons 11, 18, 22, 107
poisons, corrosive 20
poisons, selective 154
polarization of light 38
poliomyelitis 198
poppy 251
population, world 282
pork 117
post mortem examination 7–8, 15, 16, 42
Pouilly-le-Fort 40
practolol 278–9
prayer 281
pregnancy 96
preposterous proposition 104, 264
price of antibiotics 176, 181
primaquine 193
procaine 109
pro-drug 35, 150
production, scaling up 53, 175
progestogen 94
proguanil 191
promethazine 256
pronethalol 241
Prontosil 150–2, 190
propranolol 241
prostaglandin 238–9
prostate gland 215
protamine 220
protein 35, 92, 118–19, 123, 142, 220
protoplasm 58, 70, 122

Pseudomonas pyocyanea 162
psilocybe 252
psychosis 254, 259
pteroylglutamic acid 134
publication 274
Public Health Service, USA 55, 63, 192
Puerto Rico 96
purgative 9
purification 56
purine 221, 224
putrefaction 14, 41
pyocyanase 162
pyrimethamine 193, 224, 227
pyrimidine 191, 221, 224

quackery 11, 131, 210
quality control 53
Quastel, J. H. 153
Queen Charlotte's Hospital, London 152
quinine 9, 17, 20, 28–9, 190
quinine, synthesis 28–9

rabbit 31, 60, 214
rabies 41, 53, 141
racemic acid 38
radiation, measurement 213
radiation and rickets 128
radium 212
Raistrick, Harold 146, 170–1, 174
Raleigh, Sir Walter 251
Ramon, G. 144
rat 125, 128–9
Rauwolfia serpentina 239, 257
reactions to drugs 274–81
receptor 58–9, 70, 107, 240–2, 263 (table),
    265
register of births and deaths 8, 15
regulation, government 36, 276–8
Reichstein, T. 95
relaxation 257, 261
religion 251
research, interdisciplinary 202
reserpine 239, 257, 263 (table)
resin 251
resistance to drugs 176–7, 181, 182, 191, 193,
    195–6, 218, 225
Resochin 191
revolution, French 16
revolution, industrial 69, 126
Reymond, Emil du Bois 70
Rhone–Poulenc 192, 256
riboflavin 190
ribonucleic acid 199, 221
ribose 221
Richards, A. N. 175
Richardson, Sir Benjamin 31
rickets 126–9

Ridley, F. 168–70
RNA, *see* ribonucleic acid
Roberts, E. A. H. 167
Robinson, Sir Robert 167, 178
Roche U.S.A. 261
Rockefeller Foundation 104, 106, 173, 198, 212
Roentgen, W. 212
roentgen (unit of radiation) 213
rotation, optical 38
roulette, molecular 110
Rous, Peyton 211
Roux, Emile 52
Royal College of Physicians 11–12
Royal College of Surgeons 12, 19
Royal Pharmaceutical Society 19, 106
Royal Society of London 27, 147, 163, 179, 202
rust 8, 19
rye 72

safety 97, 274, 280
saffron 8
Saint Antony 72
Saint Ignatius's beans 20 (table)
Saint Mary's Hospital, London 143, 161, 163–5, 168–70, 177–9
salicylate 10
salicylic acid 37
salmon 219
salt 119, 242–3, 281
Salvarsan 60–3, 148
Sanderson, J. Burdon 161
Sandoz 111, 253
Schafer, E. A. 86
schizophrenia 254, 258, 264
Schmedieberg, Oswald 21, 103
Schneider, Jean 192
Schwartz, Robert 226
Scott, R. F. (Antarctic explorer) 120
screen 110, 181, 191, 227
scrofula 32
scurvy 6, 120, 126
secrecy 36, 86, 149
sedative, *see* hypnotic, tranquillizer
seeds 22
semen 238
Semmelweiss, Ignaz 15
sepsis, puerperal, *see* fever, childbed
serotonin 263 (table), 264
serum
    antitoxic 52, 85
    antipneumococcal 142
serum sickness 142
Sertürner, Freidrich 17
sex 62, 94–7
Shaw, G. Bernard 163

sheep 40, 56, 84
Sherrington, Sir Charles 130, 165
side-chains 58
side effect 275
siege, nutrition during 119
signs, doctrine of 8, 19
silkworms 14, 39–40
Simpson, Sir James Young 30
sleeping sickness 59
smallpox 7, 12–14, 56, 252
Smith, Kline and French 242
Snow, John 15, 209
Sobrero, Ascanio 31
Society of American Bacteriologists 172
Society, British Pharmacological 106
Society for Pharmacology and Experimental
    Therapeutics (USA) 104, 264
Society, Royal, of London 27, 147, 163, 179, 202
Society, Royal Pharmaceutical 19, 106
Sohxlet 200
Sontochin 192–3
Specia 192
spermatozoa 219
Speyer, Franziska 60
spinach 134
spirochaete 60
Squibb, E. R. 88
stability 61
standardization
    antitoxins 55–7
    drugs 107–8
    hormones 92–3
    toxins 55–7
Starling, E. H. 87
starvation 117
statisticians 143
Stephenson, Marjory 146–7, 153, 166
sterility
    microbial 61
    mammalian 96, 213
Sternbach, Leo 261–2
steroid 95
Stevenson, T. 122
stilboestrol 95, 216
Stoll, Arthur 111
stomach extracts 133–5
Stone, Edmund 10
Strasbourg (Strassburg) 21–2, 39, 55, 103
streptococcus 150–3, 178
streptomycin 180–1, 193–7
structure-activity relationship 34
structure, chemical 27–9, 34, 153
strychnine 17, 20 (table)
sulfanilamide elixir 275–6
sulfarsphenamine 63
sulphanilamide 150–3, 190, 195, 275–6

sulphapyridine 152
sulphonamide 152, 190
sulphone 195
sulphur 5
sunlight 126–9
Surgeons, Royal College 12, 19
surgery 41–3
surveillance 279
Sydenham, Thomas 7
sympathetic nerves 240
sympathin 77, 240
synthesis, lethal 35
syphilis 5, 14, 32, 60–4, 141, 148, 254

Tagamet 242
Takaki, Baron 121
tamoxifen 216
tartaric acid 38
Tartu, *see* Dorpat
tea 9
testes 85, 215
tetanus 52–3, 108
tetracycline 181
thalidomide 276–7
Theiler, Max 198
thermometers, clinical 37
thiazide 243
thymus 85
thyroid 84–5, 87–9
thyroxin, thyroxine 88–9
tissue culture 199
tobacco 251
tonic 20 (table)
Toronto 90, 133
torpedo 45
toxicity, selective 153–4
toxicity studies 36, 277–9
toxicology 18, 22, 277–8
toxin
    bacterial 51–3, 56–7, 85, 144
    food 126
toxoid 144
tranquillizer 256–8
transmission of nerve impulses 70
transmitters 70–1, 76–8, 237–42, 255, 263
    (table), 264
transplantation, organ 225–6
Treponema pallidum 60
Tréfouel, J. 150, plate 22
tretinoin 132
trial (experiment) 5, 6
trial, clinical 6, 13, 16, 61–3, 128, 133, 152,
    181, 191, 199, 243–5, 258, 259, 280–1
trichinia 117
tricyclic 260, 263 (table)
tri-iodothyronine 89
trimethoprim 227

trypaflavin 149
trypanosome 59
tryptophane 123
tuberculin 85, 194
tuberculosis 74, 75, 173, 180–1, 193–7, 276
tumour, *see* cancer
Twort, F. W. 145–6, 162
typhoid fever 143, 145
typing of organisms 145, 162
tyrocidine 172
tyrothricin 172

ulcers, gastric and duodenal 235
University College, London, 69, 72, 86, 87,
    106, 111
urea 28, 118
Ure, Alexander 34
uric acid 35, 122, 225
urine 28, 35, 94, 118, 242–3
uterus 73, 111, 239

vaccination 13, 143, 282
vaccine 13, 40, 56, 141, 193, 197–9, 227, 276
vagina, cancer 216
vagus (nerve) 71, 74, 76–7
vagusstoff 76
variolation 13
vegetables, green 134
veratrine 17
Veronal 37
vessels, blood 31
Victoria, Queen 15, 30
Vienna 15, 109
vinblastine 227
vinca alkaloids 227
vincristine 227
Virchow, Rudolf 42, 44, 210
virulence 40
virus 197–200
vision, impaired, *see* blindness
vitalism 28, 39, 45
vitamin 125–35, 147
vitamin A 125, 128, 132
vitamin B 125, 126
vitamin B_1 126, 131
vitamin B_{12} 134, 135
vitamin C 126
vitamin D 128, 129, 132
vitamin K 132
vitamine 125
vitamins, toxicity 132
vivisection 44, 54
Voegtlin, Carl 63, 147
Vogl, A. 242
Vulpian, D. 84, 86

water-soluble B 125
Waksman, S. A. 172, 180–1, plate 22

war
  Napoleonic 7
  Franco-Prussian 21, 40, 44, 119
  World, I 37, 63, 76, 127, 149
  World, II 78, 107, 142, 154, 174, 190, 192–3, 216–18
warfare, chemical 91, 216–17
Wassermann, August von 60
weight
  atomic 28
  molecular 220
Wellcome Physiological Research Laboratory 72–4, 106, 111, 144, 170
Wellcome, Sir Henry 72, 73, 201
Wellcome Trust 201–2
Wells, Horace 30
Whitby, Sir Lionel 153
Whipple, G. H. 133
willow 10, 37
Wills, Lucy 133–4
wine 8, 19, 39, 251, 276
Winthrop 191–3
Withering, William, 10, 242, plate 3

Wöhler, Freidrich 28, 35, plate 9
womb 73, 111, 239
Woods, D. D. 153–4
Worcester Foundation (USA) 96
World Health Organization 282
wounds 6, 41, 266
Wright, Almroth 61, 143, 152, 163, 178, 194

xanthine oxidase 225
X-rays 212–14
X-ray crystallography 176, 221

Yale 217
yaws 60
yeast 39, 134, 145, 153
yellow fever 144, 198
Yersin, Alexandre 52
Yorke, Warrington 190
Yule, G. Udny 143

Zaimis, E. 237
zoo 126–7

Date	Chapter 3 Antimicrobials	Chapter 4 Transmitters	Chapter 5 Hormones
1880	Cholera vaccine by Haffkine	Involuntary (autonomic) nervous system	Diabetes associa with pancreas by Mering and Minkowski
1890	Diphtheria antitoxin by Behring	Ideas of chemical transmission	Adrenaline isola by Abel
1900	Trypan Red by Laveran	Adrenaline transmitter? by Elliot	Endocrine activi of gonads by Knauer
1910	Salvarsan by Ehrlich	Biology of acetylcholine by Dale	Thyroxine isolat by Kendall
1920	Sulfarsphenamine by Voegtlin	Vagusstoff by Lœwi	Insulin by Banting an Thyroxine synthe by Harington
1930	Mapharsen by Tatum	Chemical transmission established by Dale, Feldberg	Steroid structure established Stilboestrol by Dodds

This table is a guide to the chronology of notable events described in Chapters 3 to 8, and is *
references.